```
QP           89076
501      Rickenberg, H. V.
.B527         Biochemistry of
v.8          hormones
```

MTP International Review of Science

Biochemistry
Series One

Consultant Editors
H. L. Kornberg, F.R.S. and
D. C. Phillips, F.R.S.

Publisher's Note

The MTP International Review of Science is an important new venture in scientific publishing, which is presented by Butterworths in association with MTP Medical and Technical Publishing Co. Ltd. and University Park Press, Baltimore. The basic concept of the Review is to provide regular authoritative reviews of entire disciplines. Chemistry was taken first as the problems of literature survey are probably more acute in this subject than in any other. Physiology and Biochemistry followed naturally. As a matter of policy, the authorship of the MTP Review of Science is international and distinguished, the subject coverage is extensive, systematic and critical, and most important of all, it is intended that new issues of the Review will be published at regular intervals.

In the MTP Review of Chemistry (Series One), Inorganic, Physical and Organic Chemistry are comprehensively reviewed in 33 text volumes and 3 index volumes. Physiology (Series One) consists of 8 volumes and Biochemistry (Series One) 12 volumes, each volume individually indexed. Details follow. In general, the Chemistry (Series One) reviews cover the period 1967 to 1971, and Physiology and Biochemistry (Series One) reviews up to 1972. It is planned to start in 1974 the MTP International Review of Science (Series Two), consisting of a similar set of volumes covering developments in a two year period.

The MTP International Review of Science has been conceived within a carefully organised editorial framework. The overall plan was drawn up, and the volume editors appointed by seven consultant editors. In turn, each volume editor planned the coverage of his field and appointed authors to write on subjects which were within the area of their own research experience. No geographical restriction was imposed. Hence the 500 or so contributions to the MTP Review of Science come from many countries of the world and provide an authoritative account of progress.

Butterworth & Co. (Publishers) Ltd.

BIOCHEMISTRY SERIES ONE

Consultant Editors
H. L. Kornberg, F.R.S.
Department of Biochemistry University of Leicester and
D. C. Phillips, F.R.S., *Department of Zoology, University of Oxford*

Volume titles and Editors

1. **CHEMISTRY OF MACRO-MOLECULES**
   Professor H. Gutfreund, *University of Bristol*

2. **BIOCHEMISTRY OF CELL WALLS AND MEMBRANES**
   Dr. C. F. Fox, *University of California*

3. **ENERGY TRANSDUCING MECHANISMS**
   Professor E. Racker, *Cornell University, New York*

4. **BIOCHEMISTRY OF LIPIDS**
   Professor T. W. Goodwin, F.R.S., *University of Liverpool*

5. **BIOCHEMISTRY OF CARBO-HYDRATES**
   Professor W. J. Whelan, *University of Miami*

6. **BIOCHEMISTRY OF NUCLEIC ACIDS**
   Professor K. Burton, F.R.S., *University of Newcastle upon Tyne*

7. **SYNTHESIS OF AMINO ACIDS AND PROTEINS**
   Professor H. R. V. Arnstein, *King's College, University of London*

8. **BIOCHEMISTRY OF HORMONES**
   Professor H. V. Rickenberg, *National Jewish Hospital & Research Center, Colorado*

9. **BIOCHEMISTRY OF CELL DIFFERENTIATION**
   Dr. J. Paul, *The Beatson Institute for Cancer Research, Glasgow*

10. **DEFENCE AND RECOGNITION**
    Professor R. R. Porter, F.R.S., *University of Oxford*

11. **PLANT BIOCHEMISTRY**
    Professor D. H. Northcote, F.R.S., *University of Cambridge*

12. **PHYSIOLOGICAL AND PHARMACO-LOGICAL BIOCHEMISTRY**
    Dr. H. F. K. Blaschko, F.R.S., *University of Oxford*

PHYSIOLOGY SERIES ONE

Consultant Editors
A. C. Guyton,
Department of Physiology and Biophysics, University of Mississippi Medical Center and
D. Horrobin,
Department of Medical Physiology, University College of Nairobi

Volume titles and Editors

1. **CARDIOVASCULAR PHYSIOLOGY**
   Professor A. C. Guyton and Dr. C. E. Jones, *University of Mississippi Medical Center*

2. **RESPIRATORY PHYSIOLOGY**
   Professor J. G. Widdicombe, *St. George's Hospital, London*

3. **NEUROPHYSIOLOGY**
   Professor C. C. Hunt, *Washington University School of Medicine, St. Louis*

4. **GASTROINTESTINAL PHYSIOLOGY**
   Professor E. D. Jacobson and Dr. L. L. Shanbour, *University of Texas Medical School*

5. **ENDOCRINE PHYSIOLOGY**
   Professor S. M. McCann, *University of Texas*

6. **KIDNEY AND URINARY TRACT PHYSIOLOGY**
   Professor K. Thurau, *University of Munich*

7. **ENVIRONMENTAL PHYSIOLOGY**
   Professor D. Robertshaw, *University of Nairobi*

8. **REPRODUCTIVE PHYSIOLOGY**
   Professor R. O. Greep, *Harvard Medical School*

INORGANIC CHEMISTRY SERIES ONE

Consultant Editor
H. J. Emeléus, F.R.S.
Department of Chemistry
University of Cambridge

Volume titles and Editors

1. **MAIN GROUP ELEMENTS—HYDROGEN AND GROUPS I-IV**
Professor M. F. Lappert, University of Sussex

2. **MAIN GROUP ELEMENTS—GROUPS V AND VI**
Professor C. C. Addison, F.R.S. and Dr. D. B. Sowerby, University of Nottingham

3. **MAIN GROUP ELEMENTS—GROUP VII AND NOBLE GASES**
Professor Viktor Gutmann, Technical University of Vienna

4. **ORGANOMETALLIC DERIVATIVES OF THE MAIN GROUP ELEMENTS**
Dr. B. J. Aylett, Westfield College, University of London

5. **TRANSITION METALS—PART 1**
Professor D. W. A. Sharp, University of Glasgow

6. **TRANSITION METALS—PART 2**
Dr. M. J. Mays, University of Cambridge

7. **LANTHANIDES AND ACTINIDES**
Professor K. W. Bagnall, University of Manchester

8. **RADIOCHEMISTRY**
Dr. A. G. Maddock, University of Cambridge

9. **REACTION MECHANISMS IN INORGANIC CHEMISTRY**
Professor M. L. Tobe, University College, University of London

10. **SOLID STATE CHEMISTRY**
Dr. L. E. J. Roberts, Atomic Energy Research Establishment, Harwell

INDEX VOLUME

PHYSICAL CHEMISTRY SERIES ONE

Consultant Editor
A. D. Buckingham
Department of Chemistry
University of Cambridge

Volume titles and Editors

1. **THEORETICAL CHEMISTRY**
Professor W. Byers Brown, University of Manchester

2. **MOLECULAR STRUCTURE AND PROPERTIES**
Professor G. Allen, University of Manchester

3. **SPECTROSCOPY**
Dr. D. A. Ramsay, F.R.S.C., National Research Council of Canada

4. **MAGNETIC RESONANCE**
Professor C. A. McDowell, F.R.S.C., University of British Columbia

5. **MASS SPECTROMETRY**
Professor A. Maccoll, University College, University of London

6. **ELECTROCHEMISTRY**
Professor J. O'M Bockris, University of Pennsylvania

7. **SURFACE CHEMISTRY AND COLLOIDS**
Professor M. Kerker, Clarkson College of Technology, New York

8. **MACROMOLECULAR SCIENCE**
Professor C. E. H. Bawn, F.R.S., University of Liverpool

9. **CHEMICAL KINETICS**
Professor J. C. Polanyi, F.R.S., University of Toronto

10. **THERMOCHEMISTRY AND THERMODYNAMICS**
Dr. H. A. Skinner, University of Manchester

11. **CHEMICAL CRYSTALLOGRAPHY**
Professor J. Monteath Robertson, F.R.S., University of Glasgow

12. **ANALYTICAL CHEMISTRY—PART 1**
Professor T. S. West, Imperial College, University of London

13. **ANALYTICAL CHEMISTRY—PART 2**
Professor T. S. West, Imperial College, University of London

INDEX VOLUME

ORGANIC CHEMISTRY SERIES ONE

Consultant Editor
D. H. Hey, F.R.S.,
Department of Chemistry
King's College, University of London

Volume titles and Editors

1. **STRUCTURE DETERMINATION IN ORGANIC CHEMISTRY**
Professor W. D. Ollis, F.R.S., University of Sheffield

2. **ALIPHATIC COMPOUNDS**
Professor N. B. Chapman, Hull University

3. **AROMATIC COMPOUNDS**
Professor H. Zollinger, Swiss Federal Institute of Technology

4. **HETEROCYCLIC COMPOUNDS**
Dr. K. Schofield, University of Exeter

5. **ALICYCLIC COMPOUNDS**
Professor W. Parker, University of Stirling

6. **AMINO ACIDS, PEPTIDES AND RELATED COMPOUNDS**
Professor D. H. Hey, F.R.S., and Dr. D. I. John, King's College, University of London

7. **CARBOHYDRATES**
Professor G. O. Aspinall, Trent University, Ontario

8. **STEROIDS**
Dr. W. F. Johns, G. D. Searle & Co., Chicago

9. **ALKALOIDS**
Professor K. Wiesner, F.R.S., University of New Brunswick

10. **FREE RADICAL REACTIONS**
Professor W. A. Waters, F.R.S., University of Oxford

INDEX VOLUME

MTP International Review of Science

Biochemistry
Series One

Volume 8
Biochemistry of Hormones

Edited by **H. V. Rickenberg**
National Jewish Hospital and Research Centre, Denver

Butterworths · London
University Park Press · Baltimore

THE BUTTERWORTH GROUP

ENGLAND
Butterworth & Co (Publishers) Ltd
London: 88 Kingsway, WC2B 6AB

AUSTRALIA
Butterworths Pty Ltd
Sydney: 586 Pacific Highway 2067
Melbourne: 343 Little Collins Street, 3000
Brisbane: 240 Queen Street, 4000

NEW ZEALAND
Butterworths of New Zealand Ltd
Wellington: 26–28 Waring Taylor Street, 1

SOUTH AFRICA
Butterworth & Co (South Africa) (Pty) Ltd
Durban: 152–154 Gale Street

ISBN 0 408 70502 7

UNIVERSITY PARK PRESS

U.S.A. and CANADA
University Park Press
Chamber of Commerce Building
Baltimore, Maryland, 21202

Library of Congress Cataloging in Publication Data

Rickenberg, H. V.
  Biochemistry of hormones.

  (Biochemistry, series one, v. 8) (MTP international review of science)
  1. Hormones. I. Title. II. Series.
III. Series: MTP international review of science.
[DNLM: 1. Hormones. 2. Hormones-Metabolism.
W1 BI633 ser. 1 v. 8 1974/WK100 B615 1974]
QP1. B48 vol. 8 [QP571] 599′.01′08s [599′.01′927]
ISBN 0–8391–1047–2   73–21884

First Published 1974 and © 1974
MTP MEDICAL AND TECHNICAL PUBLISHING CO LTD
St Leonard's House
St Leonardgate
Lancaster, Lancs
and
BUTTERWORTH & CO (PUBLISHERS) LTD

Typeset and printed in Great Britain by
REDWOOD BURN LIMITED
Trowbridge & Esher
and bound by R. J. Acford Ltd, Chichester, Sussex

Consultant Editors' Note

The MTP International Review of Science is designed to provide a comprehensive, critical and continuing survey of progress in research. Nowhere is such a survey needed as urgently as in those areas of knowledge that deal with the molecular aspects of biology. Both the volume of new information, and the pace at which it accrues, threaten to overwhelm the reader: it is becoming increasingly difficult for a practitioner of one branch of biochemistry to understand even the language used by specialists in another.

The present series of 12 volumes is intended to counteract this situation. It has been the aim of each Editor and the contributors to each volume not only to provide authoritative and up-to-date reviews but carefully to place these reviews into the context of existing knowledge, so that their significance to the overall advances in biochemical understanding can be understood also by advanced students and by non-specialist biochemists. It is particularly hoped that this series will benefit those colleagues to whom the whole range of scientific journals is not readily available. Inevitably, some of the information in these articles will already be out of date by the time these volumes appear: it is for that reason that further or revised volumes will be published as and when this is felt to be appropriate.

In order to give some kind of coherence to this series, we have viewed the description of biological processes in molecular terms as a progression from the properties of macromolecular cell components, through the functional interrelations of those components, to the manner in which cells, tissues and organisms respond biochemically to external changes. Although it is clear that many important topics have been ignored in a collection of articles chosen in this manner, we hope that the authority and distinction of the contributions will compensate for our shortcomings of thematic selection. We certainly welcome criticisms, and solicit suggestions for future reviews, from interested readers.

It is our pleasure to thank all who have collaborated to make this venture possible—the volume editors, the chapter authors, and the publishers.

Leicester   H. L. Kornberg

Oxford   D. C. Phillips

Preface

The title of this volume, 'Biochemistry of Hormones' is somewhat misleading. The collection of papers, written by workers in the field, is both more and less than the title implies. The majority of the contributions focus on the mode of action of hormones, although their biosynthesis and secretion, and the complex regulation of these processes, is not neglected. Clearly, it is no longer possible to survey, let alone cover, the biochemistry of hormones and their mode of action within the covers of a single volume. The editor felt therefore that the reader would be served best by a series of essays that deal with aspects of the subject in which recent progress has been rapid. Judicious use of the extensive references provided by the authors should permit the reader to expand and deepen his understanding of areas of endocrinology of interest to him, even if not covered in the present volume.

The volume deals with three classes of hormones: the polypeptide hormones, the steroid hormones, and a chemically heterogeneous group whose common denominator is that they affect the metabolism of plants. Catecholamines, which on the basis of their function meet the criterion of 'hormones', although they are frequently and with equal justification considered 'neurotransmitters' are not discussed here; they have been the subject of several recent reviews.

Findings obtained during the last few years in a number of laboratories lead to certain conclusions regarding the initial and terminal, though not intermediate, steps by which hormones act. It now appears that a majority if not all, of the polypeptide hormones interact with the surface of the cell and in some manner affect the synthesis of cyclic AMP and, perhaps, of cyclic GMP. Catecholamines probably act by a similar mechanism. It also seems that steroid hormones enter the cytoplasm of the target cell to combine with a hormone-specific proteinaceous receptor. The steroid-receptor complex then migrates to the nucleus and interacts with yet another protein which may be a component of chromatin. This interaction then, it is assumed, affects the transcription of genes which code for products ultimately relevant to the function of the hormone. The hypothesis is attractive; there are major gaps in our understanding, particularly regarding the specificity of the interaction between the hormone-protein complex and the genome. Clearly additional, or even alternative, modes of action of the hormones are not ruled out.

As stated earlier, polypeptide hormones and catecholamines seem to act by an effect on the metabolism of cyclic nucleotides. This insight by itself, however valid, leaves unresolved the question of the mode of action of the cyclic nucleotides in different types of cells and tissues. It also sheds no light on the nature of the occasional synergism observed between the actions of cyclic AMP and certain steroids.

The editor of this volume, himself a neophyte to the discipline of endocrinology and hence doubly impressed by recent progress in this field, is confident that an understanding of the precise mechanisms by which hormones exert their effects is not far distant.

Denver, Colorado H. V. Rickenberg

Contents

Mode of action of insulin 1
J. N. Fain, *Brown University, Providence R.I.*

Modes of action of ACTH 25
G. Sayers, R. J. Beall and S. Seelig, *Case Western Reserve University, Cleveland*

Adenohypophysial hormones: regulation of their secretion and mechanisms of their action 61
V. Schreiber, *Charles University, Prague*

Hormonal regulation of calcium metabolism 101
J. F. Habener and H. D. Niall, *Massachusetts General Hospital and Harvard Medical School, Boston*

Biochemical studies on the receptor mechanisms involved in androgen actions 153
S. Liao, *The University of Chicago*

The mode of action of the female sex steroids 187
B. W. O'Malley and A. R. Means, *Baylor College of Medicine, Houston*

The mode of action of glucocorticoids 211
W. D. Wicks, *University of Colorado Medical Center, Denver*

Steroid hormonal analogues 243
H. B. Anstall, *University of Utah College of Medicine*

Mode of action of plant hormones 283
R. Cleland, *University of Washington, Seattle*

Cell culture in the study of the mechanism of hormone action 305
E. H. Macintyre, *National Jewish Hospital and Research Centre, Denver, Colorado*

Index 331

1
Mode of Action of Insulin

J. N. FAIN
Brown University, Providence, R.I.

1.1	INTRODUCTION	2
1.2	INSULIN BINDING TO RECEPTORS	2
1.3	RELATIONSHIP OF INSULIN STRUCTURE TO ITS ACTIVITY	5
1.4	INSULIN ENTRY INTO CELLS	6
1.5	INSULIN ACTION ON CELL-FREE SYSTEMS	8
1.6	EFFECTS OF INSULIN ON INTACT CELLS	9
	1.6.1 *Cyclic-AMP*	9
	1.6.1.1 *Cyclic-AMP and lipolysis in fat cells*	9
	1.6.1.2 *Muscle, cyclic-AMP and glycogen metabolism*	11
	1.6.1.3 *Insulin, cyclic-AMP and hepatic glucose output*	12
	1.6.2 *Stimulation of membrane transport*	13
	1.6.2.1 *Hexoses*	13
	1.6.2.2 *Structural changes in membrane structure in relation to stimulation of hexose uptake by insulin*	14
	1.6.2.3 *Amino acid transport*	15
	1.6.3 *Effects on protein metabolism*	15
	1.6.3.1 *Protein synthesis in muscle*	15
	1.6.3.2 *Protein degradation in muscle*	16
	1.6.3.3 *Protein degradation in liver*	17
	1.6.3.4 *Insulin and protein synthesis in adipose tissue*	18
	1.6.4 *Insulin, membrane potentials and ion fluxes*	18
1.7	SUMMARY	19

1.1 INTRODUCTION

Approximately 50 years ago the discovery of insulin was announced by Banting and Best. The proceedings of at least three symposia held to celebrate the discovery of insulin[1-3] and an entire issue of the *Handbook of Physiology* on the endocrine pancreas[4] have recently been published. These provide extensive reviews of our current knowledge about insulin. Most of the earlier work on insulin action was reviewed by Krahl in 1961[5]. The aim of this article is to provide in one chapter a concise review of our current understanding of the mechanism of insulin action.

There are a multitude of effects which can be seen after the administration of insulin to intact animals or its addition to intact tissues or cells incubated *in vitro*. But despite 50 years of intensive investigation it is still impossible to pin-point in any detail the precise mode of insulin action. Many reports of insulin effects on cell-free systems have appeared from time to time but they have been most difficult to reproduce. The recent work on the characterisation of the insulin receptors has provided much information about insulin binding to tissue but we still know little about how the insulin–receptor complex is able to affect so many different processes in intact cells. The apparent multiplicity of the effects of insulin has complicated research on its mode of action. Whether these effects are independent or have a common denominator, remains to be established.

The study of insulin action appears to have been subject to the current fads and fashions in biochemical investigations to a remarkable degree. In the era of enzymology during the 1940s, the possibility that insulin would have direct effects on enzymes was studied intensively. In the 1950s, the emphasis was on membrane transport and there were reports which emphasised effects of insulin on membrane transport of glucose. In fact, it was claimed that all effects of insulin were secondary to glucose transport, but that view has proved untenable with the passage of time. In the next decade the emphasis was on molecular biology, particularly RNA and protein synthesis. It was even suggested for a time that all effects of insulin involved RNA synthesis. That theory was soon discarded, but there are marked effects of insulin on protein synthesis by ribosomes which will be discussed later. The recognition in the 1960s that cyclic-AMP was involved in the action of a large number of hormones led investigators to suggest that all effects of insulin were secondary to a depression in cyclic-AMP content. This theory has already fallen by the wayside. What the next fad will be is not yet clear. Renewed interest has been shown in the possibility that insulin has its own intracellular messenger which might be a small molecule distinct from cyclic-AMP[6].

1.2 INSULIN BINDING TO RECEPTORS

Recent work has provided support for the original suggestion by Stadie and his associates in 1949[7] that the initial step in insulin action is a firm binding to cells. Narahara[8] has shown that binding of insulin to frog sartorius muscle precedes its action on glucose transport. The binding of insulin to this muscle occurs immediately after its addition while there is a substantial lag period before transport is accelerated[8]. Insulin binds to frog muscle at 0 °C but no stimulation of glucose transport is seen at this temperature[8]. After the rapid binding of insulin to the cell, the hormone apparently causes slower reactions which alter membrane function and metabolism.

Studies on the binding of mono-iodo-insulin to a variety of receptors have been published recently[9-19]. Substantial progress has been made on the purification of the insulin receptor from liver cell membranes[14-16].

Freychet, Roth and Neville[9] found that porcine insulin labelled with not more than one ^{125}I atom per molecule by iodination at a molar ratio of iodine to insulin of 1/10 was readily bound to rat liver membrane fractions. The ability of eight different insulins and derivatives, with wholly different biological potencies, to inhibit [^{125}I] insulin binding was in direct proportion to their ability to stimulate glucose oxidation in fat cells[9]. The binding of insulin was unaffected by glucagon, the A chain or the B chain of insulin.

In fat cells, it has been estimated[10, 20] that there are some 50 000 to 160 000 potential insulin receptors per cell. However, maximum stimulation by insulin can be seen when less than 3000 receptors have combined with insulin[10]. Kono and Barham[10] concluded that the large number of spare receptors increase the sensitivity of the cells to insulin since they enable cells to bind sufficient amounts of insulin to stimulate metabolism even when the hormone concentration is very low. This conclusion was supported by the finding that trypsin treatment reduced both the maximum insulin-binding capacity of cells and the sensitivity of the cells to stimulation of glucose oxidation by insulin[10, 18].

Cuatrecasas[12] estimated that there were 11 000 insulin receptors per fat cell which were ordinarily exposed to insulin. Exposure of fat cells to high salt concentrations[17] or to phospholipase preparations[18] unmasked up to 65 000 receptors. The dissociation constant for insulin binding to fat cells (i.e. the concentration at which half-maximum binding is observed) reported initially by Cuatrecasas was around 0.05×10^{-9} mol l^{-1}, which is similar to the concentration required for half-maximum stimulation of glucose metabolism[12]. Subsequently, Cuatrecasas[14, 15] reported that the dissociation constant for insulin binding to liver membranes, fat-cell membranes and solubilised receptors prepared from these membranes was about 0.15×10^{-9} mol l^{-1}.

Other workers have reported substantially higher dissociation constants for insulin binding of around $2-7 \times 10^{-9}$ mol l^{-1} for liver membranes[13, 19] and for intact fat cells[10, 20, 21]. Kahn et al.[22] suggested that these differences might be due to the presence of two or more types of receptors for insulin. House[13] first reported the presence of two types of receptors for the binding of insulin to liver membranes. Similar results were reported using plasma membrane fractions derived from white fat cells. Hammond et al.[133] found about 3000 receptors per cell with a dissociation constant of around 0.5×10^{-9} mol l^{-1} and 30 000 insulin receptors per cell with a dissociation constant of 3×10^{-9} mol l^{-1}.

The available data suggest that there are a small number of receptors with a high affinity for insulin which may be involved in insulin action. The large number of receptors with a low affinity for insulin may have little to do with its mechanism of action. In all studies of insulin binding, one must be careful with regard to non-specific binding for insulin binds readily to a variety of substances. A further complication is the degradation of insulin by membrane preparations and intact fat cells which can affect binding measurements.

The rapid reversibility of insulin binding and action on cells is of interest. Fat cells[10] and muscle[8] retain insulin if washed at 0 °C but the bound insulin is rapidly lost if cells are rinsed at higher temperatures. Crofford[23] reported that in fat cells which were rinsed several times at 37 °C the effect of insulin was rapidly lost and no immunologically- or biologically-active insulin could be detected. His data suggested that under ordinary conditions the combined effects of insulin dissociation from the receptor and rapid inactivation may account for the short half-life of insulin action on free cells. In other tissues, and after more prolonged exposure, the effects of insulin may persist for longer periods[8].

The insulin receptors appear to be in the plasma membrane of fat cells. Crofford and Okayama[24] first reported that in sub-cellular fractions of fat-cell homogenates there was little uptake of insulin by nuclear or mitochondrial fractions while fractions containing plasma membrane vesicles rapidly took up insulin. They found that trypsin abolished both the uptake of insulin by fat cells and plasma membrane fractions isolated from trypsin-treated fat cells[24]. This was subsequently confirmed by Cuatrecasas, who also found that trypsin covalently bound to agarose was able to produce the same effects as trypsin on intact fat cells[18]. The available data indicate that insulin receptors are present exclusively in the plasma membrane of cells and that ordinarily only a small fraction are readily accessible to insulin.

Treatment of fat cells with commerical preparations of phospholipases non-specifically inhibits the response of fat cells to insulin and many other agents, but at the same time results in a three to sixfold increase in specific binding of insulin[25]. A similar dissociation was seen after treatment with neuraminidase preparations[26]. These data suggest that, unlike trypsin, these other enzymes are acting at sites beyond the binding of insulin to its receptor. Apparently there are additional receptors present in the plasma membranes of cells which are unable to bind insulin but can be unmasked by a variety of procedures[17, 18, 25, 26]. These procedures also non-specifically destroy the ability of insulin and other agents to stimulate glucose metabolism, which suggests that the perturbations of membrane structure resulting in increased insulin receptors have deleterious effects on cellular metabolism. Possibly phospholipase preparations also interfere with the ability of the insulin–receptor complex to activate cellular metabolism.

Cuatrecasas[14-16] found that insulin receptor of liver and fat-cell membranes could be solubilised with a non-ionic detergent such as Triton X-100. Further purification of the solubilised receptor from liver homogenates[16] by ammonium sulphate fractionation, DEAE cellulose chromatography and, finally, affinity chromatography using insulin–agarose derivatives resulted in an approximately 250 000-fold purification. However, the amount of receptor which could be recovered was so small that its protein content could not be accurately measured. Confirmation of these findings and description of conditions in which larger amounts of purified insulin receptor can be isolated are needed. Cuatrecasas has also claimed that the insulin receptor has a molecular weight of 300 000 and is a highly asymmetric protein which is not a lipoprotein[15], but these conclusions were based on studies with relatively impure preparations of solubilised liver membranes in which the insulin-binding protein or receptor may have accounted for less than 1% of the total protein[15, 16].

It has long been an attractive idea that the binding of insulin to its receptors involved disulphide bonds. However, recent studies on insulin binding to liver and fat-cell membranes[17, 19] have provided no support for this hypothesis. The binding of insulin was not affected by agents such as N-ethylmaleimide, arsenite, or p-chloromercuribenzoate. Similarly, the involvement of tryptophanyl or carboxyl groups in binding appears unlikely, although there is some disagreement over tryptophanyl[17, 19] involvement. Neither metal ions nor lipid appeared to be involved in the binding of insulin to its receptor[19].

The possibility that a deficiency in the number of insulin receptors might explain the insulin-insensitivity seen under a variety of conditions has been explored. Bennett and Cuatrecasas[27] reported that starvation of rats or prior treatment with prednisone or streptozotocin reduced the metabolic response of fat cells to insulin but did not affect the quantity or affinity of the receptors for insulin. Despite the decreased responsiveness to insulin of fat cells obtained from hamster, rabbits, mice and guinea-pigs as compared to rat fat cells, the binding of insulin was similar[27]. Livingstone, East and Lockwood[28] have also reported that the insulin-insensitivity of large fat cells was not due to a reduced insulin binding. There did appear to be a 65–85% reduction in the degree of insulin binding to liver plasma membranes from obese hyperglycemic mice, but whether this is the cause or the result of the decreased insulin sensitivity of these animals remains to be demonstrated[29].

1.3 RELATIONSHIP OF INSULIN STRUCTURE TO ITS ACTIVITY

The elucidation of the amino-acid sequence of insulin by Sanger and Tuppy[30] was a major achievement. More recently, Hodgkin and associates have determined the three-dimensional structure of insulin by single-crystal x-ray analysis[31-33]. Despite these remarkable advances in our understanding of insulin structure, the active site involved in insulin binding to receptors is not known.

The insulin molecule has a compact non-polar core and all the polar residues are contained in the surface structure. Any modification of insulin that seriously impairs its three-dimensional structure results in a marked loss of biological activity. Most modifications of the intrinsic structure of insulin result in a large decrease in biological activity. For example, the substitution of bulky groups on the A-1-glycine position, or deletion of this group, causes a marked drop in biological activity. On the other hand, alterations at the N-terminus of the B chain have much less effect.

Analysis of the variations in the sequence of insulins from different species has revealed that certain residues are invariant. This knowledge, combined with information on the three-dimensional structure of insulin, has led to some interesting speculations as to which residues of insulin are involved in its action. Blundell et al.[32] have suggested that the key invariant residues (Gly^{A-1}, Gln^{A-5}, Tyr^{A-19} and Asn^{A-21}) located on the A chain might be important. This is based on their location on the surface, their sensitivity to alterations in structure and the low potency of proinsulin.

The role of the thiol groups in insulin action is relatively minor. Zahn et al.[34] found that an insulin derivative with a split A-7—B-7-disulphide bridge had high biological activity. Jost et al.[35] found that [A-6—A-11-cystanthionine]-insulin was biologically active. Furthermore the A-6—A-11-disulphide is buried as part of the non-polar core of insulin[31-33]. Whether the other disulphide bridge is related to insulin action is not yet known.

Fully-iodinated insulin is inactive but mono-iodo-insulin retains biological activity[8, 36-38]. Freychet et al.[38] found that under conditions of iodination, in which the molar ratio of [I]/insulin was 1/10, only mono-iodo-insulin was formed. Its biological activity was similar to that of native insulin[38]. No data were presented as to which tyrosine was labelled. Arquilla[36, 37] isolated a preparation of mono-iodo-insulin with about 70% of the biological activity of insulin. This preparation contained about 50% of the molecules iodinated at Tyr^{A-14} and 50% at Tyr^{A-19}. It was suggested that A-19-iodo-insulin was less active than A-14[38]. Possibly the high activity of the mono-iodo insulin prepared by Freychet et al.[36] resulted from the use of carrier-free ^{125}I and conditions in which only the A-14-tyrosine of insulin was iodinated. In any case, the data suggest that A-14- and A-19-tyrosines are probably not as important for biological activity as other residues on the A chain such as Gly^{A-1}, Gln^{A-5} and Asn^{A-21}.

The sequences in insulin responsible for immunological activity are unrelated to those involved in biological activity. Fish and guinea-pig insulins have immuno reactivities that are much lower than their biological activities[9]. However porcine pro-insulin, while it has low biological activity, is almost equipotent to insulin with respect to immunological reactivity[9].

1.4 INSULIN ENTRY INTO CELLS

Cuatrecasas[39] has reported that insulin can be covalently bound through the ε-amino group of the B-29-lysyl residue to agarose beads which are as large as fat cells. The complex stimulated glucose metabolism by fat cells. Amazingly, the log(dose)–response curve for stimulation of glucose metabolism in fat cells by the insoluble insulin–agarose bead complex was almost identical to that for insulin[39]. However, no data were presented on how the activity of the insulin–agarose was determined. Cuatrecasas thought the possibility of hydrolysis of the insulin–agarose to give free insulin was unlikely since no lag period was observed before insulin–agarose was active[39]. The binding of insulin to agarose was said to be stable since washing the beads with 8 mol l^{-1} urea, sodium bicarbonate and 6 mol l^{-1} guanidine followed by washing with 0.2 mol l^{-1} ammonium bicarbonate at pH 8.8 did not result in any liberation of free insulin[39]. Insulin–agarose has been reported to activate glycogen synthase in sliced cubes of tadpole liver[40] and stimulate RNA synthesis[41] and a-amino-isobutyric acid uptake[42] by isolated mammary cells.

Oka and Topper[42] found that mammary cells in explant culture did not respond to native insulin but did respond well to insulin–agarose. The reason for this unusual finding is not known. Neither is there an explanation of how enough molecules of insulin–agarose could bind to fat cells to activate metabolic processes. Spatial considerations indicate that the number of insulin–agarose molecules which can bind at one time to a cell is very low. Cuatrecasas[43] and Oka and Topper[42] have suggested that the insulin–agarose dissociates much more rapidly from the receptor than does native insulin. The association of insulin with receptors is an extremely rapid process which appears to be limited by the rate of diffusion of insulin while insulin dissociation is slower. Insulin dissociation is markedly temperature-sensitive, and at 37 °C has a half-life of about 7 min. This hypothesis assumes that the mechanism of insulin action involves only the initial interaction between insulin and its receptor. Continued occupancy by insulin is postulated not to be required for insulin action. This is a most ingenious hypothesis which, if true, would be a remarkable new finding regarding the mode of action of hormones.

It has generally been assumed that as long as insulin was linked to its receptor a signal was transmitted into the cell and throughout the plasma membrane. This kind of model is based on an analogy to the mechanism by which cyclic-AMP activates protein kinase[134]. Cyclic-AMP binds to the receptor protein which then dissociates from the catalytic unit. When the catalytic unit is not bound to the receptor unit it is fully active, and inactivation is due to binding of the catalytic unit by free receptor. If the receptor protein binds cyclic AMP, it does not bind the catalytic unit[134]. Thus the mechanism of activation by cyclic-AMP can be considered as a release from inhibition. Possibly insulin binding to its receptor alters the conformation of this receptor which may result in reduced receptor binding to other membrane proteins. It is conceivable that such proteins are catalytically active as enzymes when no longer bound to the insulin–receptor protein.

One suggestion to explain the effects of insulin–agarose is that the insulin is hydrolysed or released during incubation with cells and active only as free insulin. Davidson et al.[44] have prepared insulin–agarose and found that it stimulated glucose oxidation by rat adipose tissue and glucose uptake by rat hemidiaphragms. However, free insulin was continuously being liberated from the preparation[44]. Fritz[45] found similar results in experiments in which insulin was reacted with Sephadex G-100. Further experimentation is thus required before we can be really sure that insulin is active when bound to agarose. No log(dose)–response curves have been published in other experiments using insulin–agarose[40-42]. Turkington[41] compared 50 µg ml^{-1} of insulin–agarose with 5 µg ml^{-1} of insulin which is a very high amount of insulin for most biological effects.

Armstrong, Noall and Stouffer[46] have reacted insulin with dextran-2000 (average molecular weight = 2 000 000) in the presence of cyanogen bromide, and obtained a product which stimulated glucose metabolism in fat cells. They found that the soluble complex of insulin–dextran-2000 was also able to lower the blood glucose of rabbits in vivo but was unable to affect amino-acid or glucose transport in intact rat diaphragm preparations. This finding is of interest since dextran-2000 should not be able to penetrate peritoneal membranes. If insulin–dextran-2000 does not exert its effect in the rat diaphragm, it is hard to visualise how the insoluble insulin–agarose complex is able to stimulate metabolism in this preparation.

Further studies have been reported on insulin covalently coupled to dextran-40 (average molecular weight = 40 000)[141,142]. This complex was water-soluble and did stimulate glycogen synthesis and sugar transport by the intact rat diaphragm preparation[141]. The insulin–dextran-40 was less effective than native insulin on the diaphragm[141], but was more effective than insulin *in vivo* or in stimulating adipose tissue slice metabolism[142]. In the latter cases, the greater effectiveness of insulin–dextran-40 was attributed to its markedly lower rate of degradation by intact rat adipose tissue, soleus muscle or homogenates of either tissue[142].

The studies with soluble insulin–dextran derivatives support the hypothesis that insulin entry into cells is not necessary in order for it to affect intracellular metabolism. The use of insoluble insulin–agarose derivatives appears to be much less suitable for such studies and there is substantial doubt that such compounds actually are effective.

1.5 INSULIN ACTION ON CELL-FREE SYSTEMS

No consistently reproducible effect of insulin in a cell-free system has yet been reported. The best current candidate is inhibition of adenylate cyclase in membrane preparations. None of the effects of insulin on cell-free systems reviewed by Krahl[5] in 1961 have withstood the test of time. Investigators have largely abandoned the search for effects of insulin on hexokinase and mitochondria. Insulin is said to increase the binding of hexokinase to mitochondria in muscle[47] and fat[48] but only after *in vivo* administration or addition to intact cells. Liver glucokinase and adipose tissue hexokinase II synthesis is also increased as a long-term adaptive response to insulin[49].

Recently, a direct stimulation by insulin of 3′,5′-AMP phosphodiesterase activity in a membrane preparation from liver has been observed[50]. House, Poulis and Weidemann[50] reported that the effect was immediate in onset (less than 30 s) and the insulin concentration required for half-maximum stimulation was 2.2×10^{-9} mol l^{-1}. The same preparation also contained most of the insulin-binding activity as well as glucagon-sensitive adenylate cyclase activity. Loten and Sneyd[51] found a stimulation of 3′,5′-AMP phosphodiesterase activity (low K_m fraction) by insulin, but only if insulin was added to fat cells prior to homogenisation. Vaughan[52] confirmed that insulin increased 3′,5′-AMP phosphodiesterase activity of the low K_m enzyme present in particulate fractions of fat-cell homogenates. Insulin had to be present prior to homogenisation, and the effect was only observed at cyclic-AMP concentrations less than 5×10^{-6} mol l^{-1}[52]. The insulin effect was abolished by freezing and thawing of the enzyme preparations. These results have been confirmed by Zinman and Hollenberg[53] and Jungas[54]. Fain and Rosenberg[55] found no effect of prior exposure to insulin on the low K_m 3′,5′-AMP phosphodiesterase activity present in the soluble supernatant of rat fat-cell homogenates. The difference is apparently due to the fact that the effects of prior exposure of fat cells to insulin on 3′,5-AMP phosphodiesterase are observed in whole homogenates[51] or in particulate preparations[52]. Ward[56] has been able in our laboratory to reproduce the findings of Vaughan[25] using particulate preparations prepared after gentle homogenisation of fat cells.

Recently, two groups have described direct inhibitory effects of insulin on the activation of adenylate cyclase by catecholamines[57, 58]. Hepp and Renner[57] found no effect of insulin on basal or fluoride-activated adenylate cyclase activity of mouse liver particulate preparations or rat fat-cell ghosts. Insulin acted as a competitive inhibitor of the stimulation by glucagon of liver and fat-cell adenylate cyclase. Insulin also inhibited the increase in adenylate cyclase due to ACTH and norepinephrine in fat-cell ghosts[57]. Hepp and Renner[57] suggested that the previous failure of attempts to observe effects of insulin on adenylate cyclase might have been due to the use of high concentrations of insulin or stimulators of adenylate cyclase, aged membrane preparations, or high concentrations of substrate. The problem is not that simple since the effects of insulin on adenylate cyclase activity are difficult to reproduce[55]. There are apparently uncontrolled variables which have not yet been elucidated. Further investigations await the description of conditions in which insulin effects can be reproducibly seen. Even if insulin does turn out to modulate adenylate cyclase, it is unlikely that this will explain many of the actions of insulin.

1.6 EFFECTS OF INSULIN ON INTACT CELLS

1.6.1 Cyclic-AMP

1.6.1.1 Cyclic-AMP and lipolysis in fat cells

At one time it was thought that all effects of insulin on fatty-acid release were due to increase re-esterification of fatty acids secondary to increased uptake of glucose. However, insulin also has a direct antilipolytic action on adipose tissue in the absence of glucose[59, 60]. The inhibitory action of insulin is overcome in the presence of high concentrations of lipolytic agents[60]. Butcher et al.[61] found that insulin also decreased the large accumulation of cyclic-AMP due to catecholamines in the presence of methylxanthines. Prior treatment of fat cells with insulin reduced the adenylate cyclase and phosphorylase activity of whole homogenates while glycogen synthase activity was increased[62]. These findings suggested that the relatively specific stimulation by insulin of glycogen deposition in adipose tissue and inhibition of lipolysis were secondary to inhibition of cyclic-AMP accumulation. This relatively simple picture has been complicated by the findings of Jarrett et al.[63], Fain[64] and Fain and Rosenberg[55]. No effect of insulin on total cyclic-AMP accumulation in fat cells was seen under conditions in which it inhibited lipolysis[55, 63].

The results shown in Figure 1.1 point out the lack of correlation between the action of insulin and adenosine on lipolysis as contrasted with that on cyclic-AMP accumulation due to norepinephrine alone. The data indicate that while adenosine is a much more effective inhibitor of cyclic-AMP accumulation than is insulin, it had little effect on lipolysis. When both insulin

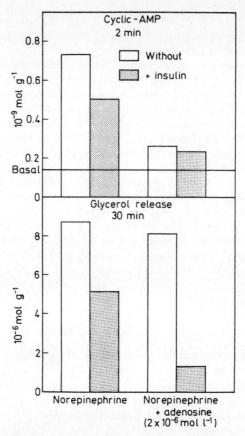

Figure 1.1 Dissociation between effects of insulin and adenosine on lipolysis as contrasted with cyclic-AMP accumulation.

The pooled white fat cells from 2–4 rats (1 g of fat per rat) were isolated and then incubated in 6–8 ml of phosphate buffer for 15 min. The cells were washed twice and then 30 mg tube^{-1} incubated for 10 min in the absence or presence of insulin (120 μU ml^{-1}). Norepinephrine (1.5×10^{-6} mol l^{-1}) was then added either without or with adenosine (2×10^{-6} mol l^{-1}). Cyclic-AMP accumulation was measured after 2 min and glycerol release after 30 min. The values shown are the means of six paired experiments. The basal value for cyclic-AMP was 0.14×10^{-9} mol g^{-1} (shown by the thin line across the figure marked basal). The basal level for glycerol release was undetectable. The inhibition by insulin of cyclic-AMP accumulation and glycerol release was statistically significant and the effect of insulin on glycerol release in the presence of adenosine was greater than the effect of insulin alone ($p < 0.05$ by paired comparisons). (J. N. Fain, unpublished experiments.)

and adenosine were present, there was no further reduction in cyclic-AMP accumulation but there was a marked drop in lipolysis. Fain, Pointer and Ward[65] have shown that adenosine and related nucleosides could completely block the small increase in cyclic-AMP accumulation due to catecholamines or theophylline without affecting lipolysis. In contrast, naphthoquinones are potent antilipolytic agents but actually increase cyclic-AMP accumulation[64]. These data suggest that there is little correlation between the effect of insulin on lipolysis and on cyclic-AMP accumulation. One possible explanation is that measurements of the total cyclic-AMP pool provide little information about the physiologically active pool of cyclic-AMP. If this is so, then the finding that under some conditions insulin can affect cyclic-AMP accumulation does not mean that this is the mechanism by which it affects lipolysis and glucose metabolism.

Possibly insulin has its own intracellular messenger which counteracts the action of cyclic-AMP. Larner[6] has mentioned this possibility, as have Robison, Butcher and Sutherland in their monograph on cyclic-AMP[66]. Perhaps such a second messenger is either a metabolite of cyclic-AMP, AMP or ATP whose formation is accelerated by insulin. The main problem is that there is no proof that such a second messenger exists to act as an intracellular mediator for insulin.

1.6.1.2 Muscle, cyclic-AMP and glycogen metabolism

In muscle as in fat cells, no effect of insulin on basal levels of cyclic-AMP has been reported. Previous exposure to insulin does decrease the maximum stimulation of cyclic-AMP accumulation by catecholamines in both muscle and fat cells[61, 67]. However, under other conditions, no effect of insulin on cyclic-AMP accumulation in rat diaphragm muscle can be seen either after *in vivo* or *in vitro* exposure in either the absence or presence of epinephrine[68].

Nuttall[69] has reviewed the mechanisms by which insulin accelerates the deposition of glycogen in the liver and Larner and Villar-Palasi[70] the factors regulating glycogen synthase. Gemmill[71, 72] first demonstrated that insulin affected the metabolism of muscle *in vitro* resulting in increased glucose uptake and glycogen deposition. Gemmill and Hamman[73] found that under certain conditions as much as 90% of the extra glucose taken up in the presence of insulin was converted to glycogen. Krahl[5] and Nuttall[69] have summarised the evidence that the effect of insulin on glycogen synthesis is not secondary to the increased transport of glucose.

Beloff-Chain *et al.*[74] were among the first to suggest that insulin specifically stimulated a pathway to glycogen. Villar-Palasi and Larner[75] in 1961 reported that insulin treatment resulted in an increase in the percentage of glycogen synthase present in the active state; insulin did not affect the total activity of the enzyme as measured in the presence of glucose-6-phosphate. The effect of insulin is seen in the complete absence of glucose[76] and in both skeletal muscle and heart[69]. The mechanism by which insulin increased glycogen synthase activity has not yet been elucidated, but the effect can be seen without any detectable change in the activities of the phosphorylase enzyme system or the concentration of cyclic-AMP[69]. Apparently, the major effect of insulin on glycogen synthase activity is via cyclic-AMP-independent pathways. The major possibilities are for insulin to reduce the activity of the protein kinase which converts the active to the inactive enzyme, or to increase the activity of the protein phosphatase which converts the inactive form of glycogen synthase to the active form. Evidence has been presented in support of both possibilities[69]. Insulin could also affect the reactivity of glycogen synthase with either the protein kinase or phosphatase. Whether insulin regulates the formation of its own hypothetical second messenger or whether it acts by some other mechanism is not clear.

1.6.1.3 Insulin, cyclic-AMP and hepatic glucose output

Convincing data which demonstrate that insulin action of muscle glycogen, fat-cell lipolysis or glucose transport in any cell is mediated via cyclic-AMP is not yet at hand. However, the strongest case for insulin action via cyclic-AMP is probably antagonism of glucagon-induced glucose release from the liver.

There is now ample evidence that insulin directly inhibits glucose output by the liver[77,78]. Glucagon, catecholamines and exogenous cyclic-AMP are potent stimulators of glucose output by the perfused rat liver and activate both glycogenolysis and gluconeogenesis[77,78]. Insulin antagonises the increase in cyclic-AMP accumulation and output of glucose brought about by low concentrations of glucagon or catecholamines in perfused livers[77,78]. Insulin also antagonises the action of added cyclic-AMP[77,78]. The available evidence suggests that the inhibition by insulin of the increased glucose production, due to agents which activate adenylate cyclase, is related to a reduction in hepatic cyclic-AMP. Whether this is due to inhibition by insulin of glucagon-activated adenylate cyclase[57,58], or activation of phosphodiesterase[50], is not yet clear. Possibly both mechanisms are involved; this would explain the ability of insulin to inhibit the stimulation of glucose output by the liver brought about by added cyclic-AMP[79]. However, insulin did not inhibit the activation of glucose output in perfused liver by dibutyryl cyclic-AMP or cyclic-3′,5′-GMP[77-79]. Whether this represents a difference between the mechanisms by which these nucleotides stimulate hepatic glucose output and that of added cyclic-AMP is not clear. A further complication is that in isolated liver cells the small rise in cyclic-AMP found after the addition of catecholamines is apparently unrelated to the stimulation of gluconeogenesis by catecholamines[135].

There are other effects of insulin on the liver besides antagonism of cyclic-AMP accumulation and glucose output. Glucagon and insulin have opposite effects on hepatic urea production and K^+ ion release which are not readily explained by their effects on cyclic-AMP. Insulin inhibits both of these processes under conditions in which glycogenolysis is unaffected by insulin[77,79]. Glinsmann and Mortimore[79] found that, in the presence of enough exogenous cyclic-AMP to completely block the inhibitory effect of insulin on glucose output by perfused livers, the inhibition by insulin of K^+ ion efflux was still seen.

In preparations of isolated rat liver parenchymal cells from starved rats, insulin increased the net accumulation of glycogen and decreased glucose output[80]. The effects of insulin could be overcome by glucagon or added dibutyryl cyclic-AMP[80].

The hepatic cyclic-AMP content is elevated in livers from alloxan-diabetic rats[81] but not in those from pancreatectomised dogs[82]. The elevated cyclic-AMP content in the livers of alloxan-diabetic animals may be due more to the presence of agents such as glucagon, which elevate cyclic-AMP, than to the absence of insulin. There is also an elevated cyclic-AMP content in livers from starved rats[83] or in rats injected with insulin antiserum[81]. To date, no clear effect of insulin on basal levels of cyclic-AMP has been seen in any tissue. However, the basal levels of cyclic-AMP may include a large component of physiologically-inactive cyclic-AMP.

1.6.2 Stimulation of membrane transport

1.6.2.1 Hexoses

Lundsgaard[84] first published studies in 1939 which suggested that transport might be the rate-limiting step in glucose uptake by perfused hind limbs, and that insulin accelerated this process. Levine and associates[85] found an increase due to insulin in the disappearance of blood galactose, a hexose which is not appreciably utilised by peripheral tissues. They suggested that insulin affected a hexose-transport system in muscle. Subsequent studies have amply confirmed the hypothesis that insulin accelerates hexose transport in a wide variety of tissues[86-91]. The localisation of the hexose-transport system in the plasma membrane may have influenced Levine[92] to propose in 1965 that insulin interacts with the membrane, and that this initiates a number of signals which lead both to the stimulation of glucose transport and the other effects of insulin which are not secondary to increased glucose uptake. This is still the most reasonable hypothesis to explain insulin action.

Morgan et al.[87] using muscle, and Crofford and Renold[88] using adipose tissue, have shown that glucose utilisation is limited by transport, and insulin increases the transport of glucose. The carrier-mediated transport system involved in hexose uptake involves passive rather than active transport. This facilitated diffusion process can be separated from simple diffusion by its saturation kinetics, stereospecificity, competition between pairs of hexoses for transport and counter-transport[90].

Very little is known about the biochemical properties of the hexose-transport system in any mammalian tissue, and even less about the mechanisms by which insulin might affect this system. Recently, Martin and associates[91,93] have been able to demonstrate effects of insulin on the rate of glucose release by plasma membrane vesicles isolated from fat-cell homogenates. Insulin facilitated both the uptake and release of D-glucose but not of L-glucose. The effect of insulin could only be seen if added to fat cells prior to homogenisation, and as yet direct effects of insulin on membrane vesicles have not been reported[91,93,138].

1.6.2.2 Structural changes in membrane structure in relation to stimulation of hexose uptake by insulin

In fat cells, a wide variety of agents which alter membrane structure and function appear to mimic the stimulation of glucose transport seen with insulin[91,94]. Most of these agents exert biphasic effects in that while low concentrations (which are rather high as compared to insulin) stimulate glucose oxidation, high concentrations generally depress glucose oxidation and cellular metabolism. Rodbell[95] first reported that commercial preparations of phospholipase C stimulated glucose metabolism in fat cells and suggested that selective hydrolysis of membrane phospholipids results in a structural change in the configuration of the plasma membrane. However, it is possible that contaminants in the phospholipase C preparations were responsible for the observed effects[96]. Subsequent studies have shown that almost any agent which alters membrane structure will increase glucose oxidation at low concentrations and inhibit at higher concentrations[91,94]. It is not clear whether any insight into insulin action will come from comparing its effects to those of other membrane-active agents.

The effects of insulin are apparently very subtle for no marked conformational changes in the proteins of plasma membrane vesicles isolated from insulin-treated, as opposed to control, adipose tissue cells could be detected by infrared spectroscopy or fluorescent probe studies[91,138]. While these procedures can often detect large-scale membrane transitions, they may not be sensitive enough to see hormone–membrane interactions.

Incubation of fat cells with neuraminidase preparations diminishes the response of fat cells to insulin[26,96]. However, the response of fat cells to cysteine was also abolished by neuraminidase but not by trypsin[96]. Rosenthal and Fain[96] suggested that the effects of neuraminidase preparations might be due to contaminants, and in any case the net result was a non-specific inhibition of metabolic responsiveness. The binding of insulin to fat cells was unaffected by neuraminidase preparations[26]. These data suggest that sialic acid residues are not required for insulin binding or specifically involved in insulin action.

Trypsin treatment of adipose tissue or fat cells selectively destroyed the response of fat cells to insulin and glucagon[25,97-99,136]. Studies by Crofford and Okayama[24], and subsequently by others[10,25], have indicated that the inhibition of insulin action by trypsin is directly proportional to its ability to reduce insulin binding. This is in contrast to studies with neuraminidase[26] or phospholipase[25] preparations where insulin binding is unaffected, or even enhanced, under conditions in which cellular metabolism is depressed. The conclusion from studies with trypsin is that it acts at the first step in insulin action to cleave exposed peptide groups of the insulin receptors on the surface of intact cells. Interestingly, trypsin treatment of intact fat cells results in a marked inhibition of the insulin-degrading activity of membrane preparations prepared from the cells[100].

The early studies of Kono[98] suggested that there was a rapid regeneration of insulin receptors after trypsin treatment. More recently, Kono and Barham[10] and El-Allawy and Gliemann[136] found smaller rates of insulin-receptor regeneration. The process of recovery is difficult to see, and in other laboratories no detectable recovery of insulin receptors was seen after trypsin treatment[25, 96]. It is hoped that conditions can be found under which recovery of insulin receptors after trypsin treatment can be readily observed. Then the question of whether recovery represents unmasking of hidden receptors or *de novo* synthesis of new receptors can be answered.

1.6.2.3 Amino acid transport

The main difference between hexose and amino acid transport in mammalian tissues is that the latter involves transport against a concentration gradient[90]. The so-called uphill transport of amino acids is accomplished by linking the uptake of sodium to that of amino acids[137].

Kipnis and Noall[101] found that insulin facilitated the accumulation of α-amino-isobutyric acid into rat diaphragm muscle. The uptake of some natural amino acids is enhanced by insulin, but that of others is unaffected[102]. There is ample evidence that the effects of insulin on protein synthesis are not secondary to increased amino-acid uptake[90, 102, 103]. The reverse is also true, for the stimulation by insulin of amino acid uptake is enhanced under conditions in which protein synthesis was blocked with puromycin[104].

In thymocytes[139] and foetal rat calvaria[140], it has been found that the stimulation of α-amino-isobutyric acid transport by insulin is unrelated to cyclic-AMP. In fact, the addition of cyclic-AMP, or agents which stimulate cyclic-AMP accumulation, mimics the action of insulin on amino acid transport. In rat thymocytes, insulin had no affect on either basal cyclic-AMP values or the increase seen after addition of prostaglandin E_1[139].

1.6.3 Effects on protein metabolism

1.6.3.1 Protein synthesis in muscle

The stimulation by insulin of protein synthesis in muscle is independent of its effects on either hexose or amino-acid uptake[103, 105]. Wool and his associates have found that insulin treatment of alloxan-diabetic rats increased muscle protein synthesis by accelerating the translation of messenger-RNA[103, 105]. They suggested that insulin altered the structure of ribosomes in a way that increased ribosomal initiation of protein synthesis. This conclusion was based on the finding that the muscle ribosomes of diabetic animals were less efficient in catalysing protein synthesis than the ribosomes of normal animals. The defect could be corrected within 5 min of the administration of insulin. No effect of insulin added to any cell-free system involved in protein synthesis could be observed. The effect of insulin was non-specific with respect to the nature of the proteins made by muscle ribosomes.

There was no readily discernible difference between ribosomes from normal and diabetic animals with respect to their protein or RNA content[103]. There was a difference in the thermal denaturation curves for normal as compared to diabetic muscle ribosomes[103]. The defective amino-acid incorporation of diabetic ribosomes could be traced to a smaller content of bound peptidyl tRNA. Whether this results from a defective capacity for initiation of endogenous protein synthesis, or simply from a decreased binding capacity of diabetic ribosomes, is not known.

Morgan et al.[106] have recently utilised perfused rat heart and muscle preparations from normally fed or starved rats to examine the effect of insulin on protein synthesis. They found that, initially, peptide-chain elongation was the rate-limiting step in protein synthesis, but that with time a block in peptide-chain initiation developed. The reduction in amino acid incorporation and the increase in ribosomal sub-units could be prevented by the presence of insulin[106]. Their measurements of intracellular amino acid accumulation suggested that the effects of insulin were not secondary to increased amino acid transport. The perfused muscle should be an excellent system for studies of insulin action on protein synthesis and it is hoped lead to further understanding of the mechanism by which insulin facilitates peptide-chain initiation.

Manchester[102] has pointed out that the defect in initiation of protein synthesis seen in skeletal muscle of alloxan-diabetic rats may not be due to a direct lack of insulin but to some indirect action of insulin. He suggested the possibility that the defective protein synthesis might be secondary to increased accumulation of intracellular free fatty acids in muscle, for lipolysis is elevated in acute alloxan-diabetic rats. Insulin is able to reduce intracellular free fatty acids both by inhibiting lipolysis and by increasing the re-esterification of fatty acids.

There is evidence that the defective protein synthesis in muscle of diabetic animals involves more than effects on ribosomes. The ability of the soluble cell sap (pH 5 fraction) to support protein synthesis in cell-free systems in the presence of ribosomes was impaired in diabetic animals[102]. This defect was corrected by prior treatment of the animals with insulin[102].

1.6.3.2 Protein degradation in muscle

Luck and associates[107, 108] first reported that insulin decreased the amino acid content in plasma as well as that in muscle. Mirsky[109] found that insulin markedly decreased the net release of amino acids by muscle in eviscerated-nephrectomised dogs, increased the uptake by muscle of exogenous amino-acids and reduced hepatic deamination of amino acids. Similar effects of insulin on amino-acid release were seen in eviscerated rats[110]. Russell[111] pointed out that insulin primarily affected nitrogen metabolism via direct effects on muscle and other insulin-responsive peripheral tissue. The early reports of an increased removal of amino acids by muscle in the presence of insulin, and of the stimulatory effects of insulin on protein synthesis, resulted in a relative neglect of the hypothesis that part of the insulin effect on amino acid release by muscle might involve inhibition of proteolysis. More recently, it has been suggested that, in muscle, insulin also inhibits proteolysis[112, 113].

1.6.3.3 Protein degradation in liver

The major acute effect of insulin on protein metabolism in the adult rat liver is an anticatabolic action. For some time it has been known that insulin increases the incorporation of label into protein from radioactive amino acids in liver slices[5]. John and Miller[114] have examined the effect of insulin on the incorporation of labelled amino acids into protein and net synthesis of certain proteins by rat livers perfused for 12 h[114]. They found that in the presence of insulin there was a reduction in urea nitrogen but no increase in synthesis of specific plasma proteins such as fibrinogen, acid glycoprotein, acute phase globulin or haptoglobin. In contrast, the incorporation of label into liver protein was enhanced.

Mortimore and Mondon[115] have pointed out that protein degradation in the perfused liver is nearly always greater than protein biosynthesis. They found that insulin produced a general reduction in perfusate amino acids, ureogenesis and a 50% drop in the release of free valine. If appropriate corrections were made for the effect of insulin on the specific activity of labelled valine introduced into the perfusion system, no effect of insulin on valine incorporation into protein was observed[115]. This study suggested that most, if not all, of the stimulation by insulin of the incorporation of labelled amino acids into liver protein was due to an increase in the specific activity of the precursor pool of valine used for protein synthesis. This amply illustrates the problems that can be involved in equating uptake of label with protein synthesis, as was pointed out by Russell[116] in 1958.

Mortimore and Mondon[115] labelled the hepatic proteins with valine and then measured the release of label by perfused rat livers. Insulin decreased the release of radioactive valine; this indicated an inhibition of proteolysis. They attributed the ability of insulin to inhibit breakdown of hepatic protein to a stabilisation of liver lysosomes[115]. There was an increase in the osmotic fragility of liver lysosomes during perfusion which could be prevented by the presence of insulin. Similar effects of insulin have also been seen on perfused hearts[106].

There are long-term effects of insulin administration to experimental animals on the activities of many hepatic enzymes. Weber[117] has pointed out that insulin increases the activity of some enzymes and decreases the activity of others. He has referred to this as a reprogramming of gene expression; but these long-term effects of insulin appear to be more of an adaptive response to effects of insulin on carbohydrate, lipid and protein metabolism. The available evidence indicates that insulin increases the biosynthesis of only certain specific hepatic proteins These may be such a small fraction of total protein synthesis in the liver that effects of insulin will have to be examined by isolating specific proteins and determining the changes in amount and incorporation of label. One protein whose activity declines in the liver during fasting is glucokinase[118]. The increase in liver glucokinase seen after re-feeding is dependent on the presence of insulin[118]. Pilkis[118] suggested that the insulin-induced increase in glucokinase was a secondary action due to effects on metabolites.

In cultured hepatoma cells, a direct effect of insulin could be seen on the synthesis of tyrosine aminotransferase (L-tyrosine: 2-oxo-glutarate aminotransferase, E.C. 2.6.1.5) which did not reuqire the presence of amino acids or glucose in the medium[119, 120]. The synthesis of this enzyme was also accelerated by glucocorticoids and blocked by inhibitors of RNA synthesis, while the increased synthesis due to insulin was unaffected by these inhibitors. The insulin effect is apparently a translational stimulation of synthesis which is unrelated to cyclic-AMP, as the hepatomas used have little adenylate cyclase or cyclic-AMP.

1.6.3.4 Insulin and protein synthesis in adipose tissue

In fat cells, insulin appears to primarily increase protein synthesis. Minemura, Lacy and Crofford[121] found no effect of insulin on the uptake of natural amino acids or of a-amino-isobutyric acid in fat cells. There was a small inhibition by insulin of fat-cell proteolysis, but this effect was not large enough to account for the stimulation by insulin of amino acid incorporation into protein[121].

The stimulation of protein synthesis by insulin was attributed to a reduction in cyclic-AMP content of fat cells[121]. However, insulin stimulates protein synthesis in the absence of agents which increase cyclic-AMP accumulation[55] and these agents do not inhibit protein synthesis except at very high concentrations. Furthermore, the increase due to insulin in labelled leucine incorporation into fat-cell protein was actually greater in the presence of low concentrations of lipolytic agents[55]. Fain and Rosenberg[55] suggested that the ability of lipolytic agents at high concentrations to inhibit insulin-mediated increases in glucose metabolism and protein synthesis was subsequent to the accumulation of intracellular fatty acids. This was based on the finding that added fatty acids mimicked the effects of high concentrations of lipolytic agents[55].

In isolated pieces of adipose tissue incubated for 16 h or more, there was an increase in hexokinase II activity in the presence of insulin[122, 123]. The effect of insulin was blocked by inhibitors of RNA and protein synthesis. In the studies on the induction of tyrosine aminotransferase, the effect of insulin is seen after only 30 min incubation and is not blocked by d actinomycin, an inhibitor of RNA synthesis. Neither are the effects of insulin on *in vitro* incorporation of label into protein by rat diaphragm blocked by dactinomycin[124]. The only effects of insulin on protein synthesis which are blocked by this agent are those which require a 16–24 h lag period[117, 123]. It may be that over such long periods of time there are other effects of dactinomycin besides inhibition of RNA synthesis or, more importantly, the turnover of certain RNA molecules may be sufficiently rapid for them to be depleted over such time periods. Apparently the stimulation by insulin of protein synthesis is not due to a transcriptional action of the hormone on DNA-dependent RNA synthesis. Rather, insulin acts at a post-transcriptional site, and even here, the stimultion due to insulin might be due to changes in metabolite levels.

1.6.4 Insulin, membrane potentials and ion fluxes

One of the earliest known effects of insulin was the reduction in plasma K^+ ion content[125]. There is ample evidence that the reduction in plasma K^+ ion due to insulin is the result of net K^+ ion uptake by muscle, adipose tissue and liver[126]. This is related to the ability of insulin to cause hyperpolarisation in these tissues even in the absence of glucose. Zierler[126] has suggested that the effect of insulin on K^+ ion uptake by cells is secondary to an insulin-induced increase in the electrical potential difference across the plasma membrane. The apparent effect of insulin is to make the membrane less permeable to sodium. A good candidate for this effect would be the Na^+-K^+-activated, ouabain-inhibited ATPase of muscle. But Rogus, Price and Zierler[127] failed to find any effect of insulin on this enzyme in broken-cell preparations. However, others have reported a small, but consistent, stimulation of this enzyme by exposure to insulin in lymphocytes[128]. How insulin action would affect ion permeability is not known, but it is possible that many other actions of insulin, perhaps even glucose transport, are secondary to changes in ion flux across the membrane.

1.7 SUMMARY

The present article has summarised some of the more recent findings with regard to the mode of action of insulin, and it is disappointing to note that

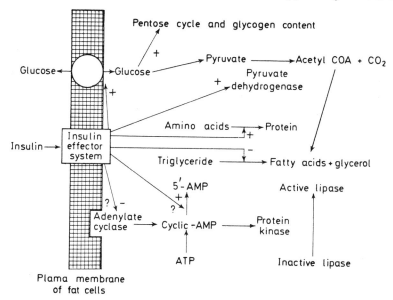

Figure 1.2 Action of insulin on fat cells. The figure summarises the major actions of insulin on fat cells with the + signs representing stimulation, the − signs inhibition by insulin and the ? mark those effects about which there are still questions. The insulin effector system is considered as the complex of insulin bound to its receptor in the plasma membrane. The model is designed to illustrate the many actions resulting from this effector system. Some appear to be associated with the membrane (adenylate cyclase and glucose transport) while others are clearly not associated with the plasma membrane (pyruvate dehydrogenase). Not all the effects of insulin are shown, e.g. such as those on membrane potential

the answer is not yet at hand. We know a considerable amount about the nature of the interaction between insulin and its receptor in the plasma membrane; the next few years should result in great progress in this area. We do not know how the insulin–receptor complex mediates the many actions of insulin. Probably these are all secondary to changes in the activity of a variety of enzymes associated with the plasma membrane.

A summary of some of the many actions of insulin on one target cell (fat cell) is shown in Figure 1.2. This figure emphasises a unitary hypothesis for insulin action involving a primary interaction of insulin with its receptor in the plasma membrane. This results in a propagated disturbance which affects the activity of many membrane-bound enzymes and also intracellular processes. The ability of insulin to stimulate pyruvate dehydrogenase[129-132] illustrates an action on mitochondrial reactions. Whether this is mediated through release of a second messenger for insulin, a reduction in intracellular free fatty acids, or some unknown process remains to be established.

References

1. (1972). Proceedings Fiftieth Anniversary Insulin Symposium, Indianapolis, Indiana, published in *Diabetes*, **21**, Suppl. 2
2. (1972). Impact of insulin on metabolic pathways, *Israel J. Med. Sci.*, **8**, Parts 3 and 6
3. (1972). *Insulin Action* (I. B. Fritz, editor) (New York: Academic Press)
4. (1972). *Handbook of Physiology*, Section 7, *Endocrinology*, Vol. 1 (*Endocrine Pancreas*) (D. F. Steiner and N. Freinkel, editors) (Washington: American Physiological Society)
5. Krahl, M. E. (1961). *The Action of Insulin on Cells* (New York: Academic Press)
6. Larner, J. (1972). *Diabetes*, **21**, 428
7. Stadie, W. C., Haugaard, N., Marsh, J. B. and Hills, A. G. (1949). *Amer. J. Med. Sci.*, **218**, 265
8. Narahara, H. T. (1972). *Handbook of Physiology*, Section 7, *Endocrinology*, Vol. 1, (*Endocrine Pancreas*), 333 (D. F. Steiner and N. Freinkel, editors) (Washington: American Physiological Society)
9. Freychet, P., Roth, J. and Neville, D. M., Jr. (1971). *Proc. Nat. Acad. Sci. USA*, **68**, 1833
10. Kono, T. and Barham, F. W. (1971). *J. Biol. Chem.*, **246**, 6210
11. Gavin, J. R., III, Roth, J., Jen, P. and Freychet, P. (1972). *Proc. Nat. Acad. Sci. USA*, **69**, 747
12. Cuatrecasas, P. (1971). *Proc. Nat. Acad. Sci. USA*, **68**, 1264
13. House, P. D. R. (1971). *FEBS Lett.*, **16**, 339
14. Cuatrecasas, P. (1972). *Proc. Nat. Acad. Sci. USA*, **69**, 318
15. Cuatrecasas, P. (1972). *J. Biol. Chem.*, **247**, 1980
16. Cuatrecasas, P. (1972). *Proc. Nat. Acad. Sci. USA*, **69**, 1277
17. Cuatrecasas, P. (1971). *J. Biol. Chem.*, **246**, 7265
18. Cuatrecasas, P. (1971). *J. Biol. Chem.*, **246**, 6522
19. Pilkis, S. J., Johnson, R. A. and Park, C. R. (1972) *Diabetes*, **21**, 335 (abstract)
20. Gliemann, J. and Gammeltoft, S. (1972). *Abstracts IV International Congress of Endocrinology*, 198 (Amsterdam: Excerpta Medica)
21. Kono, T. (1972). *Insulin Action*, 171 (I. B. Fritz, editor) (New York: Academic Press)
22. Kahn, R., Freychet, P., Neville, D. M., Jr. and Roth, J. (1972). *Diabetes*, **21**, 334 (abstract)
23. Crofford, O. B. (1968). *J. Biol. Chem.*, **243**, 362
24. Crofford, O. B. and Okayama, T. (1970). *Diabetes*, **19**, 369 (abstract)
25. Cuatrecasas, P. (1971). *J. Biol. Chem.*, **246**, 6532
26. Cuatrecasas, P. and Illiano, G. (1971). *J. Biol. Chem.*, **246**, 4938
27. Bennett, G. V. and Cuatrecasas, P. (1972). *Science*, **176**, 805

28. Livingston, J. N., East, L. E. and Lockwood, D. H. (1972). *Abstracts IV International Congress of Endocrinology*, 166 (Amsterdam: Excerpta Medica)
29. Kahn, C. R., Neville, D. M., Jr., Gorden, P., Freychet, P., and Roth, J. (1972). *Biochem. Biophys. Res. Commun.*, **48**, 135
30. Sanger, F. and Tuppy, H. (1951). *Biochem. J.*, **49**, 463 and 481
31. Hodgkin, D. C. and Mercola, D. (1972). *Handbook of Physiology-Endocrinology, Endocrine Pancreas*, 139 (D. F. Steiner and N. Freinkel, editors) (Washington: American Physiological Society)
32. Blundell, T. L., Cutfield, J. F., Cutfield, S. M., Dodson, E. J., Dodson, G. G., Hodgkin, D. C. and Mercola, D. A. (1972). *Diabetes*, **21**, 492
33. Hodgkin, D. C., Blundell, T. L., Cutfield, J. F., Cutfield, S. M., Dodson, G. G., Dodson, E. J., Mercola, D. A. and Vijayan, M. (1972). *Insulin Action*, 1 (I. B. Fritz, editor) (New York: Academic Press)
34. Zahn, H., Brandenburg, D. and Gattner, H. G. (1972). *Diabetes*, **21**, 468
35. Jost, K., Rudinger, J., Klostermeyer, H. and Zahn, H. (1968). *Z. Naturforsch.*, **23b**, 1059
36. Arquilla, E. R. and Stanford, E. J. (1972). *Insulin Action*, 29 (I. B. Fritz, editor) (New York: Academic Press)
37. Arquilla, E. R., Miles, P. V. and Morris, J. W. (1972). *Handbook of Physiology-Endocrinology, Endocrine Pancreas*, 159 (D. F. Steiner and N. Freinkel, editors) (Washington: American Physiological Society)
38. Freychet, P., Roth, J. and Neville, D. M., Jr. (1971). *Biochem. Biophys. Res. Commun.*, **43**, 400
39. Cuatrecasas, P. (1969). *Proc. Nat. Acad. Sci. USA*, **63**, 450
40. Blatt, L. M. and Kim, K. H. (1971). *J. Biol. Chem.*, **246**, 4895
41. Turkington, R. W. (1970). *Biochem. Biophys. Res. Commun.*, **41**, 1362
42. Oka, T. and Topper, Y. J. (1971). *Proc. Nat. Acad. Sci. USA*, **68**, 2066
43. Cuatrecasas, P. (1972). *Insulin Action*, 137 (I. B. Fritz, Editor) (New York: Academic Press)
44. Davidson, M. B., Gerschenson, L. E. and Van Herle, A. J. (1972). *Diabetes*, **21**, 335 (abstract)
45. Fritz, I. B. (1972). *Biochemical Actions of Hormones*, Vol. 2, 165 (G. Litwack, editor) (New York: Academic Press)
46. Armstrong, K. J., Noall, M. W. and Stouffer, J. E. (1972). *Biochem. Biophys. Res. Commun.*, **47**, 354
47. Bessman, S. P. (1972). *Israel J. Med. Sci.*, **8**, 344
48. Borrebaek, B. and Spydevold, Ø. (1969). *Diabetologia*, **5**, 42
49. Krahl, M. E. (1972). *Insulin Action*, 461 (I. B. Fritz, editor) (New York: Academic Press)
50. House, P. D. R., Poulis, P. and Weidemann, M. J. (1972). *Eur. J. Biochem.*, **24**, 429
51. Loten, E. G. and Sneyd, J. G. T. (1970). *Biochem. J.*, **120**, 187
52. Vaughan, M. (1972). *Insulin Action*, 297 (I. B. Fritz, editor) (New York: Academic Press)
53. Zinman, B. and Hollenberg, C. H. (1972). *Abstracts IV International Congress of Endocrinology*, 157 (Amsterdam: Excerpta Medica)
54. Clarke, L. J., Jungas, R. L. and Goodman, H. M. *Endocrinology*, in press.
55. Fain, J. N. and Rosenberg, L. (1972). *Diabetes*, **21**, 414
56. Ward, W. F., personal communication
57. Hepp, K. D. and Renner, R. (1972). *FEBS Lett.*, **20**, 191
58. Illiano, G. and Cuatrecasas, P. (1972). *Science*, **175**, 906
59. Jungas, R. L. and Ball, E. G. (1963). *Biochemistry*, **2**, 383
60. Fain, J. N., Kovacev, V. P. and Scow, R. O. (1966). *Endocrinology*, **78**, 773
61. Butcher, R. W., Baird, C. E. and Sutherland, E. W. (1968). *J. Biol. Chem.*, **243**, 1705
62. Jungas, R. L. (1966). *Proc. Nat. Acad. Sci. USA*, **56**, 757
63. Jarett, L., Steiner, A. L., Smith, R. M. and Kipnis, D. M. (1972). *Endocrinology*, **90**, 1277
64. Fain, J. N. (1971). *Mol. Pharmacol.*, **7**, 465
65. Fain, J. N., Pointer, R. H. and Ward, W. F. (1972). *J. Biol. Chem.*, **247**, 6866
66. Robison, G. A., Butcher, R. W. and Sutherland, E. W. (1971). *Cyclic-AMP*, 298 (New York: Academic Press)

67. Craig, J. W., Rall, T. W. and Larner, J. (1969). *Biochim. Biophys. Acta,* **177,** 213
68. Goldberg, N. D., Villar-Palasi, C., Sasko, H. and Larner, J. (1967). *Biochim. Biophys. Acta,* **148,** 665
69. Nuttall, F. Q. (1972). *Handbook of Physiology-Endocrinology, Endocrine Pancreas,* 395 (D. F. Steiner and N. Freinkel, editors) (Washington: American Physiological Society)
70. Larner, J. and Villar-Palasi, C. (1971). *Current Topics in Cellular Regulation,* Vol. 3, 195 (B. Horecker and E. W. Stadtman, editors) (New York: Academic Press)
71. Gemmill, C. L. (1940). *Bull. Johns Hopkins Hosp.,* **66,** 232
72. Gemmill, C. L. (1941). *Bull. Johns Hopkins Hosp.,* **68,** 329
73. Gemmill, C. L. and Hamman, L., Jr. (1941). *Bull. Johns Hopkins Hosp.,* **68,** 50
74. Beloff-Chain, A., Catanzaro, R., Chain, E. B., Masi, I., Pocchiari, F. and Rossi, C. (1955). *Proc. Roy. Soc. (London),* **B143,** 481
75. Villar-Palasi, C. and Larner, J. (1961). *Arch. Biochem. Biophys.,* **94,** 436
76. Craig, J. W. and Larner, J. (1964). *Nature (London),* **202,** 971
77. Mortimore, G. E. (1972). *Handbook of Physiology-Endocrinology, Endocrine Pancreas,* 495 (D. F. Steiner and N. Freinkel, editors) (Washington: American Physiological Society)
78. Exton, J. H. and Park, C. R. (1972). *Handbook of Physiology-Endocrinology, Endocrine Pancreas,* 437 (D. F. Steiner and N. Freinkel, editors) (Washington: American Physiological Society)
79. Glinsmann, W. H. and Mortimore, G. E. (1968). *Amer. J. Physiol.,* **215,** 553
80. Johnson, M. E. M., Das, N. M., Butcher, F. R. and Fain, J. N. (1972). *J. Biol. Chem.,* **247,** 3229
81. Jefferson, L. S., Exton, J. H., Butcher, R. W., Sutherland, E. W. and Park, C. R. (1968). *J. Biol. Chem.,* **243,** 1031
82. Bishop, J. S., Goldberg, N. D. and Larner, J. (1971). *Amer. J. Physiol.,* **220,** 499
83. Park, C. R., Lewis, S. B. and Exton, J. H. (1972). *Diabetes,* **21,** 439
84. Lundsgaard, E. (1939). *Uppsala Läkareforen. Forh.,* **45,** 143
85. Levine, R., Goldstein, M., Huddlestun, B. and Klein, S. (1950). *Amer. J. Physiol.,* **163,** 70
86. Levine, R. (1961). *Diabetes,* **10,** 421
87. Morgan, H. E., Henderson, M. J., Regen, D. M. and Park, C. R. (1961). *J. Biol. Chem.,* **236,** 253
88. Crofford, O. B. and Renold, A. E. (1965). *J. Biol. Chem.,* **240,** 14
89. Narahara, H. T. and Cori, C. F. (1968). *Carbohydrate Metabolism and its Disorders,* Vol. 1, 375 (F. Dickens, D. J. Randle, and W. J. Whelan, editors) (London: Academic Press)
90. Morgan, H. E. and Neely, J. R. (1972). *Handbook of Physiology-Endocrinology, Endocrine Pancreas,* 323 (D. F. Steiner, and N. Freinkel, editors) (Washington: American Physiological Society)
91. Avruch, J., Carter, J. R. and Martin, D. F. (1972). *Handbook of Physiology-Endocrinology, Endocrine Pancreas,* 545 (D. F. Steiner and N. Freinkel, editors) (Washington: American Physiological Society)
92. Levine, R. (1965). *Fed. Proc.,* **24,** 1071
93. Carter, J. R., Jr., Avruch, J. and Martin, D. B. (1972). *J. Biol. Chem.,* **247,** 2682
94. Fain, J. N. (1973). *Pharmacol. Rev.,* **25,** 67
95. Rodbell, M. (1966). *J. Biol. Chem.,* **241,** 130
96. Rosenthal, J. W. and Fain, J. N. (1971). *J. Biol. Chem.,* **246,** 5888
97. Kono, T. (1969). *J. Biol. Chem.,* **244,** 1772
98. Kono, T. (1969). *J. Biol. Chem.,* **244,** 5777
99. Fain, J. N. and Loken, S. C. (1969). *J. Biol. Chem.,* **244,** 3500
100. Crofford, O. B., Rogers, N. L. and Russell, W. G. (1972). *Diabetes,* **21,** 403
101. Kipnis, D. M. and Noall, M. W. (1958). *Biochim. Biophys. Acta,* **28,** 226
102. Manchester, K. L. (1972). *Diabetes,* **21,** 447
103. Wool, I. G., Castles, J. J., Leader, D. P. and Fox, A. (1972). *Handbook of Physiology-Endocrinology, Endocrine Pancreas,* 385 (D. F. Steiner and N. Freinkel, editors) (Washington: American Physiological Society)
104. Scharff, R. and Wool, I. G. (1965). *Biochem. J.,* **97,** 272
105. Wool, I. G., Stirewalt, W. S., Kurihara, K., Low, R. B., Bailey, P. and Oyer, D. (1968). *Rec. Prog. Horm. Res.,* **24,** 139

106. Morgan, H. E., Rannels, D. E., Wolpert, E. B., Giger, K. E., Robertson, J. W. and Jefferson, L. S. (1972). *Insulin Action*, 437 (I. B. Fritz, editor) (New York: Academic Press)
107. Luck, J. M., Morrison, G. and Wilbur, L. F. (1928). *J. Biol. Chem.*, **77,** 151
108. Luck, J. M. and Morse, S. W. (1933). *Biochem. J.*, **27,** 1648
109. Mirsky, I. A. (1938). *Amer. J. Physiol.*, **124,** 569
110. Frame, E. G. and Russell, J. A. (1946). *Endocrinology*, **39,** 420
111. Russell, J. A. (1955). *Fed. Proc.*, **14,** 696
112. Pozefsky, T., Felig, P., Tobin, J. D., Soeldner, J. S. and Cahill, G. F., Jr. (1969). *J. Clin. Invest.*, **48,** 2273
113. Cahill, G. F., Jr., Aoki, T. T. and Marliss, E. B. (1972). *Handbook of Physiology–Endocrinology, Endocrine Pancreas*, 563 (D. F. Steiner and N. Freinkel, editors) (Washington: American Physiological Society)
114. John, D. W. and Miller, L. L. (1969). *J. Biol. Chem.*, **244,** 6134
115. Mortimore, G. E. and Mondon, C. E. (1970). *J. Biol. Chem.*, **245,** 2375
116. Russell, J. A. (1958). *Perspect. Biol. Med.*, **1,** 138
117. Weber, G. (1972). *Israel J. Med. Sci.*, **8,** 325
118. Pilkis, S. J. (1970). *Biochim. Biophys. Acta*, **215,** 461
119. Gelehrter, T. D. and Tomkins, G. M. (1970). *Proc. Nat. Acad. Sci. USA*, **66,** 390
120. Lee, K.-L., Reel, J. R. and Kenney, F. T. (1970). *J. Biol. Chem.*, **245,** 5806
121. Minemura, T., Lacy, W. W. and Crofford, O. B. (1970). *J. Biol. Chem.*, **245,** 3872
122. Hansen, R. J., Pilkis, S. J. and Krahl, M. E. (1967). *Endocrinology*, **81,** 1397
123. Hansen, R. J., Pilkis, S. J. and Krahl, M. E. (1970). *Endocrinology*, **86,** 57
124. Eboue-Bonis, E., Chambaut, A. M., Volfin, P. and Clauser, H. (1963). *Nature (London)*, **199,** 1183
125. Briggs, A. P., Koechig, I., Doisy, E. A. and Weber, C. J. (1924). *J. Biol. Chem.*, **58,** 721
126. Zierler, K. L. (1972). *Handbook of Physiology–Endocrinology, Endocrine Pancreas*, 347 (D. F. Steiner and N. Freinkel, editors). (Washington: American Physiological Society)
127. Rogus, E., Price, T. and Zierler, K. L. (1969). *J. Gen. Physiol.*, **54,** 188
128. Hadden, J. W., Hadden, E. M., Wilson, E. E., Good, R. A. and Coffey, R. G. (1972). *Nature (New Biol.)*, **235,** 174
129. Jungas, R. L. (1970). *Endocrinology*, **86,** 1368
130. Coore, H. G., Denton, R. M., Martin, B. R. and Randle, P. J. (1971). *Biochem. J.*, **125,** 115
131. Weiss, L., Löffler, G., Schirmann, A. and Wieland, O. (1971). *FEBS Lett.*, **15,** 229
132. Wieland, O. H., Patzelt, C. and Löffler, G. (1972). *Eur. J. Biochem.* **26,** 426
133. Hammond, J., Jarett, L., Mariz, I. K. and Daughaday, W. H. (1972). *B.B.A.C.*, **49,** 1122
134. Krebs, E. G. (1972). Current Topics in Cellular Regulation, Vol. 5, 99
135. Tolbert, M. E. M., Butcher, F. R., and Fain, J. N. (1973). *J. Biol. Chem.*, **248,** 5686
136. El-Allawy, R. M. M. and Gliemann, J. (1972). *Biochim. Biophys. Acta*, **273,** 97
137. Schultz, S. G. and Curran, P. F. (1970). *Physiol. Rev.*, **50,** 637
138. Avruch, J., Carter, J. R. and Martin, D. B. (1972). *Biochim. Biophys. Acta*, **288,** 27
139. Goldfine, I. D. and Sherline, P. (1972). *J. Biol. Chem.*, **247,** 6927
140. Phang, J. M. and Downing, S. K. (1973). *Amer. J. Physiol.*, **224,** 191
141. Tuari, S., Saito, Y., Suzuki, F. and Takeda, Y. (1972). *Endocrinology*, **91,** 1442
142. Suzuki, F., Daikuhara, Y., Ono, M. and Takeda, Y. (1972). *Endocrinology*, **90,** 1220

2
Modes of Action of ACTH*

G. SAYERS, R. J. BEALL and S. SEELIG†
Case Western Reserve University, Cleveland

2.1	INTRODUCTION	26
2.2	INTERACTION OF ACTH WITH RECEPTORS ON THE PLASMA MEMBRANE AND/OR ACTIVATION OF INTRACELLULAR ENZYMES	28
	2.2.1 Biological activity of ACTH covalently linked to large molecules	28
	2.2.2 'Binding' of ACTH to receptors	28
	2.2.3 Nature of the receptor	29
2.3	THE EVENTS IN THE PLASMA MEMBRANE	29
2.4	CYCLIC-AMP: INTERMEDIARY OF THE ACTIONS OF ACTH	29
	2.4.1 ACTH induces cyclic-AMP accumulation	30
	2.4.1.1 Time course of cyclic-AMP accumulation	30
	2.4.1.2 Cyclic-AMP accumulation in response to various doses of ACTH	31
	2.4.1.3 Distribution of cyclic-AMP between cells and medium: suspensions of isolated adrenal cortex cells	32
	2.4.2 Cyclic nucleotides mimic the actions of ACTH	34
	2.4.2.1 Corticosterone production in response to cyclic-AMP and dibutyryl cyclic-AMP	34
	2.4.2.2 Analogues of cyclic-AMP	35
	2.4.2.3 Cyclic-GMP, cyclic-IMP and cyclic-UMP	36
	2.4.3 Methylxanthines and other inhibitors of phosphodiesterase	36
2.5	MODES OF ACTION OF CYCLIC-AMP	37
	2.5.1 Steroidogenesis	39
	2.5.1.1 Activation of a protein kinase	39
	(a) Activation of cholesterol esterase	40
	(b) Synthesis of protein (protein X)	40

* Supported in part by National Science Foundation Grant GB 27426
† Predoctoral Fellow, US Public Health Service Training Grant 5 TO1 GM-00899

26   BIOCHEMISTRY OF HORMONES

2.5.2	Cell growth and division	40

2.6 STEROID BIOSYNTHETIC PATHWAYS   41
    2.6.1 *Morphology: integration of structure and function*   41
    2.6.2 *Cholesterol sources: cholesterol synthesis; plasma cholesterol; cholesterol in lipid droplets*   41
    2.6.3 *Integration of structure and function*   41
    2.6.4 *Cytochrome P450*   45
    2.6.5 *Regulation of steroidogenesis*   46
        2.6.5.1 *Sterol-carrier protein*   46
        2.6.5.2 *Cholesterol side-chain cleavage*   46

2.7 RELATION OF STRUCTURE OF ACTH ANALOGUES TO BIOLOGICAL ACTIVITY   48
    2.7.1 *Characterisation of ACTH analogues*   49
    2.7.2 *Nature of the receptor*   51

2.8 CALCIUM   52
    2.8.1 *Calcium and the plasma membrane*   52
    2.8.2 *Calcium and steroidogenesis*   54
    2.8.3 *Calcium as a messenger*   54

2.9 'SENSITIVITY', 'AMPLIFICATION'   55

2.10 CONCLUDING REMARKS   56

ACKNOWLEDGEMENTS   57

2.1 INTRODUCTION

Adrenocorticotropic hormone (ACTH) is a polypeptide synthesised and released by the adenohypophysis for the purpose of regulating the rate of secretion of corticosteroids by the adrenal cortex. According to the prevailing point of view, ACTH interacts with a receptor on the surface of the plasma membrane resulting in increased production of cyclic-AMP; the cyclic nucleotide ('second messenger' of Sutherland) mediates the steroidogenic and growth-promoting activities of the polypeptide hormone. From this point of view, the mode of action is confined to the generation of a signal on the receptor surface. However, there are some investigators who believe that cyclic AMP is neither the sole nor the obligatory mediator of the various actions of ACTH. The evidence will be considered in the body of the text. To orientate the reader, we present an overall view of the sequence of events from the interaction of ACTH with its receptor to the secretion of steroid hormones (see Figure 2.1). ACTH interacts with a receptor on the outer surface of the plasma membrane[1] to generate a signal which is transduced and amplified as it passes through the plasma membrane[2]. Adenylate cyclase,

MODES OF ACTION OF ACTH

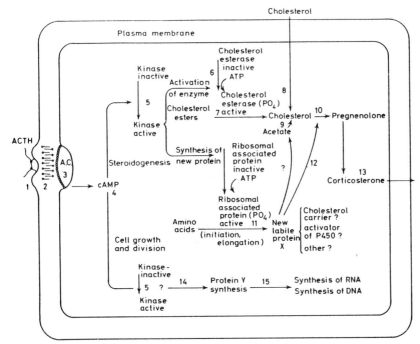

Figure 2.1 Provisional model of the action of ACTH and of cyclic-AMP in promoting steroidogenesis, cell growth and division in cells of the adrenal cortex. (See text for explanation.)

which is probably located on the inner surface of the membrane, is activated[3] and production of cyclic-AMP is enhanced[4]. Cyclic-AMP stimulates steroidogenesis and promotes cell growth and cell division by mechanisms currently under investigation. Cyclic-AMP activates a kinase or kinases[5]. Activated kinase may phosphorylate a cholesterol esterase[6] which then accelerates the rate of conversion of cholesterol fatty-acid esters in the lipid droplets of the adrenal cortex cell to free cholesterol[7]. The free cholesterol, which makes up the 'active cholesterol pool', may also come from the plasma[8] or be synthesised in the adrenal cortex cell[9]. The rate-limiting step in the biosynthetic pathway is the conversion of cholesterol to pregnenolone[10]. It is believed that cyclic-AMP induces synthesis of a labile protein (protein X) at the level of translation[11]. Protein X accelerates the conversion of cholesterol to pregnenolone possibly by binding and carrying the cholesterol to its site of side-chain cleavage on the mitochondrial P450 system. Once pregnenolone is formed, the subsequent steroid conversions occur at a rapid rate. In the case of the rat, corticosterone is formed; in other species (as in man), corticosterone and cortisol are formed. The means by which corticosterone is transported across the plasma membrane is unknown[13].

Activation of a kinase[5] is believed by certain investigators to promote cell growth and division through induction of protein (protein Y)[14], RNA and DNA[15] synthesis. This kinase may or may not be the same as that involved in steroidogenesis.

2.2 INTERACTION OF ACTH WITH RECEPTORS ON THE PLASMA MEMBRANE AND/OR ACTIVATION OF INTRACELLULAR ENZYMES

Neither ACTH nor cyclic-AMP has been shown to stimulate steroidogenesis in a broken-cell preparation of adrenal tissue. There have been claims to the contrary[1,2] but the experimental observations upon which the claims are based have not been reproduced in other laboratories. Evidence has been brought forward to support the notion that ACTH 'binds' to a receptor on the surface of the plasma membrane; the hormone does not enter the cell. The cyclic-AMP produced in response to the activation of adenylate cyclase in the plasma membrane mediates the actions of the hormone. If this is the actual state of affairs, then the mode of action of ACTH is strictly limited to the interaction with receptors on the cell surface. Subsequent events concern the varied actions of cyclic-AMP in the intracellular milieu.

2.2.1 Biological activity of ACTH covalently linked to large molecules

ACTH linked to polymers by covalent bonds exhibits steroidogenic activity when added to cultures of adrenal cortex tumour cells[3] or to suspensions of dispersed isolated adrenal cortex cells[4]. If one can be certain that the preparations contain no trace of free ACTH and that the covalent bond is stable during the course of incubation with intact cells, then the conclusion is justified that ACTH can act *without* entering the cell. The demonstration does not exclude other actions, but at least raises doubts as to the importance or even the existence of intracellular sites of action of the polypeptide hormone.

2.2.2 'Binding' of ACTH to receptors

A preparation of $ACTH_{1-39}$ labelled with radioactive iodine was shown to be biologically potent and to bind to a plasma membrane fraction of an adrenal tumour-cell homogenate[5]. The capacity of an ACTH analogue to displace labelled ACTH was related to its biological potency. Hofmann *et al.*[6] prepared $[^{14}C\text{-Phe}^9][\text{Glu}^5]ACTH_{1-20}$ in which tryptophane at position 9 has been replaced by ^{14}C-labelled phenylalanine and glutamic acid at position 5 by glutamine. The compound exhibited 2% of the steroidogenic potency of $ACTH_{1-24}$. The labelled compound was bound by a particulate fraction from beef adrenal cortical tissue. Among a series of analogues tested, ability to displace $[^{14}C\text{-Phe}^9][\text{Glu}^5]ACTH_{1-20}$ from its binding sites correlated well with biological potency.

In the opinion of the reviewers, the matter of the biological potency of iodine-labelled ACTH has not been settled. In our experience, iodination induces a drastic reduction in the potency of ACTH. However, there is the possibility that iodinated ACTH, even if biologically inactive, binds to the ACTH receptor site.

That unlabelled ACTH, but not other proteins or polypeptides, will displace iodinated ACTH from plasma membrane fragments is of considerable importance for the assay of ACTH[7].

2.2.3 Nature of the receptor

The chemical nature of the receptor in the plasma membrane remains to be defined. Isolation and characterisation will be difficult, not only on account of the chemical complexity of the entity but probably more importantly because of the loss of its 'biological namecard'.

Receptor is presently defined as an object which interacts with ACTH to induce a response. We can isolate and identify the receptors by their affinity and selectivity for ACTH molecules — their 'binding' properties. However, 'binding' properties will not provide sufficient identification. The component parts of the complex system must be reconstituted and demonstrated to respond to the addition of ACTH with a physical or chemical change of functional significance. A few hints as to certain general features of the receptor may be gained from an analysis of the relation of structure to biological activity among analogues of ACTH which enter into a complementary relationship with receptor. This subject will be taken up in greater detail in Section 2.7 of this chapter.

2.3 THE EVENTS IN THE PLASMA MEMBRANE

This is an area where our knowledge is woefully lacking and we can speak only in the broadest of terms. The evidence available suggests that adenylate cyclase is not the receptor[8]. The interaction of ACTH and a receptor is said to generate a signal which may relate to a change in the conformation of the membrane components. The signal, whatever its nature, is 'transduced' and 'amplified' as it traverses the plasma membrane to activate adenylate cyclase. Calcium may be involved in the transmission and amplification of this signal[9]. In this connection, Rodbell and associates[10, 11] have provided evidence which suggests that GTP magnifies the signal generated when glucagon interacts with its receptor on the plasma membrane of liver cells and which is characterised by the enhanced activation of adenylate cyclase.

2.4 CYCLIC-AMP: INTERMEDIARY OF THE ACTIONS OF ACTH

According to the second-messenger theory as proposed by Sutherland and Rall, a hormone induces a response in the target tissue by stimulating the production of cyclic-AMP via activation of adenylate cyclase[12]. Cyclic-AMP, formed in the cell, acts upon the metabolic machinery to elicit the biological response characteristic of the particular target tissue involved. Two monographs which describe the cyclic-AMP-mediated actions of a number of hormones have recently been published[13, 14].

Certain criteria should be met if one is to conclude that ACTH acts via the mediation of cyclic-AMP in the adrenal cortex[15]. First, ACTH should induce a measurable change in the level of cyclic nucleotide in the cell, and this increase should occur at least as soon as steroidogenesis is observed. The cyclic-AMP response should also be related to the dose of ACTH used. Second, when added to adrenocortical tissue, cyclic-AMP or its analogues should mimic the action of the first messenger, ACTH. Third, inhibitors of phosphodiesterase, the only enzyme known to be involved in the metabolism of the cyclic nucleotides, should accelerate the steroidogenic action of ACTH on the adrenal cortex. Fourth, the ability of analogues of ACTH to stimulate steroidogenesis should be reflected in their ability to increase the levels of cyclic-AMP.

2.4.1 ACTH induces cyclic-AMP accumulation

Haynes[16], using bovine adrenal cortex slices, was the first to demonstrate that ACTH causes a significant increase in tissue levels of cyclic-AMP. Subsequent studies, in which ACTH was added to incubated rat adrenal quarters[15], or to the medium perfusing cat adrenals[17], or injected into the whole animal[15,18], demonstrated that the hormone increases cyclic-AMP accumulation and that the increase precedes the biological response of the cell.

2.4.1.1 Time course of cyclic-AMP accumulation

Isolated cells, prepared by the trypsin digestion of rat adrenal quarters, are free of the diffusion barriers that are inherent in other *in vitro* preparations. It is evident from Figure 2.2 that significant accumulation of cyclic-AMP

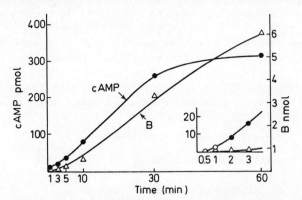

Figure 2.2 Cyclic-AMP accumulation and corticosterone (B) production with time after addition of 10 000 pg ACTH$_{1-39}$ to aliquots (0.9 ml, cell count 325 000) from a single pool of isolated adrenal cells incubated at 37°C in an atmosphere of 95% O_2 and 5% CO_2. The changes are net; quantities of cyclic AMP and of B in aliquots to which no ACTH was added (blank) have been subtracted. The insert at the lower right displays the changes for the first 3 min on an expanded scale. (From Beall and Sayers[19], by courtesy of Academic Press.)

occurred within 1 min of exposure of adrenal cells to 10 000 pg ACTH, whereas corticosterone production was not detectable until after 3 min of incubation. This observation supports the thesis that cyclic-AMP mediates the steroidogenic action of ACTH on the adrenal cortex.

2.4.1.2 Cyclic-AMP accumulation in response to various doses of ACTH

A relation between the dose of ACTH and the rates of production of cyclic-AMP and corticosteroid has been reported by Grahame-Smith et al.[15] and Beall and Sayers[19]. Figure 2.3 summarises the results obtained on studies using isolated rat adrenal cells. The alterations in the concentration of cyclic-AMP and of corticosterone in adrenal cell suspensions (cells and medium) exposed to various doses of ACTH may be described as follows: (a) low doses of ACTH (50–250 pg) stimulate steroidogenesis without causing detectable changes in the concentration of cyclic-AMP, (b) intermediate doses of ACTH

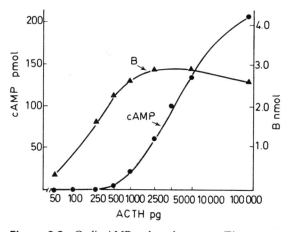

Figure 2.3 Cyclic-AMP and corticosterone (B) accumulation after addition of various doses of ACTH to aliquots (0.9 ml, cell count 300 000) from a single pool of isolated adrenal cells incubated at 37°C for 60 min. Changes are net; quantities of cyclic-AMP and of B in aliquots to which no ACTH was added (blank) have been subtracted. (After Beall and Sayers[19], by courtesy of Academic Press.)

(500–2500 pg) induce parallel increases in cyclic-AMP and corticosterone accumulation and (c) large doses of ACTH (2500–100 000 pg) cause additional increases in the concentration of cyclic-AMP without causing further increases in corticosterone accumulation.

It appears that the adenylate cyclase complex need only proceed at about 15–20% of its maximum rate to induce maximum steroidogenesis. The significance of this excess of receptor–adenylate complexes ('spare receptors') has already been discussed[15, 19, 20] and will be discussed further in Section 2.9 of this chapter.

The observation that low doses of ACTH cause steroidogenesis without detectable increases in cyclic-AMP is not unique to the adrenal cortex. In the perfused rat liver system, glucagon stimulated glycogenolysis at a concentration of 5×10^{-10} mol l^{-1}, whereas significant stimulation of intracellular cyclic-AMP accumulation required 4×10^{-10} mol l^{-1} [21]. In the epidymal fat pad, epinephrine at a dose of 0.1 µg ml^{-1} of incubation medium induced a barely significant increase in the concentration of cyclic-AMP in the medium but increased the release of free fatty acids into the medium at a rate of 30% of V_{max}[22]. One might interpret these observations to mean that low doses of ACTH act independently of cyclic-AMP production to bring about the biological response of the cell. We are inclined to the view that cyclic-AMP is an obligatory intermediary, and that at low doses of the hormone increases in cyclic-AMP do occur but at discrete loci.

Observations on perfused liver have been considered to be consistent with the view that glucagon causes an increase in 'free functionally-active' cyclic-AMP and that under basal conditions practically all the cyclic-AMP is in the inactive form (bound, sequestered or inhibited)[21]. The basal concentration of cyclic-AMP in the adrenal cortex cells is 2×10^{-6} mol l^{-1} which is at least 100 times the K_m value for the cyclic-AMP activation of the adrenal protein kinase[23].

The problem relating to the distribution of 'active' and 'inactive' forms of cyclic-AMP within the cell are complex and await resolution.

2.4.1.3 Distribution of cyclic-AMP between cells and medium: suspension of isolated adrenal cortex cells

In our original studies, cyclic-AMP was measured in the total incubation mixture, i.e. cells plus medium. We have now examined the distribution of cyclic-AMP with time following the addition of various doses of ACTH. The accumulation of cyclic-AMP in the cells and in the medium and the accumulation of corticosterone in the medium following the addition of 10^5 pg of ACTH are shown in Figure 2.4. The quantity of cyclic-AMP in the cells increased 20-fold over basal level during the first 15 min of incubation and then declined. Cyclic-AMP accumulated in the medium at a constant rate up to 30 min of incubation, then slowed to a slight extent over the next 30 min. Corticosterone production was constant from 5 to 60 min of incubation despite the intracellular decrease of cyclic-AMP after 15 min. It is evident that large quantities of the nucleotide accumulate in the medium relative to the quantity in the cells. Estimates of the concentration suggest that the intracellular concentration of cyclic-AMP is higher than the extracellular concentration even at the end of incubation when the quantity of cyclic-AMP in the incubation medium is very much greater than that in the cells. It is apparent that the efflux of the cyclic nucleotide from the cells is not against a concentration gradient. A similar relationship of ACTH on the distribution of cyclic-AMP and corticosteroid production has been shown for all doses of ACTH above 500 pg ACTH[24]. Theophylline at a concentration of 1×10^{-3} mol l^{-1} had to be present in order to demonstrate intracellular changes with 100 pg ACTH; however, the dynamics of the intra- and extra-cellular responses in the

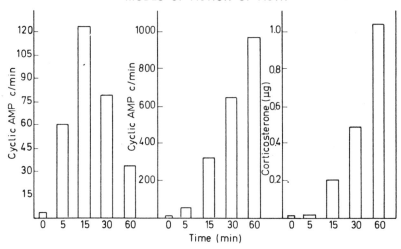

Figure 2.4 Accumulation of cyclic-[8-^{14}C]AMP in the cells, left panel; cyclic-[8-^{14}C]AMP in the medium, middle panel; and corticosterone production, right panel. These values were obtained following the addition of 1×10^5 pg ACTH$_{1-39}$ to aliquots (0.9 ml, cell count 175 000) of cell suspensions. Cells, pre-labelled with adenine, were incubated with and without ACTH$_{1-39}$ for the times shown. After incubation, the cells were separated from the incubation medium by filtration. Cyclic-[8-^{14}C]AMP and corticosterone were analysed according to the method previously described by Beall and Sayers[19]. The values shown are the quantity of cyclic-[8-^{14}C]AMP in the cells or medium and the quantity of corticosterone in a single incubate of cells. The mean values shown are net; quantities of cyclic-AMP and of B in aliquots to which no ACTH was added (blank) have been subtracted

presence of theophylline were similar to those observed with higher doses of ACTH without theophylline. The isolated adrenal cortex cells behave like isolated fat cells[25, 26] and the *in vitro* fat pad[22] in that intracellular cyclic-AMP levels increase and then decrease during the period in which the biological response (release of free fatty acids into the medium) proceeds at a constant rate. The fat cells are distinguished by the fact that relatively little cyclic-AMP appears in the medium even in the presence of methylxanthines. The addition of phosphodiesterase to the suspension of isolated adrenal cortex cells maintained the level of cyclic-AMP in the medium at near zero during the course of incubation. However, the steroidogenic action of ACTH was not modified. It appears that the nucleotide in the medium has no functional role and is simply a reflection of the rate of production. Isolated adrenal cells appear to have little, if any, phosphodiesterase activity[27]. This suggests that the extrusion of cyclic-AMP may be an important property of the cell designed to control intracellular levels of the nucleotide. Active transport does not need to be invoked in order to account for the efflux since the concentration of cyclic-AMP in the cells was always much greater than that in the medium. Furthermore, probenecid, an inhibitor of cyclic-AMP efflux from avian erythrocytes[28], does not modify the rate of efflux of cyclic-AMP in isolated adrenal cells. A factor, which has already been described, inhibits cyclic-AMP accumulation with progressively increasing intensity after addition of the hormone to isolated fat cells[25, 29]. This same factor, if present in adrenal cortex, may act to increase the permeability of the plasma membrane of the cells to the nucleotide.

2.4.2 Cyclic nucleotides mimic the actions of ACTH

2.4.2.1 *Corticosterone production in response to cyclic-AMP and dibutyryl cyclic-AMP*

The steroidogenic action of ACTH can be mimicked by the addition of cyclic-AMP or its dibutyryl derivative to bovine adrenal slices[30], rat adrenal quarters[31-34], isolated rat adrenal cells[35-38] and superfused rat adrenals[39]. The relation between the rate of secretion of corticosterone produced and the logarithm of the number of moles of dibutyryl cyclic-AMP or ACTH (pg) added to suspensions of isolated adrenal cells is shown in Figure 2.5. The

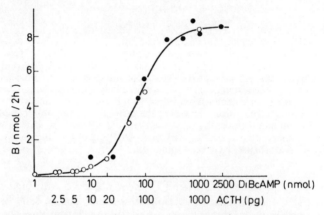

Figure 2.5 Net corticosterone (B) production (nmoles/2 h) plotted against dibutyryl cyclic-AMP (mol) or ACTH (pg) added to isolated cell suspensions. (After Sayers *et al.*[35], by courtesy of the Asscn. for Experimental and Biological Medicine.)

similarity between the log(dose)–response curves, together with the fact that the latent period of steroidogenesis in isolated adrenal cells is the same following the addition of dibutyryl cyclic-AMP or ACTH (Table 2.1), is strong evidence in support of the second-messenger theory. These data also suggest that the time before corticosterone accumulation becomes evident is to a major degree accounted for by events beyond the stimulation of adenylate cyclase.

Isolated adrenal-cell preparations are not unlike most other tissues, in that dibutyryl cyclic-AMP is clearly more potent than cyclic-AMP (Table 2.2). Dibutyryl cyclic-AMP at a concentration of 10^{-5} mol l^{-1} and cyclic-AMP at 10^{-3} mol l^{-1} induce significant increases in steroidogenesis by isolated adrenal cells. At a concentration of 10^{-3} mol l^{-1}, 5′-AMP, adenosine, or ATP induce a barely detectable increase in B production.

The increased potency of the dibutyryl derivative of cyclic-AMP over its parent compound is thought to be accounted for by differences in the rate at

Table 2.1 Time course of steroidogenesis in isolated adrenal cells in response to addition of ACTH and of dibutyryl cyclic-AMP

	\multicolumn{5}{c}{Time (min)}
	1	2	5	15	30
	\multicolumn{5}{c}{Net corticosterone (μg)}
ACTH (1×10^4 pg)	n.c.*	n.c.	0.14	0.35	0.72
Dibutyryl cyclic-AMP (1×10^{-3} mol l^{-1})	n.c.	n.c.	0.15	0.32	0.73

* n.c. = no significant change from blanks.

Table 2.2 Net corticosterone production by isolated adrenal cells in response to dibutyryl cyclic-AMP; cyclic AMP, 5′ AMP, adenosine and ATP
(After Sayers *et al.*[35], by courtesy of the Assen for Experimental and Biological Medicine.)

			Corticosterone production			
Conc. (mol l^{-1})	Dibutyryl cAMP	Conc. (mol l^{-1})	cAMP	5′-AMP	Adenosine	ATP
10^{-5}	0.54, 0.54*	10^{-5}	0.09, 0.09	0.06, 0.06	0.09, 0.06	0.06, 0.03
2×10^{-5}	1.44, 1.41	10^{-4}	0.12, 0.09	0.15, 0.15	0.12, 0.12	0.12, 0.15
4×10^{-5}	4.77, 4.83	10^{-3}	1.29, 1.23	0.12, 0.12	0.06, 0.09	0.09, 0.12

* Each value represents the quantity of corticosterone produced in a single incubate of cells, nmol/2 h

which the two compounds penetrate lipid membranes and by the differences in the rate of hydrolysis by phosphodiesterase.

Addition of cyclic-AMP to adrenal tissue not only mimics the steroidogenic effect of ACTH but other effects of the hormone as well. For example, the subcutaneous administration of dibutyryl cyclic-AMP to hypophysectomised rats resulted in the partial maintenance of adrenal weight and content of DNA, RNA and protein[40]. These observations have been confirmed by a recent ultrastructural study of the effects of cyclic-AMP and ACTH on adrenocortical cells[41]. Cyclic-AMP enhances glucose oxidation in rat adrenal quarters[42] and induces ascorbic acid depletion in the rat adrenal cortex[43]. It has also been reported that cyclic-AMP, like ACTH, will stimulate the hydrolysis of cholesterol esters to cholesterol[44].

2.4.2.2 Analogues of cyclic-AMP

The biological activities of a series of analogues of cyclic-AMP have been studied[37, 45]. Most of the 8-substituted derivatives of cyclic-AMP (e.g. 8-chloro, 8-hydroxy, 8-amino and 8-methylthio) are more potent than the parent compound in stimulating steroidogenesis in isolated adrenal cells; several of the derivatives were more potent than dibutyryl cyclic-AMP. These same 8-substituted derivatives are highly potent activators of cyclic-AMP-dependent protein kinases[46] as well as being poor substrates for phosphodiesterase[46, 47]. Substitution at the 2'–0 positions resulted in decreased potency and in some cases a complete loss of activity[45]. Complete loss of activity was observed when arabinose was substituted for ribose. The availability of these and other analogues of cylic-AMP should provide investigators with useful tools with which to study the events that take place at the molecular level.

2.4.2.3 Cyclic-GMP, cyclic-IMP and cyclic-UMP

The cyclic nucleotides known to occur in Nature are cylic-AMP and cyclic-GMP[48]. The levels of cylic-GMP are approximately one-tenth those of cyclic-AMP in most tissues studied. Cyclic-GMP has been shown to stimulate steroidogenesis in adrenal quarters[31, 33] and in isolated adrenal-cell preparations[36, 38]; potency is about one-tenth that of cyclic-AMP. In most other systems studied, cyclic-GMP is less potent than cyclic-AMP in inducing a biological response, which is in line with the fact that cyclic-GMP is a less potent activator of protein kinases[49-51].

In the light of the relatively low levels in tissues, as well as the relatively weak potency of this cyclic nucleotide, it seems highly unlikely that cyclic GMP functions in the adrenal as a physiologically-important regulator of steroidogenesis. Guanyl cyclase, the enzyme responsible for the synthesis of cyclic-GMP from GTP, has been reported to be present in adrenal tissue, but the activity of the enzyme was not affected by concentrations of ACTH known to activate adenylate cyclase[52].

Cyclic-IMP has also been shown to stimulate steroidogenesis in adrenal preparations[33, 36, 38]. However, the actions of the compound may be accounted for by conversion to cyclic-AMP. Cyclic-UMP exerts a steroidogenic effect on rat adrenal quarters[33] but not on isolated adrenal-cell preparations[36, 38].

Nucleotides containing the 2',3'-cyclic monophosphate are inactive at concentrations at which the 3',5'-cyclic monophosphates exhibit maximum activity[36, 37]. In fact, 2',3'-cyclic-GMP and 2',3'-cyclic-UMP are potent inhibitors of ACTH and cyclic-AMP-induced steroidogenesis in isolated rat adrenal-cell preparations[36].

These inhibitors may be of value in analyses directed at the site and mode of action of cyclic-AMP.

2.4.3 Methylxanthines and other inhibitors of phosphodiesterase

The common methylxanthines, theophylline or caffeine, inhibit steroidogenesis, presumably by impairing protein synthesis[53].

Figure 2.6 shows the results of experiments in which ACTH alone or ACTH with 1×10^{-3} mol l^{-1} theophylline was added to incubates of isolated adrenal cells. Theophylline will accelerate the intracellular increase of cyclic-AMP but the acceleration is only 30%. Theophylline had no effect on the accumulation of cyclic-AMP in the medium; the presence of theophylline had

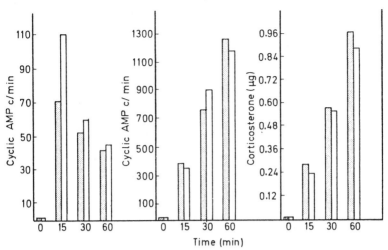

Figure 2.6 Effect of 5000 pg $ACTH_{1-39}$ in the presence of (open bars) and absence of (shaded bars) 1×10^{-3} mol l^{-1} theophylline on the cyclic-[8-^{14}C]AMP accumulation in the cells, left panel; cyclic-[8-^{14}C]AMP accumulation in the medium, middle panel; and corticosterone production, right panel. Cells, prelabelled with adenine, were incubated with and without $ACTH_{1-39}$ for the times shown. After incubation, the cells were separated from the incubation medium by filtration. Cyclic-[8-^{14}C]AMP and corticosterone were analysed according to the method previously described by Beall and Sayers[19]. Enough theophylline was dissolved in the vehicle or vehicle containing ACTH to give a final concentration of 1×10^{-3} mol l^{-1} in the incubates. The values shown are net; quantities of cyclic-[8-^{14}C]AMP and of B in aliquots over and above those in controls

a slight inhibitory effect on steroidogenesis. The small increase in intracellular cyclic-AMP observed with theophylline is compatible with the recent report that isolated adrenal cells contain little if any phosphodiesterase activity[27]. Phosphodiesterase inhibitors of greater potency than methylxanthines, and with little if any effect on protein synthesis, have recently been reported[54]. The compounds SQ 20006 and SQ 20009 were potent stimulators of steroidogenesis in ACTH- or cyclic-AMP-stimulated adrenal-cell preparations.

2.5 MODES OF ACTION OF CYCLIC-AMP

The evidence just cited suggests that cyclic-AMP is a mediator and possibly *the obligatory mediator* of the steroidogenic actions of ACTH on the adrenal cortex. We now turn to the question of the means by which cyclic-AMP stimulates steroidogenesis and promotes cell growth and division. Historically, the subject begins with the reports by Haynes and Berthet[55] and by Haynes[16] that ACTH stimulates cyclic-AMP production and activates glycogen phosphorylase when added to the medium of incubated bovine adrenal cortex slices. At this time, Rall *et al.*[56] reported that glucagon activates phosphorylase in cat liver homogenates through the mediation of cyclic-AMP. It appeared that cyclic-AMP acted on the adrenal cortex and on the liver by a common mechanism. The activation of phosphorylase was particularly attractive in the case of the adrenal cortex since it offered a means by which steroidogenesis could be regulated. Glycogenolysis would provide glucose for oxidation by the pentose shunt thus increasing the intracellular concentration of TPNH. As will be discussed more fully in the next section, steroid hydroxylations require TPNH; thus, according to Haynes and Berthet[55], cyclic-AMP increases the rate of steroidogenesis by increasing the concentration of TPNH. However, the TPNH generated would have to enter the mitochondria as well as the microsomes since TPNH requiring hydroxylations occur at both loci. Doubts as to the availability of TPNH for mitochondrial hydroxylations have been settled, at least for the bovine species, by the demonstration that a 'malate shuttle' can provide a means of transport of TPNH from the cytosol to the intramitochondrial space[57, 58]. The Haynes–Berthet thesis is an attractively simple explanation for the means by which cyclic-AMP regulates steroidogenesis. However, ACTH does not activate phosphorylase of rat adrenals[59, 60]. Furthermore, when rat adrenal quarters were incubated with ^{14}C-1-glucose, the specific activity of the $^{14}CO_2$ produced was not diminished by the addition of ACTH[61], a result to be expected if ACTH increases rate of glucose utilisation via the pentose shunt. Vance et al.[61] showed that steroidogenesis was stimulated by ACTH even after glycogen had been depleted. This difference between bovine and rat species casts doubts on the universal applicability of the Haynes–Berthet 'phosphorylase TPNH' thesis. Perhaps activation of glycogen phosphorylase can supplement supplies of TPNH under certain circumstances, but the quantitative importance of such a supplementary role is difficult to assess in the absence of information as to sources of TPNH other than that derived from glucose oxidation via the pentose shunt. At any rate, interest has turned away from activation of phosphorylase. A new development in 'mode-of-action' studies began with the demonstration by Ferguson[59] that puromycin inhibits the steroidogenic action of ACTH and of cyclic-AMP. It now appears that regulation of steroidogenesis involves *de novo* synthesis of a protein. Unfortunately, this protein has been most elusive when subjected to isolation and chemical characterisation procedures. Only a general statement can be made about its function, namely that it accelerates the conversion of cholesterol to pregnenolone. This same protein may induce cell growth and division, or perhaps another protein whose *de novo* synthesis is determined by cyclic-AMP serves these functions. Another development of some importance has been the work of Garren and Gill[23] and their colleagues on protein kinase(s) of the adrenal cortex.

The field has been moving at a rapid pace during the last 3 years, but we are not as yet at the point where a definitive mode or modes of action can be laid out step by step. Figure 2.1 is designed to assist the reader in locating individual contributions on a map of adrenal cortex functions. ACTH (steps 1, 2 and 3 of Figure 2.1) through the mediation of cyclic-AMP (step 4, Figure 2.1) regulates steroidogenesis (Section 2.5.1) and cell growth and division (Section 2.5.2). We present provisional models; each of which involves the conversion of inactive to active protein kinase (step 5, Figure 2.1). Whether one or more kinases subserve the various functions remains unresolved. Steroidogenesis is considered to be regulated by activation of an enzyme (cholesterol esterase) (step 6, Figure 2.1) and by synthesis of protein (Protein X, step 11, Figure 2.1). These models are discussed in Sections 2.5.1.1(a) and 2.5.1.1.(b) respectively.

Activation of a kinase is proposed as a means by which cell growth and division are regulated (steps 14 and 15, Figure 2.1) (Section 2.5.2).

2.5.1 Steroidogenesis

Of the mechanisms proposed to explain the steroidogenic action of cyclic-AMP at the molecular level, activation of protein kinase has received the greatest attention in recent years.

2.5.1.1 Activation of a protein kinase

Much of the current research in this field has been prompted by the discovery of a cyclic-AMP-dependent protein kinase in muscle tissue[62] which initiates a cascade of activation steps ending in the conversion of phosphorylase b to phosphorylase a[63]. Cyclic-AMP-dependent protein kinases appear to exist in all mammalian tissues, possibly in the tissues of all phyla of the animal kingdom and in bacteria[64-66]. The diverse effects of cylic-AMP may have a common mechanism, the activation of protein kinases[64, 65]. Specificity of the action would be determined by cellular substrate. For example, activation of a kinase increases the quantity of phosphorylase a[63] and decreases the activity of glycogen synthetase in muscle tissue[67-69]. A cyclic-AMP-dependent protein kinase is believed to be responsible for the phosphorylation and subsequent activation of lipases associated with adipose tissue[70, 71]. Garren and his co-workers have identified a protein-like substance in the cytosol and in the microsome fractions of an adrenal cortex homogenate that binds ^{3}H-cyclic-AMP[72]. The binding to this protein receptor appears to be specific. The only compound other than cold cyclic-AMP that displaces ^{3}H-cyclic-AMP from the receptor protein is cyclic-GMP, and this nucleotide is only 1% as efficient as cyclic-AMP. The binding constant, determined by equilibrium dialysis, and equal to the concentration of cyclic-AMP at which bound equals free, was reported to be 3×10^{-8} mol l^{-1}. Incidentally, the K_m value for cyclic-AMP activation of muscle protein kinase is equal to 1.7×10^{-8} mol l^{-1} [63]. Gill and Garren have identified a protein kinase in the adrenal cortex with a Km value for cyclic-AMP activation of histone phosphorylation equal to 1.4×10^{-8} mol l^{-1} [73]. The binding protein and the protein kinase had similar sub-cellular distributions in adrenal tissue fractions, most of the activity being associated with cytosol and microsomal fractions. Binding protein and kinase exhibited similar nucleotide specificity. Partial separation of binding protein and protein kinase activity has been accomplished by chromatography on DEAE; binding protein appears to form a complex with protein kinase[73].

Cyclic-AMP-dependent protein kinase of the adrenal cortex consists of a regulatory protein and a catalytic kinase unit[73, 74]. The regulatory protein bound to the catalytic kinase unit inhibits protein kinase activity. The activation of the protein kinase results from the interaction of cyclic-AMP with the regulatory protein. When cyclic-AMP binds to the regulatory protein, the regulatory protein and protein kinase dissociate. The dissociation results in an increase in protein kinase activity. The situation in the adrenal cortex is similar to that in skeletal muscle. Kinase and regulatory protein of that tissue have been separated chromatographically on casein–Sepharose columns in the presence of cyclic-AMP[75]. The following is a model for the adrenal cortex and for skeletal muscle:

Cyclic-AMP + regulatory protein · inactive catalytic unit ⇌ active catalytic unit + cyclic-AMP · regulatory protein

Reviews of the cyclic-AMP activation of protein kinases of adrenocortical tissue have appeared recently[23, 76, 77].

Casein, histone and protamine are substrates for the protein kinase(s) of the adrenal cortex. A key problem has been the identification of substrates having functional significance. Cholesterol esterase and proteins associated with ribosomes are likely candidates.

(a) *Activation of cholesterol esterase*—Simpson et al.[78], in a preliminary communication, report that the cholesterol esterase activity of a supernatant fraction (105 000 × g) increases following the addition of cyclic-AMP and ATP, as reflected in the increased rate of release of free cholesterol from labelled cholesteryl oleate. Prior cycloheximide treatment of the rat did not inhibit the cholesterol esterase activation by the cyclic nucleotide. Here for the first time is the demonstration of an action of cyclic-AMP in the adrenal cortex that does not not involve the synthesis of protein. The ATP requirement suggests that the protein kinase phosphorylates cholesterol esterase, as is true of the protein kinase which activates adipose tissue lipase[71].

The mobilisation of free cholesterol from stored cholesterol esters probably is not an obligatory step in the conversion of cholesterol to pregnenolone. This particular action of cyclic-AMP is probably merely supportive since free cholesterol can come from the plasma or be synthesised in the adrenal cortex.

(b) *Synthesis of protein (protein X)*—Cycloheximide and puromycin, but not actinomycin, inhibit steroidogenesis; from this it has been concluded that the action of cyclic-AMP in promoting synthesis is at the level of translation[59, 79, 80]. Protein X, the protein synthesised in response to cyclic-AMP, probably regulates steroidogenesis by accelerating the conversion of cholesterol to pregnenolone[23, 76, 81, 82]. However, no definite evidence is yet available to establish this point. We still await the isolation and chemical characterisation of the moiety.

Walton et al.[83] demonstrated that partially purified protein kinase from the cytosol fraction of adrenocortical tissue phosphorylates purified preparations of ribosomes. The degree of phosphorylation depended upon both cyclic-AMP and protein kinase. The K_m value for this activation (4×10^{-8} mol l^{-1} cyclic-AMP) was about the concentration required to induce the half-maximum rate of protamine phosphorylation.

We await the characterisation of the protein substrate associated with the ribosomes and definite evidence that this substrate is intimately involved in the synthesis of protein X.

2.5.2 Cell growth and division

In addition to its effect on steroidogenesis, ACTH plays an important role in maintaining adrenal weight. Thus, the hormone increases the protein and nucleic acid content of the gland. Cyclic-AMP has been shown to mimic the effect of ACTH in maintaining the DNA, RNA and protein content of the adrenals of hypophysectomised rats[40]. The increase in RNA[84] precedes the increase in DNA content[80]. Inhibitors of protein synthesis block the increase in adrenal RNA in response to ACTH stimulation[85]. A provisional model for the action of cyclic-AMP has been presented in Figure 2.1. Activation of kinase (step 5) promotes synthesis of protein Y which in turn enhances synthesis of RNA and of DNA[23, 77].

2.6 STEROID BIOSYNTHETIC PATHWAYS

Cholesterol is the precursor of the corticosteroids. The conversion involves a complex sequence of transformations and translocations of the steroid molecule. We shall attempt here to relate existing knowledge about function and ultrastructure. Special attention will be directed toward potentially significant sites or mechanisms of regulatory control in this complex array.

2.6.1 Morphology: integration of structure and function

For excellent reviews on the special morphology of steroid hormone secreting cells, the reader is referred to Christensen and Gillim[86] and to Fawcett et al.[87].

Steroid hormone-secreting cells, including the cells of the adrenal cortex, are distinguished by an abundance of smooth (agranular) endoplasmic reticulum, by large, oval, or rounded mitochondria which contain tubular or vesicular cristae rather than lamellar cristae and by a large number of lipid droplets whose chief component is cholesterol in the form of fatty-acid esters. Electron micrographs of cells of the adrenal cortex are presented in Figure 2.7(a), (b) and (c).

2.6.2 Cholesterol sources: cholesterol synthesis; plasma cholesterol; cholesterol in lipid droplets

The cholesterol pool which provides the mitochondrion with the 'active' precursor of pregnenolone may arise from (a) cholesterol synthesised from acetate in the smooth endoplasmic reticulum[90], (b) plasma cholesterol[91, 92], or (c) cholesterol stored in lipid droplets as ester (see Figure 2.1, step 7; Figure 2.7; Figure 2.8). The quantitative importance of these three sources is difficult to judge, particularly since the relative contributions may vary with the intensity and duration of stimulation of secretory activity by ACTH.

2.6.3 Integration of structure and function

It appears that steroid biosynthesis involves a complex of enzymes with distinct cellular loci, and that a steroid molecule during the course of its transformation must travel a tortuous course, back and forth among cellular

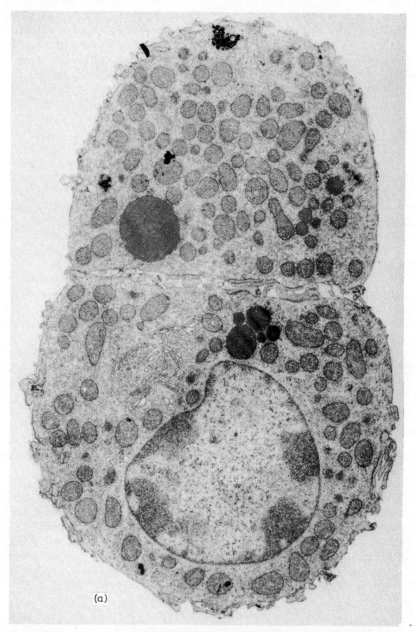

Figure 2.7 (a) Electron micrograph (× 12 000, reduced ⅔rds on reproduction) of a pair of adrenal cortex cells of the rat. The pair was in a suspension of cells dispersed by the trypsin technique of Sayers et al.[88]. Note the plasma membrane with microvilli, the extensive mass of smooth endoplasmic reticulum, the mitochondria with vesicular cristae, the lipid droplets, the Golgi apparatus and the nucleus

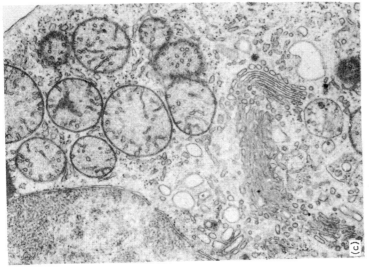

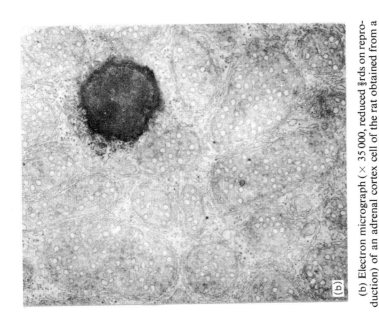

(b) Electron micrograph (× 35 000, reduced ⅔rds on reproduction) of an adrenal cortex cell of the rat obtained from a suspension of isolated adrenal cortex cells. Note the vesicular mitochondria surrounded by smooth endoplasmic reticulum and the free ribosomes dotted throughout the smooth endoplasmic reticulum. (From Malamed et al.[89] by courtesy of Springer, Berlin.)

(c) Electron micrograph (× 35 000, reduced ⅔rds on reproduction) of an isolated adrenal cortex cell of the rat. Note the tubular mitochondria, the Golgi apparatus and the free ribosomes dotted throughout the smooth endoplasmic reticulum organelles (see Figure 2.8). Cell fractionation suggests that the enzymes involved in the conversion of cholesterol to pregnenolone are located in the mitochondria.

Pregnenolone is transported out of the mitochondrion into the smooth endoplasmic reticulum to be converted to progesterone, 17α-hydroxyprogesterone, 11-desoxycorticosterone or 11-desoxy-17α-hydroxycorticosterone. Hydroxylations at positions 17 and 21, conversion of hydroxyl at 3 to a ketone and the shift of the double bond from the 5–6 to the 4–5 position are confined

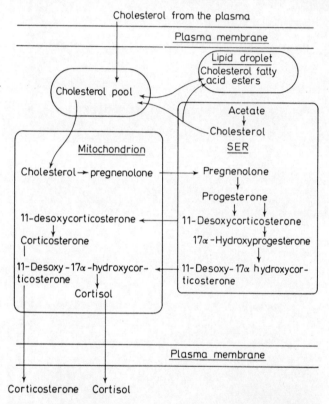

Figure 2.8 Synthesis and secretion of the corticosteroids. A 'cholesterol pool' which is most likely a small fraction of the total free cholesterol in the adrenal cortex cell is considered the immediate precursor of corticosterone and cortisol. This 'cholesterol pool' may be located within the mitochondrion, or it may be cholesterol in transit from the plasma, from the smooth endoplasmic reticulum (SER) or from the lipid droplets into the mitochondrion. Cholesterol in transit may be bound to a protein (cholesterol-binding protein). Within the mitochondrion, cholesterol is converted to pregnenolone which moves into the smooth endoplasmic reticulum wherein it is converted to 11-desoxycorticosterone or to 11-desoxy-17-hydroxycorticosterone. These intermediates move from the confines of the smooth endoplasmic reticulum back to the mitochondrion where the final, 11β-hydroxylation occurs, with the synthesis and secretion of corticosterone and of cortisol

to the smooth endoplasmic reticulum (the 'microsome' fraction of homogenate centrifugation). Hydroxylation at position 11 takes place in the mitochondrion. For an excellent review of the steroid transformations, see Samuels and Uchikawa[90]. In one of these forms, the steroid molecule re-enters the mitochondrion to be turned into the finished products corticosterone and cortisol*.

The cavity enclosed by the endoplasmic reticulum is a separate compartment and distinct from the 'cytosol'. The smooth endoplasmic reticulum may be the conduit for the transport of steroid molecules and enzyme activators. If so, the sequence of events associated with secretory activity may involve not only enzyme-catalysed conversions but a dynamic flow within the endoplasmic reticulum.

2.6.4 Cytochrome P450

The numerous steroid hydroxylations which occur in the course of the transformation of cholesterol to corticosterone and cortisol are catalysed by *mixed function oxidases*. The oxidases of the adrenal cortex are made up of a flavoprotein, a non-heme iron protein (adrenodoxin) and cytochrome P450. In the liver, these oxidases are confined to the microsomes, whereas in the adrenal cortex they are located in the mitochondria and the microsomes. An example of hydroxylation catalysed by mixed function oxidase is presented in Figure 2.9. The electron flow is from TPNH to flavoprotein, to adrenodoxin

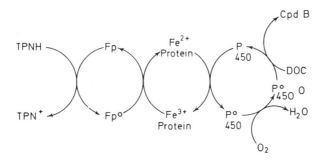

Figure 2.9 Electron transport sequence associated with the 11β-hydroxylation of deoxycorticosterone. The symbols Fp, Fe protein and P450 represent flavoprotein, non-haem iron protein (adrenodoxin) and cytochrome P450. (After Simpson and Estabrook[58].

and to P450. The cytochrome P450 is a terminal oxidase, one atom of oxygen of the medium entering the steroid nucleus the other being reduced to H_2O.

It is presumed that the number of hydroxylations are catalysed by an equal number of mixed function oxidases. However, purification of the mixed

* A word of caution is appropriate. Homogenisation may rupture organelles whose contents are now free to associate with various particles. An enzyme may be bound by an organelle with which it had no association in the living cell[93].

function oxidases is still at an elementary stage of development. Jefcoate and Boyd[94] have isolated from mitochondria two mixed function oxidases, one which catalyses 11β-hydroxylation, another which catalyses side-chain cleavage of cholesterol.

2.6.5 Regulation of steroidogenesis

2.6.5.1 Sterol-carrier protein

One possible mechanism of regulatory control over steroidogenesis may involve the synthesis of a cholesterol-binding protein[95] which transports cholesterol to the site of side-chain cleavage in the mitochondrion[78].

2.6.5.2 Cholesterol side-chain cleavage

The rate-limiting step in the conversion of cholesterol to the secretory products of the adrenal cortex is the biosynthesis of pregnenolone. The implication is that once pregnenolone production is increased the subsequent steps proceed at an accelerated rate simply on the basis of increase in the concentration of substrate intermediaries. The stimulatory action of ACTH and of cyclic-AMP has been localised to the conversion of cholesterol to pregnenolone by analyses of the rates of conversion of labelled precursors[96]. For this reason, the mixed function oxidase which catalyses side-chain cleavage of cholesterol is of particular interest to those workers interested in the regulation of steroidogenesis by ACTH. Side-chain cleavage is not a simple process. On the contrary, it involves two hydroxylation steps, one at position 20 the other at 22, and a cleavage by a desmolase to yield the 20-carbon pregnenolone. The overall reaction is designated side-chain cleavage.

Mitochondrial P450 binds cholesterol and pregnenolone. A characteristic spectral change occurs (type I difference spectrum) when cholesterol binds to the cytochrome[97]. When an excess of pregnenolone is now added, a type II difference spectrum develops which is interpreted as due to the displacement of bound cholesterol by pregnenolone.

Boyd and his co-workers[98] have isolated mitochondria from the adrenals of quiescent and from stressed rats. A comparison reveals that the mitochondria from stressed rats produce pregnenolone from endogenous cholesterol at an accelerated rate and contain P450 with a relatively high content of cholesterol. These changes do not occur if the rats are first treated with cycloheximide. Boyd and his co-workers[98] interpret these observations to mean that ACTH through the mediation of cyclic-AMP induces *de novo* synthesis of a protein whose function is to increase the rate of delivery of cholesterol to a locus on the P450 system at which side-chain cleavage occurs. Further developments of this fascinating work will be followed with interest. A hypothetical model of ACTH action incorporating the concepts of Boyd and his colleagues is presented in Figure 2.10.

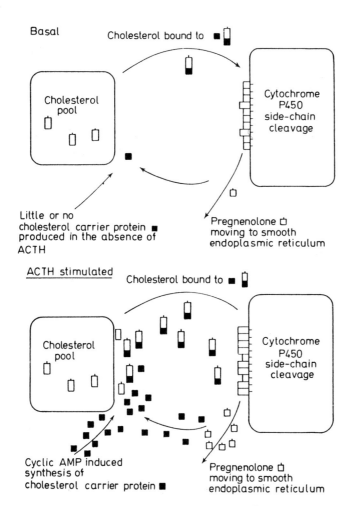

Figure 2.10 Hypothetical model for the accelerated rate of conversion of cholesterol to pregnenolone induced by ACTH via cyclic-AMP. The model considers the rate of side-chain cleavage of cholesterol to be determined by the rate at which cholesterol is transported from a cholesterol pool to a catalytic site on cytochrome P450 by a cholesterol carrier protein. Synthesis of this cholesterol carrier protein is induced by cyclic-AMP

2.7 RELATION OF STRUCTURE OF ACTH ANALOGUES TO BIOLOGICAL ACTIVITY

Porcine adrenocorticotropin is an unbranched polypeptide consisting of 39 amino acids (Figure 2.11). The primary structure of porcine, ovine, bovine and human ACTH are identical for the first 24 amino acids of the N-terminus. Species variations in primary structure occur only between amino acids 25 and 33 of the hormone[99-103]. Circular dichroism reveals little or no secondary structure[104] and fluorescent probe studies on [21-lysine (5-dimethylamino-1-naphthalenesulphonyl)]ACTH$_{1-24}$ suggest that ACTH exists as a random coil in solution[105]. The relation of the primary structure of ACTH to biological activity has been investigated by a variety of techniques: adrenal ascorbic acid depletion and steroidogenesis in hypophysectomised rats, steroidogenesis in incubated adrenal quarters, activation of adenylate cyclase in adrenal membrane fragments and cyclic-AMP accumulation in adrenals of hypophysectomised rats[15, 106-110]. Almost invariably, the potency of an ACTH analogue has been determined by comparison of the responses induced by one or two doses of an analogue and of ACTH$_{1-39}$.

H-Ser-Tyr-Ser-Met-Glu-His-Phe-Arg-Trp-Gly-Lys-Pro-Val-Gly-Lys-Lys-Arg-Arg-Pro-
  1   2   3   4   5   6   7   8   9 10 11 12 13 14 15 16 17 18 19

$$NH_2$$
|
Val-Lys-Val-Tyr-Pro-Asp-Gly-Ala-Glu-Asp-Glu-Leu-Ala-Glu-Ala-
 20 21 22 23 24 25 26 27 28 29 30 31 32 33 34

Phe-Pro-Leu-Glu-Phe-OH
 35 36 37 38 39

Figure 2.11 Primary structure of α-porcine ACTH$_{1-39}$. (After Riniker *et al.*[103].)

From these studies, several important observations were made about the relation of primary structure of ACTH to biological activity. ACTH$_{1-24}$ retained full biological activity[111-113], and further shortening to ACTH$_{1-18}$ resulted in only a slight loss of potency[106, 112, 114, 115]. The dibasic amino acids Lys[15], Lys[16], Arg[17], Arg[18] have an important influence on the potency of biological activity[106, 112, 113, 116]. However, they are not essential since ACTH$_{1-10}$ has been demonstrated to have weak but nevertheless significant biological activity[113]. Recently, Fujino *et al.*[117] have examined the effect of shortening the polypeptide chain at the N-terminus. They found that ACTH$_{4-23}$ amide retained significant steroidogenic activity and that further shortening of the N-terminus to 5-23, 6-24 and 7-23 resulted in a progressive loss of activity. These authors suggested that the sequence His^6-Phe^7-Arg^8-Trp^9 is an important region of the molecule for steroidogenic action.

Introduction of the isolated adrenal cortex-cell technique has provided new insight into the relation between structure and biological activity of ACTH and analogues. Suspensions of isolated adrenal cortex cells of the rat respond to ACTH with increased production of cyclic-AMP and corticosterone. The cells are highly selective; a number of hormones including insulin, vasopressin, oxytocin and angiotensin induce no effect. Most importantly, complete log(dose)–response curves can be constructed even for analogues of low potency. Analogues can be characterised by two parameters: first, capacity to induce a maximum biological response and second, the dose which induces one-half of the maximum response. In general terms, the maximum response induced is a measure of capacity to excite the receptor ('intrinsic activity' efficiency); the dose necessary to induce one-half of the maximum response is a measure of the affinity of the analogue for the receptor[118-121].

2.7.1 Characterisation of ACTH analogues

The capacity of various ACTH analogues to induce maximum cyclic-AMP response and their molar potency based on cyclic-AMP production have been determined (Table 2.3). $ACTH_{1-39}$, $ACTH_{1-24}$, $[Lys^{17}, Lys^{18}]ACTH_{1-18}$ amide and $ACTH_{1-16}$ amide induce the same maximum rate of cyclic-AMP production. $ACTH_{5-24}$ induces a maximum rate of cyclic AMP-production equal to 0.45 of that of $ACTH_{1-39}$, whereas the value for $[Trp(Nps)^9]$ $ACTH_{1-39}$ is 0.01. These analogues have different potencies.

Table 2.3 Cyclic AMP molar potencies and 'intrinsic activities' of ACTH analogues

	a*	Molar potency†
$ACTH_{1-39}$	1	100
$ACTH_{1-24}$	1	143
$[Lys^{17}, Lys^{18}]ACTH_{1-18}$ amide	1	24
$ACTH_{1-16}$ amide	1	0.12
$ACTH_{5-24}$	0.45	0.04
$[Trp(Nps)^9]ACTH_{1-39}$	<0.01	—‡

* The intrinsic activity (a) of an analogue equals the maximum rate of cyclic-AMP production for the analogue divided by the maximum rate of cyclic-AMP production for $ACTH_{1-39}$
† The cyclic-AMP molar potency of an analogue equals [dose of $ACTH_{1-39}$ in moles required to induce one-half of the maximum rate of cyclic-AMP production] × 100 divided by [dose of analogue in moles required to induce one-half of the maximum rate of cyclic-AMP production]
‡ Due to the exceedingly low intrinsic activity of this analogue, the cyclic-AMP molar potency could not be determined

We interpret these results to mean that $ACTH_{1-39}$, $ACTH_{1-24}$, $[Lys^{17}, Lys^{18}]ACTH_{1-18}$ amide and $ACTH_{1-16}$ amide have the same capacity to activate the receptor. $ACTH_{5-24}$ and $[Trp(Nps)^9]ACTH_{1-39}$ activate the receptor but with a reduced capacity. The reduction in potency with the loss of the dibasic amino acids at positions 15, 16, 17 and 18 is ascribed to a loss of affinity of the polypeptide for the receptor. The reduction in potency with the loss of amino acids 1–4 is also ascribed to a loss of affinity of the hormone for the receptor and, in addition, to a loss of capacity of the hormone to activate the receptor.

It was of interest to determine the capacity to induce a maximum corticosterone response and molar potency based on corticosterone production for the same analogues (Table 2.4). $ACTH_{1-39}$, $ACTH_{1-24}$, $[Lys^{17}, Lys^{18}]$ $ACTH_{1-18}$ amide, $ACTH_{1-16}$ amide and $ACTH_{5-24}$ induce the same maximum rate of corticosterone production. $[Trp(Nps)^9]ACTH_{1-39}$ induces a maximum rate of corticosterone production equal to 0.77 of that of $ACTH_{1-39}$. The corticosterone molar potencies are similar to the cyclic-AMP molar potencies.

Table 2.4 Steroidogenic molar potencies and 'intrinsic activities' of ACTH analogues

	a*	Molar potency†
$ACTH_{1-39}$	1	100
$ACTH_{1-24}$	1	140
$[Lys^{17}, Lys^{18}]ACTH_{1-18}$ amide	1	22
$ACTH_{1-16}$ amide	1	0.05
$ACTH_{5-24}$	1	0.03
$[Trp(Nps)^9]ACTH_{1-39}$	0.77	1.6

* The intrinsic activity (a) of an analogue equals the maximum rate of corticosterone production of the analogue divided by the maximum rate of corticosterone production for $ACTH_{1-39}$.
† The steroidogenic molar potency of an analogue equals [dose of $ACTH_{1-39}$ in moles required to induce one-half of the maximum rate of corticosterone production] × 100 divided by [dose of analogue in moles required to induce one-half of the maximum rate of corticosterone production.]

These observations have important implications for the delineation of the 'active centre' of the ACTH molecule as determined by the biological response, namely corticosterone production, or by cyclic-AMP production, an early event in the sequence. $ACTH_{5-24}$ is clearly less effective in activating the receptor than $ACTH_{1-39}$, as measured by the maximum rate of cyclic-AMP production. Measurement of corticosterone production obviously fails to reveal the loss of effectiveness when $ACTH_{1-39}$ is shortened to $ACTH_{5-24}$.

The reason for this apparent discrepancy is revealed by an examination of the relation between cyclic-AMP and corticosterone production. The quantity of cyclic-AMP which induces near-maximum rate of production of corticosterone is produced when less than 20% of the ACTH receptor · adenylate cyclase complexes are activated[19]. Hence $ACTH_{5-24}$, which exhibits an efficiency of 0.45 in activating the ACTH receptor · adenylate cyclase complexes, is capable of inducing a maximum rate of corticosterone production equal to that induced by $ACTH_{1-39}$.

Potencies of $ACTH_{1-39}$, $ACTH_{1-24}$, $[Lys^{17}, Lys^{18}]ACTH_{1-18}$ amide and $ACTH_{1-16}$ amide are remarkably similar as determined by cyclic-AMP production and by corticosterone production. The differences among these polypeptides are strictly based on their affinity for receptor. However, in the case of $ACTH_{5-24}$, potency as judged by corticosterone production will be determined not only by affinity for receptor but also by capacity to excite the receptor. The generalisation that 'active centre' and regions providing affinity can be distinguished does not hold for corticosterone production.

We have taken the position that cyclic-AMP production reflects receptor activation with fidelity. However, we are aware of the possibility that an analogue of ACTH may have a capacity less than that of $ACTH_{1-39}$ to generate a signal when it interacts with receptor, but because of transduction and amplification in the plasma membrane is capable of inducing maximum activation of adenylate cyclase. Obviously, present knowledge of the events

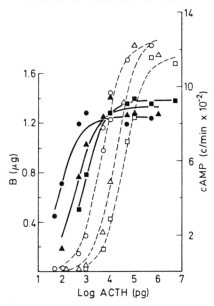

Figure 2.12 Log(dose)–response curves for cyclic-AMP production (- - - -) and corticosterone production (———) by aliquots of a suspension of isolated adrenal cells in response to $ACTH_{1-39}$ alone (● or ○); in combination with 50 μg $ACTH_{11-24}$ (▲ or △) and with 100 μg $ACTH_{11-24}$ (■ or □). The points represent the means of analyses on two aliquots of cell suspension; the lines are least-square best fits. The abscissa represents log(picogram dose of $ACTH_{1-39}$); the ordinates are μg corticosterone/60 min and c.p.m. cyclic-$[8-^{14}C]$ AMP/60 min.
(From Seelig, S. and Sayers, G. (1973). *Arch. Biochem. Biophys.*, **154**, 230 by courtesy of Academic Press.)

which take place in the membrane is at a most elementary state and certainly quite inadequate to resolve the problem posed.

The fragment $ACTH_{11-24}$ exhibits no steroidogenic activity nor does it increase cyclic-AMP production when added to suspensions of adrenal cortex cells at high concentration (about 1 mg ml^{-1}). This polypeptide reversibly inhibits cyclic-AMP production and corticosterone production when added in combination with $ACTH_{1-39}$ (Figure 2.12). We conclude that the major role of amino acids in the sequence 11-24 is to enhance the affinity of the hormone for the receptor.

2.7.2 Nature of the receptor

Provisionally, we present the following picture of the ACTH receptor. The receptor surface has regions complementary to the amino-acid residues of the

ACTH molecule. A small region of the receptor surface comes into intimate contact with the sequence 4–10 of ACTH and a signal is generated. The sequence 5–10 is capable of generating a signal but it is relatively weak compared to that generated by 4–10. Other regions of the receptor have charges which attract; for example, the sequence 15–18 of the ACTH molecule. Still others have lipophilic groups which attract such amino acids as tyrosine and tryptophan.

2.8 CALCIUM

Calcium exerts important and multifarious actions on cell functions. It has been implicated in muscle contraction, exocrine and endocrine gland secretion, metabolic processes and in the regulation of the excitability and the permeability of plasma membranes. Usually, calcium exerts a number of separate and distinct functions within a single cell type. This, together with the fact that the Ca^{2+} ion is not distributed evenly throughout the cell but characteristically is concentrated in special compartments, has made for technical difficulties in approaching the modes of action of calcium.

An important role for calcium in adrenal cortex function was first revealed by Birmingham and her colleagues[122] who showed that omission of calcium from the incubation medium reduced the steroidogenic response of rat adrenal quarters to ACTH.

2.8.1 Calcium and the plasma membrane

Haksar and Péron[123] examined corticosterone production by isolated adrenal cortex cells of the rat in response to ACTH and dibutyryl cyclic-AMP at various concentrations of calcium in the suspension medium. They concluded that the events preceding cyclic-AMP production are much more sensitive to calcium than those that follow the synthesis of the nucleotide.

Sayers et al.[9] examined accumulation of cyclic-AMP by isolated adrenal cortex cells of the rat in response to ACTH at various calcium concentrations (Figure 2.13). The authors concluded that the effect of calcium on ACTH-induced steroidogenesis is in large measure a consequence of the reduction in cyclic-AMP production.

Two observations with cellular fractions appear to be in conflict with the studies conducted on intact cells. Concentrations of calcium greater than 1.0×10^{-3} mol l^{-1} suppress the binding of ACTH to receptor and inhibit activation of adenylate cyclase by ACTH when measured in 'sub-cellular membrane particles' of mouse adrenal tumour tissue[124].

These apparent discrepant observations relating to intact cell versus subcellular fractions are brought into harmony with the model in Figure 2.1.

Increased biological activity of ACTH may well be associated with decreased occupancy of receptor site, an interpretation in line with the findings reported here on isolated adrenal cells. According to the rate theory proposed by Paton[125], high efficacy of ACTH would be associated with high dissociation rate, low efficacy with low dissociation rate. In this connection, Rodbel et al.[10,11] have demonstrated that guanyl nucleotides play a specific and

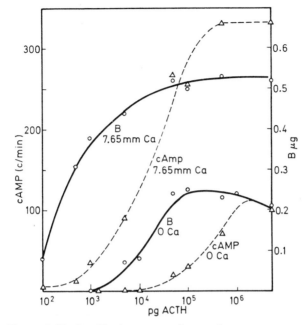

Figure 2.13 Log(dose)–response for corticosterone production (μg B/30 min) and accumulation of cyclic-[8-^{14}C] AMP (count min^{-1}) by isolated adrenal cells in response to increasing concentrations of ACTH in picograms. A single pool of adrenal cells from rats was divided into two parts and centrifuged. One pellet was suspended in a buffer which contained 7.65×10^{-3} mol l^{-1} calcium, the other in a buffer which contained no calcium. Portions of cells (0.9 ml; 300 000 cells) plus ACTH in 0.1 ml of vehicle were incubated for 30 min at 37 °C in 95% O_2, 5% CO_2. (After Sayers et al.[9] by courtesy of the American Assen for Advancement of Science.)

obligatory role in glucagon activation of adenylate cyclase located in rat liver plasma membrane fragments and decrease affinity of the binding sites for glucagon. Calcium may exert similar effects on the ACTH receptor · adenylate cyclase complex on adrenal cell plasma membrane.

We propose that the strength of the signal generated at (1) and transmitted (2) through the membrane to the adenylate cyclase compartment is related to the concentration of calcium ion on the outside of the plasma membrane. On the other hand, activation by ACTH of adenylate cyclase of membrane fragments is optimal at a concentration of about 10^{-7} mol l^{-1} of calcium ion[126] and markedly inhibited at concentrations of the cation greater than 10^{-3} mol l^{-1} [124, 126]. We further propose that adenylate cyclase in the intact plasma membrane is confined to a compartment [(3) of Figure 2.1] where the concentration of calcium ion approximates to that of the cytosol (about 10^{-7} mol l^{-1} and remains relatively fixed in the face of wide fluctuations in the concentration of the cation on the outside of the cell. Evidence has been brought forth to support the notion that adenylate cyclase itself is not the receptor for ACTH[8]. If adenylate cyclase is located on the inner surface of the plasma membrane as depicted at (3), then the enzyme of the membrane fragments is exposed to abnormal ('outside') concentrations of calcium ion. This would explain why Lefkowitz et al.[124] observed inhibition of ACTH activation of adenylate cyclase of membrane fragments exposed to high concentrations of calcium ion ($>1 \times 10^{-3}$ mol l^{-1}). The signal arising from the interaction of ACTH and its receptor was strong, but the abnormally high concentration of calcium ion inhibited adenylate cyclase. Furthermore, the model explains why Lefkowitz et al.[124] were obliged to use very high concentrations of ACTH (10^{-6} mol l^{-1}) to activate adenylate cyclase when the membrane fragments were suspended in a medium containing a low concentration of calcium ion. In this instance, adenylate cyclase was responsive to the signal, but in order to overcome the adverse effect of low calcium ion concentration on the signal strength high concentrations of ACTH had to be employed.

2.8.2 Calcium and steroidogenesis

The absence of calcium also impairs the events following cyclic-AMP production. Maximum corticosterone production of isolated adrenal cortex cells in response to dibutyryl cyclic-AMP was reduced by 50% with the removal of calcium from the medium. In contrast, the dose of dibutyryl cyclic-AMP which induced one-half of the maximum rate of corticosterone production was not altered by various calcium ion concentrations in the medium[9].

The results obtained on omission of calcium from the incubation medium of isolated adrenal cortex cells are interpreted as indicating that the marked increase in the dose of ACTH required to induce corticosterone production is a result of a weakening of the signal at (2), and the reduction in the maximum rate of corticosterone production is primarily the result of impairment of reactions subsequent to cyclic-AMP production [(4) of Figure 2.1].

2.8.3 Calcium as a messenger

Rasmussen[127] has proposed that polypeptide hormones lead to an increase in calcium ion entry or intracellular translocation and an activation of adenylate cyclase, both of which mediate hormone action. The important, but as yet unresolved, problem is whether the polypeptide hormone influences these two 'messengers' by separate mechanisms.

It has been suggested that ACTH shifts calcium from a rapidly exchanging to a more slowly exchanging pool in adrenal cortex cells[128]. Rubin et al.[129] have recently examined the actions of ACTH on the cat adrenal gland perfused with regular Locke's solution or with calcium-free Locke's solution. They claim that steroid synthesis and steroid release are differentially influenced by cyclic-AMP and by calcium. They further claim that an increase in adrenal cyclic-AMP is not sufficient to initiate the release of corticosteroids. They propose that ACTH, in addition to activating adenylate cyclase, also brings about a re-distribution of calcium to some active site—possibly the endoplasmic reticulum or mitochondria. These two actions must proceed in parallel in order for steroid production to be coupled with steroid release.

The functional relations among ACTH, cyclic-AMP and calcium are of great importance to the understanding of the steroidogenic action of the polypeptide hormone. We are obviously not, as yet, in a position to present the final chapter of this fascinating problem which has broad impilcations for the regulation of cell function.

2.9 'SENSITIVITY', 'AMPLIFICATION'

The 'sensitivity' of the adrenal cortex cell may be defined in terms of the increment in the concentration of ACTH which will induce a given increment in the rate of production of corticosterone. The more 'sensitive' the cell, the less the concentration of ACTH required to induce a given increment in steroid production.

The relation between the dose of ACTH and the cyclic-AMP production fits the equation

$$cAMP = \frac{(cAMP_{max})A}{A + k_A} \quad (2.1)$$

where $cAMP$ is the rate of cyclic-AMP production, $cAMP_{max}$ is the maximum rate of cyclic-AMP production, A is the concentration of $ACTH_{1-39}$ and k_A is the concentration of $ACTH_{1-39}$ required to induce $1/2cAMP_{max}$.

Furthermore, cyclic-AMP and corticosterone are related by

$$\frac{B}{B_{max}} = \frac{cAMP}{cAMP + k_c} \quad (2.2)$$

where B is the rate of corticosterone production, B_{max} is the maximum rate of corticosterone production, $cAMP$ is the concentration of cyclic-AMP and k_c is the concentration of cyclic-AMP required to induce $1/2B_{max}$[35].

We have found it useful to express 'sensitivity' in terms of the parameters of equations (2.1) and (2.2). If these two equations are combined, equation (2.3) is obtained

$$\frac{B}{B_{max}} = \frac{A}{A(1 + k_c/cAMP_{max}) + (k_c/cAMP_{max})k_A} \quad (2.3)$$

Factors which determine 'sensitivity' include:
(a) The number of receptors on the cell surface. The greater the number of receptors, the greater the probability that an ACTH molecule will be captured. Maximum cyclic-AMP production reflects the number of these receptors. Hence, the greater the value of $cAMP_{max}$ the more 'sensitive' the system. The fact that the system has the capacity to produce cyclic-AMP at a rate far in excess of that required to induce maximum corticosterone production has been referred to as 'receptor reserve' or 'spare receptors'[19]. The significant functional meaning of this excess is revealed by equation (2.3) from which it becomes clear that this excess endows the adrenal cortex cell with a high degree of 'sensitivity' to ACTH. Cyclic-AMP not only mediates the steroidogenic action of ACTH but perhaps, more importantly, amplifies the signal generated by the interaction of the hormone with its receptor. In our opinion,

it is quite likely that another amplification step occurs in the plasma membrane. The signal generated by the interaction of ACTH and receptor is converted and amplified during its passage through the membrane to the locus of the adenylate cyclase. This amplification step in the membrane provides another means by which the 'sensitivity' of the adrenal cortex cell is enhanced[130].

(b) The concentration of cyclic-AMP which induces one-half of the maximum corticosterone production (k_c). The smaller the value of k_c, the more 'sensitive' the system.

(c) The attractive force exerted by the receptor on the ACTH molecule. This will be determined by the number of charges and of lipophilic points arranged on the receptor surface in an array complementary to the ACTH molecule. This force of attraction is measured as k_A, the concentration of $ACTH_{1-39}$ which induces one-half of the maximum cyclic-AMP production. The smaller the value of k_A, the greater the 'sensitivity'.

2.10 CONCLUDING REMARKS

ACTH increases the secretory activity of the adrenal cortex. The site of action of the polypeptide hormone is probably at the surface of the adrenal cortex cell. Interaction of ACTH and receptor on the plasma membrane initiates the sequence of events ending with increased synthesis of corticosterone and cortisol. It is generally accepted that cyclic-AMP is the intermediary, the 'second messenger'. The molecular events in the plasma membrane which intervene between ACTH–receptor interaction and activation of adenylate cyclase remain unclear. The means by which cyclic-AMP accelerates the conversion of cholesterol to pregnenolone and thereby increases the secretory activity is also an important problem awaiting definition. It is presumed that the interaction of hormone and receptor generates a signal which is transduced and transmitted to adenylate cyclase. Calcium ions appear to play an important role in this 'coupling' process.

Isolation of the receptors poses technical difficulties and our meagre knowledge of the chemical nature of these components of the membrane is based on the relation of the structure of ACTH analogues to their biological actions. Complementary affinity and active sites on hormone and receptor have been attributed to lipophilic and charged areas on the receptor surface. Attempts have been made to quantify affinity by determining the concentration of radioactive iodine-labelled ACTH in the medium associated with half-saturation of receptors. In the opinion of the reviewers, the assumptions that iodine-labelled ACTH interacts with receptor and exerts biological actions essentially similar to unlabelled ACTH are not justified. The kinetics of association and dissociation of hormone and receptor remain to be established.

Of considerable interest is the fact that the number of receptors on the plasma membrane are far in excess of the number required to induce maximum rate of production of corticosteroid. The designation 'spare receptors' is inappropriate since the relatively large population of receptors serves to increase the sensitivity of the adrenal cortex cell to ACTH; the larger the receptor population, the smaller the concentration of ACTH required to induce a given steroidogenic response.

Cyclic-AMP accelerates the conversion of cholesterol to pregnenolone, the rate-limiting step in the pathway leading from cholesterol to corticosterone and cortisol. The cyclic nucleotide activates cholesterol esterase and thus mobilises free cholesterol from cholesterol ester stores, a mode of action analogous to the activation of lipase in adipocytes. Another, and most likely the most important mode of action of cyclic-AMP in determining the rate of conversion of cholesterol to pregnenolone, involves the *de novo* synthesis of a labile protein at the level of translation. The labile protein remains to be isolated and characterised.

The steroid intermediates in the pathway of conversions from cholesterol to corticosterone to cortisol have been isolated. The majority of these conversions involve a mixed function oxidase system composed of cytochrome P450, a flavoprotein and a non-heme iron containing protein, adrenodoxin.

ACKNOWLEDGEMENTS

The authors express their gratitude to Professor Robert Schwyzer, Institute of Molecular Biology and Biophysics, Swiss Federal Institute of Technology, Zurich, Switzerland, who provided a number of analogues of ACTH utilised in the studies on relation of structure to biological activity and who contributed to the interpretation of the data.

We thank Drs. Rittel and Desaulles of Ciba–Geigy and Dr. Strade of Organon for supplies of ACTH and ACTH analogues.

The expertise of Mrs. Brady and Mrs. Vegh has been invaluable to the conduct of the experiments on isolated adrenal cortex cells.

To Ms. Sasko, Ms. Poplawski and Mrs. Smith, we express our thanks for help in the preparation of the manuscript.

References

1. Roberts, S., Creange, J. E. and Young, P. L. (1965). *Biochem. Biophys. Res. Commun.*, **20,** 446
2. McKerns, K. W. (1968). *Functions of the Adrenal Cortex*, Vol. 2, 479 (New York: Appleton-Century-Crofts)
3. Schimmer, B. P., Ueda, K. and Sato, G. H. (1968). *Biochem. Biophys. Res. Commun.*, **32,** 806
4. Selinger, R. C. L. and Civen, M. (1971). *Biochem. Biophys. Res. Commun.*, **43,** 793
5. Lefkowitz, R. J., Roth, J., Pricer, W. and Pastan, I. (1970). *Proc. Nat. Acad. Sci. USA*, **65,** 745
6. Hofmann, K., Wingender, W. and Finn, F. M. (1970). *Proc. Nat. Acad. Sci. USA*, **67,** 829
7. Lefkowitz, R. J., Roth, J. and Pastan, I. (1970). *Science*, **170,** 633
8. Rodbell, M. (1971). *Karolinska Symp. No. 3*, In Vitro *Methods in Reproductive Cell Biology*, 337 (Bogtrykkeriet Forum 1971: Copenhagen)
9. Sayers, G., Beall, R. J. and Seelig, S. (1972). *Science*, **175,** 1131
10. Rodbell, M., Krans, H. M. J., Pohl, S. L. and Birnbaumer, L. (1971). *J. Biol. Chem.*, **246,** 1872
11. Rodbell, M., Birnbaumer, L., Pohl, S. L. and Krans, H. M. J. (1971). *J. Biol. Chem.*, **246,** 1877

12. Sutherland, E. W. and Rall, T. W (1960). *Pharmacol. Rev.*, **12**, 265
13. Robison, G. A., Butcher, R. W. and Sutherland, E. W. (1971). *Cyclic AMP* (New York: Academic Press)
14. Robison, G. A., Nahas, G. G. and Triner, L. (1971). *Cyclic AMP and Cell Function* (New York: Ann. N.Y. Acad. Sci.)
15. Grahame-Smith, D. G., Butcher, R. W., Ney, R. L. and Sutherland, E. W. (1967). *J. Biol. Chem.*, **242**, 5535
16. Haynes, R. C. (1958). *J. Biol. Chem.*, **233**, 1220
17. Carchman, R. A., Jaanus, S. D. and Rubin, R. P. (1971). *Mol. Pharmacol*, **7**, 491
18. Bell, J., Brooker, G. and Harding, B. W. (1970). *Biochem. Biophys. Res. Commun.*, **41**, 938
19. Beall, R. J. and Sayers, G. (1972). *Arch. Biochem. Biophys.*, **148**, 70
20. Robison, G. A., Butcher, R. W. and Sutherland, E. W. (1971). *Cyclic AMP*, 236 (New York: Academic Press)
21. Exton, J. H., Lewis, S. B., Ho, R. J., Robison, G. A. and Park, C. R. (1971). *Ann. N.Y. Acad. Sci.*, **185**, 85
22. Butcher, R. W., Ho, R. J., Meng, H. C. and Sutherland, E. W. (1965). *J. Biol. Chem.*, **240**, 4515
23. Garren, L. D., Gill, G. N., Masui, H. and Walton, G. M. (1971). *Rec. Prog. Horm. Res.*, **27**, 433
24. Beall, R. J. and Sayers, G. (1972). *Fed. Proc. (Fed. Amer. Soc. Exp. Biol.)*, **31**, 293 Abs.
25. Manganiello, V. C., Murad, F. and Vaughan, M. (1971). *J. Biol. Chem.*, **246**, 2195
26. Kuo, J. F. and DeRenzo, E. C. (1969). *J. Biol. Chem.*, **244**, 2252
27. Kitabchi, A. E., Wilson, D. P. and Sharma, R. K. (1971). *Biochem. Biophys. Res. Commun.*, **44**, 898
28. Davoren, P. R. and Sutherland, E. W. (1963). *J. Biol. Chem.*, **238**, 3009
29. Ho, R. J. and Sutherland, E. W. (1971). *J. Biol. Chem.*, **246**, 6822
30. Haynes, R. C., Koritz, S. B. and Péron, F. G. (1959). *J. Biol. Chem.*, **234**, 1421
31. Glinsmann, W. H., Hern, E. P., Linarelli, L. G. and Farese, R. V. (1969). *Endocrinology*, **85**, 711
32. Farese, R. V., Linarelli, L. G., Glinsmann, W. H., Ditzion, B. R., Paul, M. I. and Park, G. L. (1969). *Endocrinology*, **85**, 867
33. Mahaffee, D. and Ney, R. L. (1970). *Metab. Clin. Exp.*, **19**, 1104
34. Tsang, C. P. W. and Péron, F. G. (1971). *Steroids*, **17**, 453
35. Sayers, G., Ma, R-M. and Giordano, N. (1971). *Proc. Soc. Exp. Biol. Med.*, **136**, 619
36. Rivkin, I. and Chasin, M. (1971). *Endocrinology*, **88**, 664
37. Free, C. A., Chasin, M., Paik, V. S. and Hess, S. M. (1971). *Biochemistry*, **10**, 3785
38. Kitabchi, A. E. and Sharma, R. K. (1971). *Endocrinology*, **88**, 1109
39. Schulster, D., Tait, S. A. S., Tait, J. F. and Mrotek, J. (1970). *Endocrinology*, **86**, 487
40. Ney, R. L. (1969). *Endocrinology*, **84**, 168
41. Nussdorfer, G. G. and Mazzocchi, L. D. (1972). *Lab. Invest.*, **26**, 45
42. Jones, D. J., Nicholson, W. E., Liddle, G. W. and Finnegan, A. W. (1970). *Proc. Soc. Exp. Biol. Med.*, **133**, 764
43. Earp, H. S., Watson, B. S. and Ney, R. L. (1970). *Endocrinology*, **87**, 118
44. Davis, W. W. (1969). *Fed. Proc. (Fed. Amer. Soc. Exp. Biol.)*, **28**, 701 Abs.
45. Free, C. A., Chasin, M., Paik, V. S. and Hess, S. M. (1972). *Fed. Proc. (Fed. Amer. Soc. Exp. Biol.)*, **31**, 555 Abs.
46. Muneyama, K., Bauer, R. J., Shuman, D. A., Robins, R. K. and Simon, L. (1971). *Biochemistry*, **10**, 2390
47. Michal, G., Nelböck, M. and Weimann, G. (1970). *Fresenius' Z. Anal. Chem.*, **252**, 189
48. Ishikawa, E., Ishikawa, S., Davis, J. W. and Sutherland, E. W. (1969). *J. Biol. Chem.*, **244**, 6371
49. Exton, J. H., Hardman, J. G., Williams, T. F., Sutherland, E. W. and Park, C. R. (1971). *J. Biol. Chem.*, **246**, 2658
50. Levine, R. A. (1969). *Fed. Proc. (Fed. Amer. Soc. Exp. Biol.)*, **28**, 707 Abs.
51. Braun, T., Hechter, O. and Bär, H. P. (1969). *Proc. Soc. Exp. Biol. Med.*, **132**, 233
52. McMillan, B. H., Ney, R. L. and Schorr, I. (1971). *Endocrinology*, **89**, 281
53. Halkerston, I. D. K., Feinstein, M. and Hechter, O. (1966). *Proc. Soc. Exp. Biol. Med.*, **122**, 896

54. Free, C. A., Chasin, M., Paik, V. S. and Hess, S. M. (1971). *Fed. Proc. (Fed. Amer. Soc. Exp. Biol.)*, **30,** 1268 Abs.
55. Haynes, R. C. and Berthet, L. (1957). *J. Biol. Chem.*, **225,** 115
56. Rall, T. W., Sutherland, E. W. and Berthet, J. (1957). *J. Biol. Chem.*, **224,** 463
57. Simpson, E. R., Cammer, W. and Estabrook, R. W. (1968). *Biochem. Biophys. Res. Commun.*, **31,** 113
58. Simpson, E. R. and Estabrook, R. W. (1968). *Arch. Biochem. Biophys.*, **126,** 977
59. Ferguson, J. J., Jr. (1963). *J. Biol. Chem.*, **238,** 2754
60. Kobayashi, S., Nagasumi, Y., Morisaki, M. and Ichii, S. (1963). *Steroids*, **2,** 167
61. Vance, V. K., Girard, F. and Cahill, G., Jr. (1962). *Endocrinology*, **71,** 113
62. Walsh, D. A., Perkins, J. P. and Krebs, E. G. (1968). *J. Biol. Chem.*, **243,** 3763
63. Walsh, D. A., Krebs, E. G., Reiman, E. M., Brostrom, M. A., Corbin, J. D., Hickenbottom, J. H., Soderling, T. R. and Perkins, J. P. (1970). *Advances in Biochemical Psychopharmacology*, Vol. 3, 265 (P. Greengard and E. Costa, editors) (New York: Raven Press)
64. Kuo, J. F. and Greengard, P. (1969). *J. Biol. Chem.*, **244,** 3417
65. Kuo, J. F. and Greengard, P. (1969). *Proc. Nat. Acad. Sci. USA*, **64,** 1349
66. Kuo, J. F., Krueger, B. K., Sanes, J. R. and Greengard, P. (1970). *Biochim. Biophys. Acta*, **212,** 79
67. Schlender, K. K., Wei, S. H. and Villar-Palasi, C. (1969). *Biochim. Biophys. Acta*, **191,** 272
68. Soderling, T. R. and Hickenbottom, J. P. (1970). *Fed. Proc. (Fed. Amer. Soc. Exp. Biol.)*, **29,** 601 Abs.
69. Reimann, E. M. and Walsh, D. A. (1970). *Fed. Proc. (Fed. Amer. Soc. Exp. Biol.)*, **29,** 601 Abs.
70. Corbin, J. D., Reimann, E. M., Walsh, D. A. and Krebs, E. G. (1970). *J. Biol. Chem.*, **245,** 4849
71. Huttunen, J. K. and Steinberg, D. (1971). *Biochim. Biophys. Acta*, **239,** 411
72. Gill, G. N. and Garren, L. D. (1969). *Proc. Nat. Acad. Sci. USA*, **63,** 512
73. Gill, G. N. and Garren, L. D. (1970). *Biochem. Biophys. Res. Commun.*, **39,** 335
74. Gill, G. N. and Garren, L. D. (1971). *Proc. Nat. Acad. Sci. USA*, **68,** 786
75. Reimann, E. M., Brostrom, C. O., Corbin, J. D., King, C. A. and Krebs, E. G. (1971) *Biochem. Biophys. Res. Commun.*, **42,** 187
76. Garren, L. D. (1968). *Vitam. Horm.*, **26,** 119
77. Gill, G. N. (1972). *Metab., Clin. Exp.*, **21,** 571
78. Simpson, E. R., Trzeciak, W. H., McCarthy, J. L., Jefcoate, C. R. and Boyd, G. S. (1972). *Proc. 527th Mtg. Biochem. Soc., Factors Affecting Cholesterol Esterase and Cholesterol Side-Chain-Cleavage Activities in Rat Adrenal* (University of Aberdeen, Scotland)
79. Farese, R. V. (1964). *Biochim. Biophys. Acta*, **87,** 699
80. Garren, L. D., Ney, R. L. and Davis, W. W. (1965). *Proc. Nat. Acad. Sci. USA*, **53,** 1443
81. Bransome, E. D., Jr. (1968). *Annual Review of Physiology, Adrenal Cortex*, Vol. 30, 171 (Palo Alto: Annual Reviews, Inc.)
82. Mulrow, P. J. (1972). *Annual Review of Physiology, Adrenal Cortex*, Vol. 34, 409 (Palo Alto: Annual Reviews, Inc.)
83. Walton, G. M., Gill, G. N., Abrass, I. B. and Garren, L. D. (1971). *Proc. Nat. Acad. Sci. USA*, **68,** 880
84. Fiala, S., Sproul, E. E. and Fiala, A. (1956). *J. Biophys. Biochem. Cytol.*, **2,** 115
85. Farese, R. V. and Schnure, J. J. (1967). *Endocrinology*, **80,** 872
86. Christensen, A. K. and Gillim, S. W. (1969). *The Gonads*, 415 (K. W. McKerns, editor) (New York: Appleton-Century-Crofts)
87. Fawcett, D. W., Long, J. A. and Jones, A. L. (1969). *Rec. Prog. Horm. Res.*, **25,** 315
88. Sayers, G., Swallow, R. L. and Giordano, N. D. (1971). *Endocrinology*, **88,** 1063
89. Malamed, S., Sayers, G. and Swallow, R. L. (1970). *Z. Zellforsch.*, **107,** 447
90. Samuels, L. T. and Uchikawa, T. (1967). *The Adrenal Cortex*, 61 (A. B. Eisenstein, editor) (Boston: Little, Brown and Company)
91. Matsuba, M., Ichii, S. and Kabayashi, S. (1966). *Steroid Dynamics*, 357 (G. Pincus, T. Nakao, and J. F. Tait, editors) (New York: Academic Press)

92. Borkowski, A. J., Levin, S., Delcroix, C., Mahler, A. and Verhas, V. (1967). *J. Clin. Invest.*, **46**, 797
93. Siekevitz, P. (1963). *Ann. Rev. Physiol.*, **25**, 15
94. Jefcoate, C. R. and Boyd, G. S. (1971). *FEBS Lett.*, **12**, 279
95. Kan, K. W. and Ungar, F. (1972). *Fed. Proc.* (*Fed. Amer. Soc. Exp. Biol.*), **31**, 429 Abs.
96. Karaboyas, G. C. and Koritz, S. B. (1965). *Biochemistry*, **4**, 462
97. Harding, B. W. (1971). Personal communication in Brownie, A. C., Simpson, E. R., Jefcoate, C. R. and Boyd, G. S. (1972). *Biochem. Biophys. Res. Commun.*, **46**, 483
98. Boyd, G. S., Brownie, A. C., Jefcoate, C. R. and Simpson, E. R. (1971). *Biochem. J.*, **125**, 1P
99. Howard, K. S., Shepherd, R. G., Signer, E. A., Davies, D. S. and Bell, P. H. (1955). *J. Amer. Chem. Soc.*, **77**, 3419
100. Shepherd, R. G., Howard, K. S., Bell, P. H., Cacciola, A. R., Child, R. G., Davies, M. C., English, J. P., Finn, B. M., Meisenbelder, J. H., Moyar, A. W. and van der Scheer, J. (1956). *J. Amer. Chem. Soc.*, **78**, 5051
101. Li, C. H., Geschwind, I. I., Cole, R. D., Raacke, I. D., Harris, J. I. and Dixon, J. S. (1955). *Nature* (*London*), **176**, 687
102. Li, C. H., Dixon, J. S. and Chung, D. (1958). *J. Amer. Chem. Soc.*, **80**, 2587
103. Riniker, B., Sieber, P. and Rittel, W. (1972). *Nature* (*New Biol.*), **235**, 114
104. Edelhoch, H. and Lippoldt, R. E. (1969). *J. Biol. Chem.*, **244**, 3876
105. Schiller, P. W. (1972). *Proc. Nat. Acad. Sci. USA*, **69**, 975
106. Hofmann, K. (1960). *Brookhaven Symp. Biol.*, **13**, 184
107. Ramachandran, J., Chung, D. and Li, C. H. (1965). *J. Amer. Chem. Soc.*, **87**, 2696
108. Schwyzer, R. (1964). *Ann. Rev. Biochem.*, **33**, 259
109. Desaulles, P. A., Barthe, P., Schar, B. and Staehelin, M. (1966). *Acta Endocrinol.*, **51**, 609
110. Ide, M., Tanaka, A., Nakamura, M. and Okabayashi, T. (1972). *Arch. Biochem. Biophys.*, **149**, 189
111. Guttmann, St., Pless, J. and Boissonnas, R. A. (1967). *Peptides*, 221 (Amsterdam: North-Holland Publishing Company)
112. Hofmann, K., Andreatta, R., Bohn, H. and Moroder, L. (1970). *J. Med. Chem.*, **13**, 339
113. Ney, R. L., Ogata, E., Schimizu, N., Nicholson, W. E. and Liddle, G. W. (1964). *Proc. Sec. Int. Cong. Endocr.*, Part II, 1184
114. Li, C. H., Meienhofer, J., Schnabel, E., Chung, D., Lo, T. B. and Ramachandran, J. (1961). *J. Amer. Chem. Soc.*, **83**, 4449
115. Hofmann, K., Yanaihara, N., Lande, S. and Yajima, H. (1962). *J. Amer. Chem. Soc.*, **84**, 4470
116. Bajusz, S. and Medzihradszky, K. (1967). *Peptides*, 209 (Amsterdam: North-Holland Publishing Company)
117. Fujino, M., Hatanaka, C. and Nishimura, O. (1971). *Chem. Pharm. Bull.*, **19**, 1066
118. Rudinger, J. and Krejčí, I. (1962). *Experientia*, **18**, 585
119. Stephenson, R. P. (1956). *Brit. J. Pharmacol.*, **11**, 379
120. Ariëns, E. J. and Simonis, A. M. (1964). *J. Pharm. Pharmacol.*, **16**, 137
121. Hofmann, K., Wingender, W. and Finn, F. M (1970). *Proc. Nat. Acad. Sci. USA*, **67**, 829
122. Birmingham, M. K., Elliott, F. H. and Valere, P. H.-L. (1953). *Endocrinology*, **53**, 687
123. Haksar, A. and Péron, F. G. (1972). *Biochem. Biophys. Res. Commun.*, **47**, 445
124. Lefkowitz, R. J., Roth, J. and Pastan, I. (1970). *Nature* (*London*), **228**, 864
125. Paton, W. D M. (1970). *Molecular Properties of Drug Receptors*, 3 (R. Porter and M. O'Connor, editors) (London: Churchill)
126. Bär, H.-P. and Hechter, O. (1969). *Biochem. Biophys. Res. Commun.*, **35**, 681
127. Rasmussen, H. (1970). *Science*, **170**, 404
128. Jaanus, J. D. and Rubin, R. P. (1971). *J. Physiol.*, **213**, 581
129. Rubin, R. P., Carchman, R. A. and Jaanus, S. D. (1972). *Biochem. Biophys. Res. Commun.*, **47**, 1492
130. Mayer, S. (1972). *J. Pharmacol. Exp. Ther.*, **181**, 116

3
Adenohypophysial Hormones: Regulation of their Secretion and Mechanisms of their Action

V. SCHREIBER
Charles University, Prague

3.1	INTRODUCTION	62
	3.1.1 *Biochemistry and enzymology of the adenohypophysis*	62
	3.1.2 *Source of hormones: adenohypophysial cell types*	64
	3.1.3 *General regulation*	66
	3.1.3.1 *Neurovascular theory*	66
	3.1.3.2 *Feed-back regulation*	66
3.2	THYROTROPHIC HORMONE (THYROTROPHIN, TSH)	67
	3.2.1 *Biochemistry of TSH*	67
	3.2.2 *Regulation of TSH secretion*	67
	3.2.2.1 *Thyrotrophin-releasing hormone (TRH)*	69
	3.2.2.2 *Feed-back regulation of TSH secretion*	70
	3.2.3 *Mechanism of TSH action*	70
3.3	ADRENOCORTICOTROPHIC HORMONE (ADRENOCORTICOTROPHIN, ACTH)	71
	3.3.1 *Biochemistry of ACTH*	72
	3.3.2 *Regulation of ACTH secretion*	73
	3.3.2.1 *Corticotrophin-releasing hormone (CRH)*	73
	3.3.2.2 *Feed-back regulation of ACTH secretion*	74
	3.3.3 *Mechanism of ACTH action*	75
3.4	THE GONADOTROPHINS	77
	3.4.1 *Biochemistry of the gonadotrophins*	77
	3.4.2 *Regulation of gonadotrophin secretion*	78
	3.4.2.1 *Gonadotrophin-releasing hormone (GnRH, LH–RH = FSH–RH)*	79
	3.4.2.2 *Feed-back regulation of gonadotrophin secretion*	80
	3.4.3 *Mechanism of gonadotrophin action*	81

3.5	GROWTH HORMONE (GH, SOMATOTROPHIC HORMONE, STH)	82
	3.5.1 Biochemistry of GH	83
	3.5.2 Regulation of GH secretion	85
	3.5.2.1 Growth hormone–releasing hormone (GH–RH)	85
	3.5.2.2 Other factors influencing GH secretion	85
	3.5.3 Mechanism of GH action	86
3.6	PROLACTIN (LUTEOTROPHIC HORMONE, LTH)	86
	3.6.1 Biochemistry of prolactin	87
	3.6.2 Regulation of prolactin secretion	87
	3.6.2.1 Prolactin-inhibiting hormone (PIH)	87
	3.6.2.2 Other factors influencing prolactin secretion	89
	3.6.3 Mechanism of prolactin action	89

3.1 INTRODUCTION

The adenohypophysis is a complex organ composed of at least six types of cells and producing at least six hormones. Three of the hormones are polypeptides or proteins and three are glycoproteins. The latter may reflect surviving ancient biochemical characteristics, such as the adenohypophysis probably developed from the gastrointestinal mucosa[1]. In phylogenetically recent animals also, the adenohypophysis still has some characteristics in common with the salivary glands[2]. The present review deals with the biochemistry of the adenohypophysial hormones and with biochemical aspects of the regulation of their secretion and the mechanism of their action. In addition to its role in hormone biosynthesis, however, the adenohypophysis also has interesting biochemical and enzymological characteristics which merit brief discussion.

3.1.1 Biochemistry and enzymology of the adenohypophysis

The data on different substances present in the adenohypophysis are summarised in Table 3.1, together with information on certain metabolic activities. Next to the adrenal cortex and corpus luteum, the adenohypophysis is the organ richest in ascorbic acid, although we still know nothing of the significance of this substance in the gland. Total protein synthetic activity in the adenohypophysis is in an inverse relationship to the blood level of the peripheral gland hormones; it is highest under conditions of thyroxine deficiency (after thyroidectomy), is also raised under conditions of sex hormone deficiency (after castration) and rises little or not at all with corticoid deficiency (after adrenalectomy). The reactions of protein synthesis to thyroxine deficiency or substitution are particularly interesting, as in the adenohypophysis they are the reverse of those in other tissues, e.g. the liver. The same applies to the reaction of respiration under conditions of thyroxine deficiency: respiration is stimulated in the adenohypophysis, but not in other tissues.

Table 3.1 Biochemical composition and some metabolic activities of the adenohypophysis (selected references)

Ascorbic acid[3-9]
Histamine[10]
Lipids, phospholipids[11,12]
Glycogen[13,14]
Sialic acid[15,16]
Free nucleotides[17]
^{35}S-incorporation[18]
^{32}P-incorporation[19-21]
DNA, RNA[22-40]
Amino acid incorporation[41-52]
Glucose utilisation, respiration[53-57]
Hexose monophosphate shunt[58-62]

Table 3.2 Enzymatic activities in the adenohypophysis (demonstrated by both histo- and bio-chemical methods)

Oxidoreductases
Cytochrome oxidase[53]
NADPH-cytochrome C-reductase[66]
NADH-diaphorases[68-71,74,77,79]
NADPH-diaphorases[66,68-71,74,77,79]
Lactic dehydrogenase[72,74,76-78]
Succinic dehydrogenase[53,74,75,77,79]
Malic dehydrogenase[65,73,74,77]
Glutamic dehydrogenase[74]
β-Hydroxybutyric dehydrogenase[74]
Glucose-6-P-dehydrogenase[58,59,66,74,77,80,81]
6-Phosphogluconate dehydrogenase[58,59]
a-Glycerophosphate dehydrogenase[73,79]

Hydrolytic enzymes
Alkaline phosphatase[68,82-91]
Acid phosphatase[68-71,91-108]
ATPase[68-71,96,109]
Thiamine pyrophosphatase[71,96,110]
Nucleoside diphosphatase[71,111]
β-Glucuronidase[86,95,97,101]
β-Glucosaminidase[101]
Sulphatase[97,101]
Non-specific esterases[68-71,95,108,112-114]
E-600 resistant n.e.[68-71,95,112-114]
Aminopeptidases[115-126]
Proteases[114,127-129]
Insulinase[134]
Phosphodiesterase[137,138]

Miscellaneous enzymes
Deiodases[130,131]
Methanol-forming enzyme[132]
Testosterone-5a-reductase[133]
Adenyl cyclase[137,138]
Glycogen phosphorylase[67]

According to the data of different authors, the hexose monophosphate shunt accounts for 5–25% of glucose metabolism.

The literature on adenohypophysial enzymatic activities is very extensive (for reviews see Refs. 63 and 64). Selected references are given in Table 3.2. In general, it can be claimed that the activity of oxidative enzymes is greatest in the basophils, moderate in the acidophils and lowest in the chromophobes[65]. The same applies to hydrolytic enzymes. It is also known[63] that enzymatic activation is minimal or absent after adrenalectomy, moderate after castration and highest after thyroidectomy or a goitrogen block of the thyroid. After castration and thyroidectomy, hydrolytic enzymes are activated to a greater extent than oxidative enzymes, and it can thus be assumed that the activity of the hydrolytic enzymes is more closely related to the specific secretory functions of the adenohypophysis than is the activity of the oxidative enzymes. A semi-quantitative comparison of enzymatic activities in the rat adenohypophysis after adrenalectomy, castration and thyroidectomy is submitted in Table 3.3. The reaction of enzymatic activities in the adenohypophysis to thyroxine deficiency is again the reverse of that in other tissues. For instance, glucose-6-P-dehydrogenase activity in the adenohypophysis rises after thyroidectomy and falls after thyroxine substitution, while in the liver it falls after thyroidectomy and rises after the administration of thyroxine. At present we know very little about the mechanisms of the regulation of adenohypophysial enzyme activities and the metabolic and trophic reactions of the adenohypophysis to different hormone deficiencies, although it seems highly probable that the hypothalamic-releasing hormones participate.

3.1.2 Source of hormones: adenohypophysial cell types

Although unanimity has not yet been achieved in the histological classification of the types of adenohypophysial cells (reviewed in Refs. 139 and 140), it can be generally claimed, with some simplification, that the adenohypophysis contains three basic cell-types: chromophobes (with no staining granules), basophils (with basophilic granules) and acidophils (with acidophilic granules). This general classification is based on optical microscopy, but study of the structure of the adenohypophysial cells in the electron microscope usually also demonstrates the presence of secretory granules in some of the 'chromophobe' cells, which are often regarded as hormonally inactive or as precursors of the other cell-types. The glycoprotein adenohypophysial hormones (TSH, LH, FSH*) and the peptide ACTH are synthesised and secreted by the basophils, while the protein hormones STH (GH) and prolactin (LTH) are produced by the acidophils. In the electron microscope, the individual types of cells can be identified by the size of their secretory granules[141,142]. They can also be identified by immunohistochemical methods[143,144]. Dispersed adenohypophysial cells can likewise be identified[145]. The various types of adenohypophysial cells are reviewed in Table 3.4 where they are classified from the functional aspect, i.e. according to the hormone they produce.

* TSH = thyrotrophic hormone, LH = luteinising hormone, FSH = follicle-stimulating hormone, STH (GH) = somatotrophic hormone (growth hormone), LTH = luteotrophic hormone, ACTH = adrenocorticotrophic hormone.

Table 3.3 Enzymatic activities in the adenohypophysis of male control rats, adrenalectomised rats (AE), castrated rats (Cast.) and thyroidectomised (TE) or goitrogen-treated (MTU) rats[71]

Enzyme	Cont.	AE	Cast.	TE/MTU
Alkaline phosphatase	+	+	+	++
Acid phosphatase	+	+	++	+++
ATPase	+	+	+	+/++
5-Nucleotidase	+	+	+	+/++
Thiamine pyrophosphatase	+	+	++	++/+++
Nucleoside diphosphatase	+	+	++	++/+++
Non-specific esterase	+	+	+++	+++
E-600 resistant non-specific esterases	+	+	++	+++
NAD-diaphorase	+	+	+/++	++
NADP-diaphorase	+	+	+/++	++
Succinic dehydrogenase	+	+	+/++	++
NAD-dependent dehydrogenases (lactic, malic, α-glycerophosphate)	+	+	+/++	++
NADP-dependent dehydrogenases (isocitric, glucose-6-phosphate)	+	+	+/++	++

Table 3.4 Adenohypophysial cell-types, the maximum diameter of their granules (after Ref. 145) and the hormones they produce

General type and percentage	Name	Maximum diameter of granules (nm)	Hormone
Chromophobes (50%)	—	—	—
Basophils (15%)	Thyrotrophs	150	TSH
	Gonadotrophs	200	FSH, LH = ICSH
	Corticotrophs (?)	200	ACTH
Acidophils (35%)	Somatotrophs	350	GH (STH)
	Lactotrophs	600	LTH (prolactin)

3.1.3 General regulation

Secretion of the adenohypophysial hormones is regulated by two main mechanisms—by the hypothalamic releasing hormones (RH) and by feedback action of the peripheral gland hormones. Transport of the hypothalamic RH to the adenohypophysis via the hypophysial stalk (see next section) is the final common pathway by which the modulating influences of the higher parts of the central nervous system are also mediated.

3.1.3.1 Neurovascular theory

The adenohypophysis is controlled by the brain, but it lacks secretomotor innervation and is connected to the brain by a system of blood vessels known as the portal vessels. The neurovascular theory[146] (also reviewed in Ref. 147) explains the functional relationship between the brain and the adenohypophysis as follows. In the area of the median eminence, regulatory factors are released from the hypothalamic neurones into the primary portal plexus; from here they are transported by the blood to the adenohypophysis where they specifically influence secretion of the individual adenohypophysial hormones. For some time past, these factors have been subjected to intensive study, their existence has been reliably demonstrated, in some cases their chemical structure has been determined and they have been prepared synthetically. They were at first termed 'releasing factors', but are now known as 'hypothalamic-releasing hormones' (RH) or 'hypothalamic adenohypophysiotrophic hormones'. The literature describing findings on these hormones is very extensive, but here only a few reviews have been referred (Refs. 148–154).

Most reports show that the hypothalamic RH promote release of the adenohypophysial hormones; the number of reports demonstrating their influence on the synthesis of these hormones is smaller and even less is known of how adenohypophysial growth and cell differentiation are regulated. Here again, however, the most feasible explanation is the regulatory action of factors present in the portal blood.

3.1.3.2 Feed-back regulation

A high level of the peripheral gland hormones inhibits secretion of the corresponding adenohypophysial hormones, e.g. a high thyroxine level inhibits TSH secretion and a high cortisol level inhibits ACTH secretion. Although it is not always clear whether inhibition occurs directly in the adenohypophysis (e.g. as desensitisation to the effect of tonic hypothalamic RH secretion) or in the hypothalamus (inhibition of RH secretion), it seems likely that at least part of the effect occurs in the adenohypophysis. This is known as *negative feed-back*. In some cases peripheral gland hormones stimulate adenohypophysial hormone secretion and in this case we refer to *positive feed-back*. According to recent findings, a feed-back between adenohypophysial hormones and hypothalamic RH also exists; this is known as *internal feed-back*.

3.2 THYROTROPHIC HORMONE (THYROPROPHIN, TSH)

TSH is an adenohypophysial glandotrophic hormone and its physiological action is mediated by the thyroid. It has a stimulant effect on thyroid function and structure. TSH is formed in characteristic polygonal basophils of the adenohypophysis; its secretion is regulated both by the hypothalamic thyrotrophin-releasing hormone (TRH) and by feed-back in the sense that the thyroid hormones thyroxine and tri-iodothyronine act on TSH secretion.

3.2.1 Biochemistry of TSH

Like other hormones produced by the adenohypophysial basophils, TSH is a glycoprotein. It contains about 15% carbohydrate and its molecular weight varies from 25 000 to 28 000. Bates and Condliffe[155] analysed the history of the identification of TSH and Fontaine[156] described interspecies differences. Recent studies, especially by Pierce's team[157], showed that TSH, like the other glycoprotein hormones, was composed of two sub-units—TSH-α and TSH-β. By themselves, these sub-units are both inactive, but when recombined they restore thyrotrophic activity. Most TSH-α molecules contain[96] amino-acid residues and have carbohydrate moieties at positions 56 and 82. TSH-β is composed of 113 amino-acid residues and contains one carbohydrate moiety at position 23. The amino-acid sequence in both sub-units is given in Table 3.5.

The two sub-units forming the TSH molecule are not absolutely homogeneous. Some TSH-α sub-units lack the terminal dipeptide Phe-Pro and about 70% of TSH-β sub-units lack the terminal methionine and end with tyrosine[158]. The physiological significance of the presence of two sub-units in the TSH molecule is unknown. It is remarkable, however, that the TSH-α sub-unit is identical with, and interchangeable for, the α sub-unit of the luteinising hormone (LH = ICSH, see Section 3.4) and the α sub-unit of human chorionic gonadotrophin (HCG)[159]. Pierce[158] suggests that a sub-unit (α) common to all the glycoprotein hormones is responsible for activation of the adenyl cyclase system and that specific sub-units (β) are responsible for the selective binding with the receptors of the relevant target cells, i.e. of the thyroid cells in the case of TSH. The exact localisation of the disulphide bonds is still unknown, so that suggestions of the three-dimensional structure of TSH are, for the time being, purely hypothetical.

3.2.2 Regulation of TSH secretion

TSH secretion is regulated so as to saturate thyroid hormone requirements in the tissues. Consequently, secretion of TSH rises under conditions of thyroid hormone deficiency (after thyroidectomy or a goitrogen block of thyroid hormonogenesis), under conditions of enhanced inactivation of the thyroid

Table 3.5 The amino acid sequence of bovine TSH-α and TSH-β, the hormone-specific chain. (Simplified from Pierce[158])

TSH-α	TSH-β
NH$_2$-Phe-Pro-Asp-Gly-Glu-Phe-Thr-Met-Glx-Gly-^{10}	NH$_2$-Phe-Cys-Ile-Pro-Thr-Glu-Tyr-Met-Met-His-^{10}
Cys-Pro-Glx-Cys-Lys-Leu-Lys-Glu-Asn-Lys-^{20}	Val-Glu-Arg-Lys-Glu-Cys-Ala-Tyr-Cys-Leu-^{20}
Tyr-Phe-Ser-Lys-Pro-Asx-Ala-Pro-Ile-Tyr-^{30}	Thr-Ile-Asn-Thr-Thr-Val-Cys-Ala-Gly-Tyr-^{30}
Gln-Cys-Met-Gly-Cys-Cys-Phe-Ser-Arg-Ala-^{40}	Cys-Met-Thr-Arg-Asx-Val-Asx-Gly-Lys-Leu-^{40}
Tyr-Pro-Thr-Pro-Ala-Arg-Ser-Lys-Lys-Thr-^{50}	Phe-Leu-Pro-Lys-Tyr-Ala-Leu-Ser-Gln-Asp-^{50}
Met-Leu-Val-Pro-Lys-Asn-Ile-Thr-Ser-Glx-^{60}	Val-Cys-Thr-Tyr-Arg-Asp-Phe-Met-Tyr-Lys-^{60}
Ala-Thr-Cys-Cys-Val-Ala-Lys-Ala-Phe-Thr-^{70}	Thr-Ala-Glu-Ile-Pro-Gly-Cys-Pro-Arg-His-^{70}
Lys-Ala-Thr-Val-Met-Gly-Asn-Val-Arg-Val-^{80}	Val-Thr-Pro-Tyr-Phe-Ser-Tyr-Pro-Val-Ala-^{80}
Glx-Asn-His-Thr-Glu-Cys-His-Cys-Ser-Thr-^{90}	Ile-Ser-Cys-Lys-Cys-Gly-Lys-Cys-Asx-Thr-^{90}
Cys-Tyr-Tyr-His-Lys-Ser-COOH	Asx-Tyr-Ser-Asx-Cys-Ile-Hig-Glu-Ala-Ile-^{100}
	Lys-Thr-Asn-Tyr-Cys-Thr-Lys-Pro-Gln-Lys-^{110}
	Ser-Tyr-Met-COOH

3.2.2.1 Thyrotrophin-releasing hormone (TRH)

The TRH activity of hypothalamic extracts was demonstrated in the adenohypophysis *in vitro*[160] and *in vivo*[161] in the early 1960s; further findings on TRH are described in various reviews[162-164]. The work reached a critical phase in 1965–1966, when it looked as though TRH was not a peptide at all. In 1966, however, Schally *et al.*[165] distinguished three amino acids (Glu, His and Pro) in the TRH molecule. Different tripeptides synthesised from these three amino acids were inactive and it was not until the Glu-His-Pro tripeptide molecule was modified to (pyro)Glu-His-Pro-NH$_2$ that a compound with TRH activity was obtained. This structure was demonstrated in porcine[166] and ovine[167] TRH and is probably also present in bovine[168] and human[169] TRH. The introduction of a methyl group into the imidazole nucleus of the TRH molecule made the (pyro)Glu-N^{3im}Met-His-Pro-NH$_2$ compound eight times more active than natural TRH[170].

Synthetic and natural TRH are both active in ng concentrations *in vitro*, and in the same amounts when administered intravenously to experimental animals[164]. An increase in the blood TSH level was repeatedly demonstrated after injections of 100–800 μg of TRH in man. In experiments dealing with the diagnostic uses of TRH (to differentiate a hypothalamic from a hypophysial disturbance of TSH secretion), TRH can also be administered perorally or sublingually in doses of the order of 0.01 g.

The mechanism of action of TRH in the adenohypophysis has not been elucidated definitively. Although preliminary evidence[171,172] indicated that TRH stimulated both TSH release and synthesis in the adenohypophysis, most later observations showed that it was the release process which was primarily influenced[173]. TRH-induced TSH secretion does not require either new protein synthesis or DNA-dependent RNA synthesis. Measurement of the incorporation of radioactive amino acids into the TSH of adenohypophyses incubated with TRH showed no increase in TSH synthesis. The reaction of TSH secretion to TRH is energy-dependent, however, and according to some findings, TRH causes activation of the adenyl cyclase system in the adenohypophysis[163,174]. TSH secretion is stimulated, and the action of TRH is potentiated, if the K$^+$ ion concentration in the medium is high[175], while in the absence of Ca^{2+} ion they are inhibited[174,176]; ionic shifts on the membranes of the adenohypophysial cells and their depolarisation, followed by Ca^{2+} ion uptake and by the release of secretory granules, thus evidently form part of the action of TRH. The displacement of thyroxine from its bond with the adenohypophysial proteins may be a part of the effect of TRH within the adenohypophysis[177]. TRH also stimulated the oxidation of 6-^{14}C-, but not of 1-^{14}C-labelled glucose to ^{14}CO$_2$[178].

Although the first studies of TRH repeatedly demonstrated that its effect was specific, i.e. that it stimulated only the secretion of TSH, but not of ACTH or of LH, a few recent findings show that TRH stimulates prolactin secretion from the adenohypophysis *in vitro*[179, 180] and *in vivo*[181]. The significance of this prolactin-releasing activity of TRH is not known. It is tempting to assume that this activity is related to the marked growth of the adenohypophysis under conditions of thyroid hormone deficiency. Since release of the prolactin-inhibiting hormone (PIH, see Section 3.6.2.1 and 3.6.2.2) possibly has a growth-inhibiting effect in the adenohypophysis[182], the prolactin-releasing effect of TRH could be a manifestation of its growth-promoting activity.

3.2.2.2 Feed-back regulation of TSH secretion

The reciprocal relationship between the blood level of thyroid hormones and TSH secretion has been known for a long time[183] and recent findings have been analysed in detail in various reviews[184, 185]. A high thyroxine and tri-iodothyronine level causes inhibition directly in the adenohypophysis. The intrahypophysial injection of thyroxine inhibits TSH secretion[186] and the adenohypophysis reacts to thyroxine deficiency by hypersecretion of TSH, even when implanted in the anterior chamber of the eye[187]. This would seem to indicate that thyroxine acts directly on the adenohypophysis, independently of TRH. On the other hand, thyroxine indubitably desensitises the adenohypophysis to the stimulant action of TRH[172, 188]. This inhibitory effect is blocked by inhibitors of protein synthesis[189, 190], showing that thyroxine does not cause this inhibition directly, but rather via synthesis of a proteinaceous inhibitor.

Despite the data demonstrating the direct effect of thyroxine in the adenohypophysis, physiological evidence shows that thyroxine also probably acts at the hypothalamic level, i.e. on TRH secretion. According to some reports, TRH secretion is raised in the presence of thyroxine deficiency. Studies of histological and trophic reactions of the adenohypophysis to thyroxine deficiency likewise indicate either that TRH also has a trophic and differentiative effect in the adenohypophysis, or that other growth factors with a trophic and differentiative effect act on the adenohypophysis under conditions of thyroxine deficiency.

TSH also lowers the TRH content of the hypothalamus in thyroidectomised animals[191]. This seems to testify to the existence of an internal feed-back between TSH and TRH secretion. Contrary evidence is provided by the extreme and prolonged increase in TSH secretion after thyroidectomy[192] and the evidently unabated hypothalamic stimulation of the adenohypophysis in thyroidectomised animals, culminating in the formation of adenohypophysial tumours[193].

3.2.3 Mechanism of TSH action

Although large doses of TSH can have an extrathyroidal lipolytic effect[194, 195] and high blood TSH concentration seem able to cause an increase in the mucopolysaccharide content of the retrobulbar tissue of the orbit and the skin (the existence of a separate exophthalmus-producing substance, EPS, is sometimes presumed[195]), the main effect of TSH is activation of the thyroid[155]. TSH has an activating effect both on the structure of the thyroid follicules (the flat epithelial cells become cuboidal and the epithelium:colloid ratio alters in favour of the epithelium) and on various phases of biosynthesis of the thyroid hormones (thyroxine and tri-iodothyronine). TSH stimulates iodide uptake, organic iodine binding and actual hormone secretion, i.e. release of the hormones from their bond with thyroglobulin. TSH likewise stimulates the metabolism of the thyroid, including oxygen and glucose consumption and phospholipid, electrolyte and water metabolism.

Much light has been thrown on the mechanism of TSH action in the thyroid in recent years (reviewed in Ref. 196). In the first place, it was shown that TSH is specifically bound to receptors in the wall of the follicular cells[197]. Metabolic changes can be induced in the thyroid simply by incubating slices for 1 min in a medium containing TSH and then thoroughly washing them; this effect is blocked by incubation in a solution of anti-TSH antibodies. Since the antibodies do not actually penetrate the cells, it is obvious that adsorption of TSH to their surface is sufficient to activate them. Another step in the effect of TSH is the activation of plasma membrane adenyl cyclase[198]. This reaction of the thyroid cells is highly specific for TSH. Activation of adenyl cyclase can also be demonstrated in the isolated plasma membrane fraction of the thyroid cells[199]. The cyclic-3′,5′-adenosine monophosphate content of the thyroid cells increases after exposing them to TSH[200, 201]. The effect of TSH on thyroid hormone secretion can be imitated by dibutyryl-3′,5′-adenosine monophosphate[202, 203] and also by 3′,5′-adenosine monophosphate[204]. Further evidence is the blocking effect of propranolol on the stimulation of thyroid function and on the activation of thyroid cell adenyl cyclase[205]. Pastan[196] concluded that, after TSH binding to receptors in the plasma membrane and activation of adenyl cyclase, the increase in the cyclic-3′,5′-adenosine monophosphate level is the factor which leads to stimulation of glucose and phospholipid metabolism in the thyroid cells, to thyroxine secretion, to RNA and protein synthesis and to increased iodide uptake. Organic iodide binding is blocked by inhibitors of protein synthesis and is thus evidently dependent on RNA and protein synthesis.

3.3 ADRENOCORTICOTROPHIC HORMONE (ADRENOCORTICOTROPHIN, ACTH)

ACTH is another adenohypophysial glandotrophic hormone, its main effect being stimulation of the adrenal cortex and glucocorticoid secretion (cortisol in man and corticosterone in certain other species). The site of ACTH synthesis is known with less certainty than that of the other adenohypophysial hormones. It was variously thought to be the chromophobes, the acidophils and the basophils (for reviews see Refs. 139, 206), but the latest immunohistochemical studies localised it in basophils[143] known in functional terminology as corticotrophs. ACTH secretion is regulated by the hypothalamic corticotrophin-releasing hormone (CRH) and by feed-back in which glucocorticoids, especially cortisol or corticosterone, act on ACTH secretion.

3.3.1 Biochemistry of ACTH

ACTH is a single-chain polypeptide containing 39 amino acids[207, 208]. The amino-acid sequence is given in Table 3.6.

The first 24 amino acids form the main component of the molecule and are the same in all species. The amino acids in positions 25–33 determine the species specificity of the molecule, while those in positions 34–39 assure its stability and are again identical in all species. Schwyzer[209] summed up the significance of the individual parts of the ACTH molecule as follows:

Amino acids 5–10 ('Word' 1): bind with receptors of amphibian melanophores; amino acids 1–4 ('Word' 2) and amino acids 11–14 ('Word' 3): enhance this activity thousandfold; amino acids 15–18 ('Word' 4): introduce adrenocorticotrophic and reduce melanophorotropic activity; amino acids 19–24 ('Word' 5) and amino acids 34–39 ('Word' 7): stabilise the molecule; amino acids 25–33 ('Word' 6): species specific structure: introduce antibody production in other species (together with 'Word' 7).

Table 3.6 Structure of the ACTH molecule (Taken from Evans et al.[208])

Human ACTH

1   10
H-Ser-Tyr-Ser-Met-Glu-His-Phe-Arg-Try-Gly
                                    20
Lys-Pro-Val-Gly-Lys-Lys-Arg-Arg-Pro-Val-
Lys-Val-Tyr-Pro-
25   33
Asp-Ala-Gly-Glu-Asp-Gln-Ser-Ala-Glu-
34   39
Ala-Phe-Pro-Leu-Glu-Phe-OH

Bovine ACTH

25   33
Asp-Gly-Glu-Ala-Glu-Asp-Ser-Ala-Gln

Ovine ACTH

Ala-Gly-Glu-Asp-Asp-Glu-Ala-Ser-Gln

Porcine ACTH

Asp-Gly-Ala-Glu-Asp-Gln-Leu-Ala-Glu

Dozens of peptides with a structure corresponding to different parts of the ACTH molecule and structural analogues of ACTH have been synthesised (for reviews see Refs. 210, 211). The activity of some of these analogues is actually far greater than that of natural ACTH. For instance, D-Ser^1-Nl$_e^{*4}$-Val^{25}-NH$_2$-ACTH$_{1-25}$ has between six and eight times greater depletive effect on the ascorbic acid content of the adrenals, three to five times greater effect on corticosterone production, only slightly greater lipolytic activity, but over 13 times greater melanophorotropic activity than natural ACTH[210]. The substitution of lysine or ornithine for the amino acids in positions 17–18 likewise potentiates adrenocorticotrophic activity.

As indicated above, ACTH has melanophorotropic activity and causes expansion of the pigment granules in the melanophores of amphibian skin. In some species, the intermediate lobe of the hypophysis produces a separate peptide with a melanophorotropic action, known as the melanocyte-stimulating hormone (MSH, intermedin). The function of this hormone in mammals, and particularly in man, has not been reliably demonstrated. MSH occurs in two forms[212]: α-MSH and β-MSH. There are interspecies differences in the structure of the molecules, but, in general, α-MSH is identical with the first

* Norleucine

ADENOHYPOPHYSIAL HORMONES

13 amino acids of ACTH, to which the melanophorotropic activity of ACTH is bound. The structure of bovine α-MSH and β-MSH is shown in Table 3.7. Similarity to the ACTH molecule can likewise be seen in the β-MSH molecule,

Table 3.7 Structure of bovine α-MSH and β-MSH (After Novales[211])

α-MSH

CH_3.CO-Ser-Tyr-Ser-Met-Glu-His-Phe-Arg-Try-$\overset{10}{Gly}$-Lys-Pro-Val-NH_2

β-MSH

Asp-Ser-Gly-Pro-Tyr-Lys-Met-Glu-His-$\overset{10}{Phe}$-Arg-Try-Gly-Ser-Pro-Pro-Lys-Asp

---

as the β-MSH amino acids 9, 11–17 and 19 are identical with the ACTH amino acids 2, 4–10 and 12. Structural analogues of MSH have also been synthesised[213]. Comparison of the activity of different ACTH and MSH analogues showed further similarity in their lipolytic effect[214].

3.3.2 Regulation of ACTH secretion

ACTH secretion is regulated in two ways (for reviews see Refs. 215, 216). The first is homoeostatic regulation, the purpose of which is to keep the blood concentration of adrenal cortex hormones at a constant level. It takes effect mainly under conditions of a diminished blood corticoid level, ACTH secretion being stimulated after adrenalectomy and after a chemical block or enzymatic disturbance of hormone biosynthesis in the cortex. The other type is reflex regulation, when ACTH secretion rapidly increases during exposure to the most diverse forms of stress (heat, cold, trauma, emotions, etc.). In these situations, ACTH secretion rises to some extent independently of the blood corticoid level. Both types of regulation use a single final common pathway, i.e. secretion of hypothalamic CRH into the adenohypophysial portal blood.

3.3.2.1 Corticotrophin-releasing hormone (CRH)

Study of CRH started with tests of neurohypophysial extracts (reviewed in Ref. 151). Several compounds which stimulate ACTH secretion under experimental conditions can be isolated from the neurohypophysis. These are: (a) Lys-vasopressin and Arg-vasopressin; (b) the substance β-CRF (for neurohypophysial factors we have kept the shortened form of the corticotrophin-releasing factor, CRF), which has the structure acetyl-Ser-Tyr-Cys-Phe-His-Asn-Glu-Cys-Pro-Val-Lys-Gly-NH_2[217]; (c) the substance α_2-CRF, which has the same amino-acid sequence as α-MSH, but with a blocked group on the N-terminal Ser[218]; and (d) the substance α_1-CRF, the structure of which has not been determined, but which likewise resembles α-MSH. β-CRF possesses greater activity than α_1- and α_2-CRF; for example, 0.06 µg β-CRF gave the same stimulation of ACTH secretion as 1.5 μg a_1-CRF or a_2-CRF[150]. Neurohypophysial extracts contain far more a_1-CRF and a_2-CRF than β-CRF, however. A number of structural analogues of vasopressin and both CRFs, with different degrees of corticotrophin-releasing activity *in vitro* and *in vivo*, have also been synthesised. According to the existing physiological evidence, probably none of these substances is natural CRH, but they can be of significance as supplementary regulators.

In the search for CRH in hypothalamic extracts (reviewed in Refs. 151, 219), attempts to isolate and identify the active hormone have so far (at the time this review was written) failed. CRH activity similar to vasopressin, β-CRF and a_1- and a_2-CRF has been found repeatedly in hypothalamic extracts. Since the activity of hypothalamic extracts can be destroyed by different proteolytic enzymes, a peptide compound again seems to be involved. The study of hypothalamic CRH is hampered by the low concentration of the active substance in hypothalamic extracts and also by the apparent instability of CRH.

The mechanisms of the action of CRH in the adenohypophysis has likewise not been elucidated. Indirect evidence obtained by studying ACTH secretion during stress (reviewed in Refs. 173, 220) indicates that ACTH secretion can, under certain conditions, be blocked by inhibitors of protein synthesis. In some experiments, hypothalamic extracts causing ACTH release also produced an increase in the ACTH content of the adenohypophysis *in vivo*[221] and *in vitro*[222]. If these findings are confirmed in experiments using pure CRH, it would mean that CRH acts on both the release and synthesis of ACTH. ACTH secretion is also stimulated by K^+ ion in large excess, while corticosterone inhibits this reaction[223]. Incubation of the adenohypophysis in the absence of Ca^{2+} ion inhibited the stimulant effect of hypothalamic extract, as well as of vasopressin, dibutyryl cyclic-3′,5′-adenosine monophosphate and theophylline[224]; the release of ACTH from unstimulated adenohypophyses *in vitro* was not inhibited by Ca^{2+} ion deficiency, however. This indicates that the mechanism of the action of CRH may consist of depolarisation of the adenohypophysial cells, as appears to be the case with TRH. Membrane depolarisation may be associated with the uptake of Ca^{2+} ion which activates the release of storage granules. These findings also suggest that the adenyl cyclase system participates in the reaction of the adenohypophysis to CRH.

3.3.2.2 Feed-back regulation of ACTH secretion

The feed-back regulation of ACTH secretion by the blood cortisol or corticosterone level (reviewed in Refs. 225, 226) mainly concerns the relatively slow adjustment of ACTH secretion to changed blood levels of the adrenal cortical hormones. Rapid (reflex) ACTH secretory reactions (stress) are relatively independent of the corticoid level; they can be only partly inhibited, and only by large doses of corticosteroids. The inhibitory effect of excess corticoids on ACTH secretion (development of atrophy of the adrenal cortex) was described by Ingle as early as 1938[227]. Conversely, increased ACTH secretion in the presence of corticoid deficiency is given as the explanation for the regeneration of the adrenals after their partial removal and for the compensatory hypertrophy of the remaining adrenal after unilateral adrenalectomy.

High ACTH levels can be demonstrated in the blood of adrenalectomised animals and in patients with adrenal insufficiency (Addison's disease); in the latter case, excess ACTH actually leads to the development of typical forms of pigmentation.

Diminished ACTH secretion after large doses of corticoids can also be accompanied by a decrease in the ACTH content of the adenohypophysis[228]. Conversely, the blood ACTH concentration and the amount of ACTH in the adenohypophysis both rise after adrenalectomy. This indicates that the blood corticoid level simultaneously influences both ACTH secretion and synthesis It also shows that CRH likewise exerts both types of action (see Section 3.2.2.1).

The mechanisms of the feed-back action of the adrenal cortical hormones evidently involve effects on hypothalamic CRH secretion. Study of the possible site of action of corticoids (the adenohypophysis or the hypothalamus) shows rather conclusively that the site of the inhibitory action of corticoids is the hypothalamus, and possibly other structures in the higher parts of the central nervous system as well. Implantation of a corticoid into the hypothalamus reduces ACTH secretion, while implantation into the adenohypophysis has no effect[230]. Further feed-back receptors for corticoids evidently exist in the anterior forebrain[231]. Findings showing that large doses of corticoids do not inhibit the adenohypophysial reaction to CRH[232] are in agreement with the concept that the receptors for corticoids are localised in the hypothalamus and in higher parts of the central nervous system. If corticoids acted in the adenohypophysis, we should expect the latter's reaction to CRH to be depressed. That is not the case, however, and animals in which endogenous ACTH secretion has been blocked by large doses of corticoids are actually used for the assay of CRH[233]. The feed-back action of the cortical hormones and thyroxine (see Section 3.2.2.2) thus evidently takes effect at different sites.

There is fairly convincing evidence for the existence of an internal feed-back between ACTH and CRH secretion (reviewed in Ref. 191). Stressful stimuli cause a marked increase in ACTH secretion in adrenalectomised animals if the initial ACTH level is low, and a small increase if the initial level is high. The administration of exogenous ACTH produces an increase in the ACTH content of the adenohypophysis and reduces the stress-induced decrease in the amount of ACTH in the gland. Implantation of the adenohypophysis into the hypothalamus leads to a decrease in adrenal cortical function[234]. ACTH implanted in the hypothalamus has the same effect, but not ACTH implanted in the adenohypophysis[235]. If adrenalectomy, which raises the CRH content of the hypothalamus, is followed by hypophysectomy, the amount of CRH in the hypothalamus rises still further[236]. These and other findings indicate that there is probably an internal feed-back between ACTH and CRH. In addition, evidence that ACTH also has a direct effect (not mediated by the adrenals) on higher nervous structures exists[191].

3.3.3 Mechanism of ACTH action

The main physiological stimulatory effect of ACTH (reviewed in Ref. 208) is on the structure, metabolism and function of the adrenal cortex (especially the zona fasciculata and the zona reticularis). The main adrenal cortical product secreted through the action of ACTH in man (and in dogs and guinea-pigs) is cortisol, and in various other species (rats, mice, cats) corticosterone. ACTH reduces the cholesterol content of the adrenals (consumption for corticoid biosynthesis) and lowers the ascorbic acid content. Nothing is known of either the mechanism or the significance of ACTH-induced ascorbic acid depletion in the adrenals. The extra-adrenal effects of ACTH are expansion of the pigment granules in the melanophores of amphibian skin and pigmentation of the skin in the presence of a high blood ACTH level in man; in this respect, ACTH has the same effect as MSH. Its extra-adrenal lipolytic effect[194, 195] and its other extra-adrenal effects (changes in adipose tissue metabolism, hypoglycaemic effect, diabetogenic effect, etc.) are all of a pharmacological nature and their physiological role in the presence of normal blood ACTH levels is uncertain.

The main adrenal effect of ACTH is stimulation of the utilisation of cholesterol for corticoid biosynthesis, the first intermediary product being Δ-5-pregnenolone. The cholesterol content of the adrenals consequently falls in response to ACTH. In some situations, however (namely a block of steroidogenesis between cholesterol and Δ-5-pregnenolone), ACTH raises the amount of cholesterol in the adrenals, i.e. it evidently also stimulates cholesterol uptake and biosynthesis in the adrenals. The enzymatic system transforming cholesterol to Δ-5-pregnenolone is localised in the mitochondria of the cortical cells[237]; Δ-5-pregnenolone leaves the mitochondria and is converted in the cytoplasm to progesterone. The enzymatic system of the microsomes transforms progesterone to deoxycortisol or deoxycorticosterone, which re-enter the mitochondria where cortisol or corticosterone is finally produced.

The mechanism of the action of ACTH in the adrenals initially involves interaction with receptors in the plasma membranes of the cortical cells[238]. There are two types of receptors, with different association constants for ACTH, in the cortical cells. The high-order sites ($K_A = 9 \times 10^{11}$, 60 sites per cell) bind about half of the molecules bound in the presence of physiological levels of ACTH, while the low-order sites ($K_A = 3 \times 10^7$, 600 000 sites per cell) bind the other half[239]. The specificity and mechanism of the interaction of ACTH and its analogues with cell receptors have been the subject of a series of studies[214, 240, 241]. Another phase of the action of ACTH is activation of cell membrane adenyl cyclase and increased formation of 3',5'-adenosine monophosphate in the cells[238, 242, 243]. The resultant nucleotide is assumed to induce the synthesis of a regulatory protein by acting at the level of mRNA translation[244]. Formation of this regulatory protein is blocked by inhibitors of protein synthesis and this would also explain the effect of inhibitors of protein synthesis on the general steroidogenic action of ACTH in the adrenals. The regulatory protein activates cortical steroidogenesis, starting with the transformation of cholesterol to Δ-5-pregnenolone (activation of desmolase, the mitochondrial enzyme splitting off the side-chain of cholesterol). A dynamic model was also formulated to explain the reaction of adrenal steroidogenesis to ACTH[246]. The mechanism of the adrenal action of ACTH has been investigated in further studies[245, 247-250], in some cases with special reference to the role of the intermediary reaction requiring protein synthesis[251, 252].

Activation of the enzyme protein kinase, which catalyses phosphorylation of ribosomal protein and stimulates protein synthesis, is evidently a significant factor (reviewed in Ref. 264). Calcium ions also participate in the action of ACTH in the cells of the adrenal cortex[264]. In the presence of Ca^{2+} ions, the strength of the signal generated by the interaction of ACTH with its receptor and transmitted to the adenyl cyclase compartment is proportionately increased.

The lipolytic action of ACTH also involves binding to receptors in the wall of the fat cells and activation of the adenyl cyclase system[253-255]. Since several hormones induce a similar lipolytic reaction in the fat cells (e.g. adrenalin, glucagon, ACTH, LH) the question arises as to whether the fat cells have one type of receptor with relatively low specificity (capable of activation by different hormones), or more than one type, each reacting to a different lipolytic hormone. The second hypothesis was recently found to be the more likely[256].

3.4 THE GONADOTROPHINS

Gonadotrophins are glandotrophic hormones of the adenohypophysis and act via the gonads. They have a stimulant effect on the structure and function of the gonads, both with respect to endocrine function (production of the sex steroid hormones) and with respect to the production of germ cells. Gonadotrophins are synthesised in oval basophils of the adenohypophysis and evidently both gonadotrophins originate in the same type of cell (or in two very similar cells). Their secretion is probably regulated by a common hypothalamic regulatory hormone whose effect on secretion of the individual gonadotrophins is co-determined by the sex hormone level in blood. In man and other mammals there are two gonadotrophins—the follicle-stimulating hormone, FSH (reviewed in Ref. 257) and the luteinising or interstitial cell-stimulating hormone, LH, ICSH (reviewed in Ref. 258). In certain other species (rat, sheep) there is a third gonadotrophin, the luteotrophic hormone or prolactin, LTH, which maintains the function of the corpus luteum. Since prolactin probably does not have a luteotrophic effect in man it will be discussed separately; furthermore, prolactin has a different chemical nature (protein) and site of synthesis (acidophils) from the other gonadotrophins.

In addition to the adenohypophysis, the placenta also produces a gonadotrophin (human chorionic gonadotrophin, HCG), which resembles the adenohypophysial gonadotrophins in structure and effect[259]. Human menopausal gonadotrophin (HMG) and pregnant mare serum gonadotrophin (PMSG)[260] are mixed gonadotrophin preparations extracted from urine or blood.

3.4.1 Biochemistry of the gonadotrophins

The adenohypophysial gonadotrophins (FSH and LH = ICSH) are glycoproteins[261]; their molecular weight is given as about 30 000 and they contain about 20% carbohydrate. They are both formed of two sub-units, a and β, the a sub-unit being identical with the TSH-a sub-unit and evidently with the HCG-a sub-unit also. The β sub-units are specific components of the hormone molecule[261-263]. The structure of the a sub-unit has already been described in the case of TSH (Table 3.5) and the structure of the LH-β sub-unit is given in Table 3.8. A number of positions in the structure of LH-α have been defined more exactly than in the case of TSH-α[266]: Gln^9, Glu^{13}, Asp^{26}, Glu^{60}, Glu^{81}. Both gonadotrophins are glycoproteins and the carbohydrate moieties in their molecules are fucose, galactose, mannose, glucosamine and galactosamine (LH)[267] and fucose, galactose, mannose, acetylglucosamine and N-acetylneuraminic acid (FSH)[262].

Table 3.8 Amino acid sequence in the sub-unit LH-β. (After Papkoff[265].)

```
                            10
      H-Ser-Arg-Gly-Pro-Leu-Arg-Pro-Leu-Cys-Glu-
                            20
        Pro-Ile-Asn-Ala-Thr-Leu-Ala-Ala-Glu-Lys-
                            30
        Glu-Ala-Cys-Pro-Val-Cys-Ile-Thr-Phe-Thr-
                            40
        Thr-Ser-Ile-Gly-Ala-Tyr-Cys-Cys-Pro-Ser-
                            50
        Met-Lys-Arg-Val-Leu-Pro-Val-Pro-Pro-Leu-
                            60
        Ile-Pro-Met-Pro-Gln-Arg-Val-Cys-Thr-Tyr-
                            70
        His-Gln-Leu-Arg-Phe-Ala-Ser-Val-Arg-Leu-
                            80
        Pro-Gly-Pro-Cys-Pro-Val-Asp-Pro-Gly-Met-
                            90
        Val-Ser-Phe-Pro-Val-Ala-Leu-Ser-Cys-His-
                           100
        Gly-Pro-Cys-Cys-Arg-Leu-Ser-Ser-Thr-Asp-
                           110
        Cys-Gly-Pro-Gly-Arg-Thr-Glu-Pro-Leu-Ala-
                           120
        Cys-Asp-His-Pro-Pro-Leu-Pro-Asp-Ile-Leu-OH
```

Unlike the sub-units TSH-α and TSH-β, which are inactive by themselves (see Section 3.2.1), the isolated LH-β sub-unit produces ovulation in female hamsters[286]. A dose of 20 µg LH-β is required to produce ovulation in 100% of the animals, while only 5 µg of non-dissociated LH is needed. On the other hand, both isolated sub-units are inactive in the ovarian ascorbic acid test and only the whole LH molecule is active[287]. The two sub-units are bound in the non-dissociated LH molecule by non-covalent linkages, i.e. by secondary forces such as hydrogen, hydrophobic or electrostatic bonds[268]. There are probably structural differences between the adenohypophysial gonadotrophins and those extracted from urine[269]. Crooke[270] has reviewed the isolation and characterisation of human gonadotrophins.

3.4.2 Regulation of gonadotrophin secretion

The regulation of gonadotrophin secretion is a far more complex problem than the regulation of the secretion of the adenohypophysial hormones already discussed above. In the first place, there is the sex difference. In women and all female mammals, gonadotrophins are secreted cyclically, whereas in men and male mammals they are secreted tonically, i.e. without a marked short-term rhythmic pattern (of a few days or weeks). Long-term rhythms of a seasonal character can also occur in males, however. Another complicating factor is the absence of a simple feed-back between the gonadal hormones and gonadotrophins. Feed-back action of the gonadal hormones does exist, but it is complex and is not only negative in character. In some circumstances, oestrogens stimulate LH secretion, i.e. by a positive feed-back.

Lastly, the situation is complicated by the fact that neither of the two gonadotrophins has a single target or regulates the secretion of only one hormone; in females, for example, oestrogen secretion is influenced by both FSH and LH. The gametogenic effect of gonadotrophins and, in some species, the presence of a third gonadotrophin, the luteotrophic hormone, also makes matters more complicated.

Gonadotrophin secretion is likewise controlled by a hypothalamic regulatory hormone, which was recently shown to be the same for both LH = ICSH and FSH (LH–RH, FSH–RH) and ought therefore be re-named the gonadotrophin-releasing hormone (GnRH). The nature of its action in the adenohypophysis (LH or FSH secretion) is determined by the blood level of the gonadal steroids. This effect of the gonadal steroids can be regarded as part of their feed-back action. In addition, the blood level of the gonadal steroids influences the rate of GnRH secretion. Additional hypothalamic RH influencing separately LH and/or FSH secretion may also exist.

3.4.2.1 Gonadotrophin-releasing hormone (GnRH, LH–RH = FSH–RH)

It was originally thought that separate hypothalamic regulatory hormones existed for FSH and LH = ICSH (reviewed in Refs. 148–154). After the first demonstration of the LH–RH[271] and FSH–RH[272] activity of hypothalamic extracts, various laboratories tried for a whole decade to isolate and determine the structure of these two hormones. For some time it was actually doubted, as in the case of TRH, whether they were of peptide nature. Finally, however, it was found that the two adenohypophysial gonadotrophins probably have one common regulatory hormone, with a simple decapeptide structure[273]. The suggested structure of GnRH is shown in Table 3.9. Additional separate LH–RH and/or FSH–RH may exist.

In the first studies of the LH–RH and FSH–RH activity of hypothalamic extracts, it was stated several times that the two activities could be isolated independently by separation methods. It has since been concluded, however[274–276], that the two activities are inseparable, even after using 12 purification steps and applying a variety of separation methods. A synthetic

Table 3.9 Structure of GnRH (LH–RH = FSH–RH). (From Matsuo et al.[273])

(pyro)Glu-His-Trp-Ser-Tyr-Gly-Leu-Arg-Pro-Gly-NH$_2$

GnRH decapeptide has also been prepared, but so far it has raised only the LH, but not the FSH, level[277]. The reason seems to be inadequate dosage, as the sensitivity of LH secretion to natural GnRH is much greater (i.e. smaller doses are needed) than FSH secretion.

The mechanism of action of GnRH in the adenohypophysis has been only partly elucidated. From the very first studies, GnRH was found, much more distinctly than in the case of the other RH (for review see Ref. 173), to act on both gonadotrophin secretion and synthesis. The same was recently found to apply to the decapeptide mentioned above[278]. GnRH likewise appeared to influence the structural differentiation of the adenohypophysial gonadotrophic basophils[279]. LH secretion is also stimulated by a high K^+ ion concentration in the medium[280,281] and is inhibited by the absence of Ca^{2+} ion, while FSH secretion is inhibited by omitting K^+ ion[282]. Depolarisation of the membranes of the adenohypophysial gonadotrophs and Ca^{2+} ion uptake by the cells, followed by the release of secretory granules, may thus be a factor also in the action of GnRH in the adenohypophysis. Participation by the adenyl cyclase system is another probability. The possibility of a difference in the mechanism of action of GnRH on LH and FSH secretion should likewise not be ignored. Oligomycin-induced inhibition of LH release[280] indicates that a protein-synthetic process may participate in the overall reaction. The effect of GnRH on FSH secretion is also blocked by inhibitors of protein synthesis[283,284]. McCann et al.[285] have submitted a survey of these and other characteristics of the action of GnRH in the adenohypophysis.

3.4.2.2 Feed-back regulation of gonadotrophin secretion

The position of the feed-back action of the gonadal hormones in the overall regulation of gonadotrophin secretion has been reviewed by Harris and Campbell[288] for LH, by Donovan[289] for FSH and by Davidson[290] for both gonadotrophins. For the reasons given in Section 3.4.2, it seems appropriate to discuss the feed-back action of the gonadal steroids in males and females separately.

The feed-back action of the gonadal steroids *in males* has been reviewed by Davidson[291]. Castration of males led to an increase in the blood FSH level, but an increase in the FSH content of the adenohypophysis was found only in certain studies. On the other hand, the LH content of the adenohypophysis rose markedly within a few weeks after castration, as well as blood levels. Unlike the case of females, unilateral castration was not followed by pronounced compensatory hypertrophy of the remaining gonad in males.

Androgens in excess generally cause inhibition of LH secretion, but evidence of a 'positive feed-back', i.e. stimulation of LH secretion by testosterone, was also found in males. Since it is still not clear whether FSH (apart from its gametogenic function) stimulates the production of some hormone (local androgens, hypothetical 'inhibin') in the testes, there is likewise no reliable information on the feed-back action of the testes on FSH secretion. Some findings indicate that testicular oestrogens may have a feed-back action.

Determination of the site of action of the male gonadal steroids is a very difficult problem and it is not even clear whether the same receptors are involved in the reaction to either a decrease or an increase in the gonadal steroid level[291]. Most of the evidence shows, however, that the main site of feed-back action is the hypothalamus. There is no reliable information on the existence of an internal feed-back between gonadotrophins and hypothalamic GnRH although, according to some reports, intrahypothalamic LH implants have an inhibitory effect on the LH content of the adenohypophysis (reviewed in Ref. 191).

The feed-back action of the gonadal steroids *in females* has been reviewed by Flerkó[292] and Davidson[290]. The situation here is more complex than in males. Ovariectomy is followed by an increase in both the blood FSH level and the FSH content of the adenohypophysis[289] and the same applies to LH[288]. Oestrogen inhibits the post-castration increase, but progesterone inhibits it only when combined with oestrogen. This is the negative feed-back action of oestrogen. As early as 1934, Hohlweg[293] discovered that oestrogen stimulated LH secretion, i.e. the induction of ovulation. This positive feed-back action of oestrogen has since been confirmed many times over and is a significant part of the regulation of adenohypophysial LH secretion. Under certain conditions, however, progesterone can also induce LH release and ovulation. Harris and Campbell[288] sum this up as follows: 'It seems clear that a raised level of progesterone in the blood can both excite and inhibit ovulation. The effect seen following a single dose of progesterone probably depends on the timing of the injection, as well as on the dose, state of the ovary and previous priming action of oestrogen'. Findings on the effect of oestrogens could be similarly formulated. Oestrogens further exert a biphasic action on the gonadotrophin-releasing mechanism, gonadotrophin secretion being first of all inhibited (FSH more than LH) and then stimulated (LH more than FSH)[294].

Here it would be appropriate to discuss the participation of feed-back in the general regulation of gonadotrophin secretion, especially in the cyclic regulation of gonadotrophin secretion in females. This falls outside the scope of the present review, however, and the reader is therefore referred to the relevant authors[288-291]. The relationship between gonadotrophin, oestrogen and gestagen secretion in the individual phases of the cycle in women has been analysed by Harris and Naftolin[295].

3.4.3 Mechanism of gonadotrophin action

Gonadotrophins act mainly on the structure and function of the gonads (ovaries and testes). Their physiological extragonadal effect is uncertain but some information on the peripheral lipolytic action of LH exists. The main effect of FSH[257, 288] in the ovaries is stimulation of the follicles to an advanced stage of maturity, but not to complete maturity. Views on the isolated effect of FSH on oestrogen secretion vary, but it is definitely known that oestrogen secretion, under physiological conditions, is simultaneously influenced by both FSH and LH. In males, FSH induces development of the seminiferous tubules and maturation of the spermatozoa, but ICSH (LH) also participates in the final phase of maturation.

In females, LH[258, 288] participates mainly in final maturation of the ovarian follicles and in stimulation of oestrogen secretion. An abrupt increase in the LH level during the cycle induces ovulation and conversion of the follicle to corpus luteum secreting progesterone. In man and certain other species, LH also has a luteotrophic effect, i.e. it maintains the function of corpus luteum. In other species (rat, sheep), prolactin is responsible for the luteotrophic effect. In males, ICSH (LH) completes maturation of the spermatozoa and, in particular, stimulates development of the interstitial cells of the testis (Leydig cells) and testosterone secretion.

Little is known of the mechanism of FSH action in the ovaries[297]. *In vitro*, FSH stimulates amino acid uptake and amino acid incorporation into protein; it also stimulates ovarian glycolysis. These two effects of FSH are both produced in the thecal and interstitial cells of the ovary, but not in the granulosa cells. There is an interesting difference between the two effects (i.e. on protein synthesis and ovarian glycolysis). FSH influences proteosynthesis only when administered to experimental animals *in vivo*, prior to killing, but not when added to incubation medium; this effect is blocked by puromycin. Ovarian glycolysis, however, is influenced both when FSH is administered *in vivo*, prior to killing the animal, and when it is added to incubation medium *in vitro*; this effect is not blocked by puromycin. The above effects can also be demonstrated *in vitro* in isolated cells from the individual components of ovarian tissue. It has been demonstrated[298] that the first phase of the mechanism of FSH action in the testes is the binding of FSH to specific binding sites. It was therefore postulated that FSH may initiate a biochemical sequence of events by binding to the cell membrane. Rapid stimulation of testicular nuclear-RNA synthesis was observed in immature rats injected with FSH and specific stimulation of nuclear transcription is regarded as an early manifestation of the effect of FSH[299]. The complex action of FSH on spermatogenesis has not yet been explained at the molecular level.

More reports are available on LH action in the ovaries. First of all, it was found that LH is specifically bound to ovarian cell receptors[300, 301], especially in the particulate fraction, presumably in the plasma membrane fraction of the theca cell layer[301]. Here, the binding of ^{125}I-LH is blocked by unlabelled LH, HCG and anti-LH serum. Binding to receptors is followed by activation of plasma membrane adenyl cyclase and by enhanced intracellular synthesis of cyclic-3′,5′-adenosine monophosphate[248, 301, 303]. In the subsequent phases, the conditions can be presumed to be similar to those after stimulation of other steroidogenic cells, i.e. synthesis of regulatory protein, etc.

The mechanism of action of ICSH (LH) in the testes has also been partly elucidated. ICSH was found to stimulate testicular protein[304, 305]. The administration of ICSH (and phospholipid[306] synthesis) to hypophysectomised rats was followed by raised incorporation of labelled amino acids into the protein of isolated Leydig cells[305]. Dibutyryl cyclic-3′,5′-adenosine monophosphate likewise raises amino acid incorporation, so that activation of the adenyl cyclase system evidently also forms part of ICSH action on Leydig cells.

3.5 GROWTH HORMONE (GH, SOMATOTROPHIC HORMONE, STH)

GH is an adenohypophysial hormone with direct tissue action, possibly mediated by a plasma protein, somatomedin. GH produces protein-synthetic and growth reactions in tissues and its effect, especially at high concentrations, is accompanied by complex metabolic changes. It is formed in adenohypophysial acidophils known as somatotrophs. Its secretion is not regulated by any clearly defined feed-back (blood level of peripheral gland hormones), although there are signs that it may be influenced by feed-back action of the blood glucose level. GH secretion is regulated by a hypothalamic hormone (GH–RH) and is stimulated by a whole series of different non-specific factors.

3.5.1 Biochemistry of GH

Human GH is a polypeptide with a molecular weight of 21 500[307], and consists of a single chain of 188 amino acids whose sequence is shown in Table 3.10[308]. Synthetic GH has also been prepared. There are two disulphide bridges in the molecule. The GH of other species[310] differs quite markedly from human GH (HGH) as regards molecular weight, the shape of the molecule and the number of disulphide bridges. For instance, bovine GH contains 400 amino acids, has a molecular weight of 46 000 and the molecule is Y-shaped. This great variability of GH structure is evidently the cause of the inactivity of different animal GHs (except primate) in man. The reason for this biological variability, which is most unusual for an adenohypophysial hormone, seems to be phylogenetic age.

Another curious feature is that the whole GH molecule is not needed for a biological effect and that biological activity is preserved in fragments of the native molecule. A thorough study has been made of the products of partial degradation of bovine GH by chymotrypsin[311] and carbopeptidase[312]. Removal of the phenylalanine residue from the C-terminal does not destroy biological activity and even the removal of 25% of the amino acids from the molecule does not destroy or lessen it[313]. The hormone molecule thus evidently has an active core, termed the a-core. The biological activity of fragments of human and monkey GH[314] and porcine GH[315] was also found to be unaffected after partial degradation by chymotrypsin. GH further undergoes a reversible association reaction and may exist in monomeric and dimeric form or aggregates[316].

The biological variability of GH structure results in immunological variability, which is no doubt one of the causes of the interspecies inefficacy of GH. Different GHs can be classified according to their reaction with anti-human GH antibodies, when marked relationships to the phylogenetic position of the different species are found[317].

The similarity of the structure (and effect, see Section 3.5.3) of GH to prolactin is another interesting characteristic. All GH preparations (including the synthetic hormone[309]) have prolactin activity[318]. The possibility that human GH and prolactin may be the same substance has even been discussed. In a comparison of human GH and ovine prolactin, three homologous fragments, containing in all about 45% of the amino acids, were found in each molecule[319]. One possible explanation of the relationships between GH and prolactin structure and action is that the two hormones (both of which are phylogenetically very old) have developed from a common precursor.

Table 3.10 Sequence of amino acids in human GH. (Taken from Li et al.[308].)*

NH$_2$-Phe-Pro-Thr-Ile-Pro-Leu-Ser-Arg-Leu-Phe-^{10}
Asp-Asn-Ala-Met-Leu-Arg-Ile-Ser-Leu-Leu-^{20}
Leu-Ile-Gln-Ser-Try-Leu-Glu-Pro-Val-Glu-^{30}
Phe-Ala-His-Arg-Leu-His-Gln-Leu-Ala-Phe-^{40}
Asp-Thr-Tyr-Glu-Glu-Phe-Glu-Glu-Ala-Tyr-^{50}
Ile-Pro-Lys-Glu-Gln-Lys-Tyr-Ser-Phe-Leu-^{60}
Gln-Asp-Pro-Glu-Thr-Ser-Leu-Cys-Phe-Ser-^{70}
Glu-Ser-Ile-Pro-Thr-Pro-Ser-Asn-Arg-Glu-^{80}
Glu-Thr-Gln-Lys-Ser-Asp-Leu-Glu-Leu-Leu-^{90}
Arg-Ser-Val-Phe-Ala-Asn-Ser-Leu-Val-Tyr-^{100}
Gly-Ala-Ser-Asn-Ser-Asp-Val-Tyr-Asp-Leu-^{110}
Leu-Lys-Asp-Leu-Glu-Glu-Gly-Ile-Glu-Thr-^{120}
Leu-Met-Gly-Arg-Leu-Glu-Asp-Pro-Ser-Gly-^{130}
Arg-Thr-Gly-Gln-Ile-Phe-Lys-Glu-Thr-Tyr-^{140}
Ser-Lys-Phe-Asp-Thr-Asn-Ser-His-Asn-Asp-^{150}
Asp-Ala-Leu-Leu-Lys-Asp-Tyr-Gly-Leu-Leu-^{160}
Tyr-Cys-Phe-Arg-Lys-Asp-Met-Asp-Lys-Val-^{170}
Glu-Thr-Phe-Leu-Arg-Ile-Val-Gln-Cys-Arg-^{180}
Ser-Val-Glu-Gly-Ser-Cys-Gly-Phe-COOH

* There are two disulphide bridges between the Cys residues 68–162 and 179–186

3.5.2 Regulation of GH secretion

GH secretion from the adenohypophysis is stimulated by a hypothalamic regulatory hormone, the growth hormone-releasing hormone (GH–RH). Its activity is related to the blood sugar level and is also influenced by a series of non-specific factors—chemical (arginine), physical (stress) and physiological (sleep)—which will be discussed in Section 3.5.2.2.

3.5.2.1 *Growth hormone–releasing hormone (GH–RH)*

After the first reports on the stimulant effect of hypothalamic extracts on GH secretion from the adenohypophysis[320, 321], the existence of GH–RH was demonstrated in several laboratories (for reviews see Refs. 322–324). Reports on the existence of a growth hormone inhibiting factor, GIF[325], also appeared. From the outset, it was generally agreed that GH–RH is a peptide. In 1971, it was found to be a decapeptide and its structure was determined[326]. The amino-acid sequence in GH–RH is shown in Table 3.11. A synthetic GH–RH decapeptide was also prepared[327]. Synthetic GH–RH induces GH secretion from the adenohypophysis. This can be demonstrated only by biological assay of GH and not by radioimmunological methods; natural GH–RH behaves in the same way[328].

Table 3.11   Supposed structure of GH–RH. (From Schally et al.[326])

Val-His-Leu-Ser-Ala-Glu-Glu-Lys-Glu-Ala

A number of reports on the mechanism of GH–RH action in the adenohypophysis have been published. Right at the outset it was repeatedly found that GH–RH stimulated both GH release and synthesis[329-331]. After intracarotid injections of GH–RH, structural changes were also found in the adenohypophysial somatotrophs, in particular extrusion of their granules[332-334]. Some authors found that GH–RH caused activation of adenohypophysial adenyl cyclase[335-337]. Depolarisation of the adenohypophysial cells and Ca^{2+} ions also evidently play a role in the action of GH–RH[388, 339].

3.5.2.2 *Other factors influencing GH secretion*

Most of the factors influencing GH secretion probably do so by way of changes in GH–RH secretion (for reviews see Refs. 340–342). GH secretion rises under conditions of absolute, relative, actual or impending shortage of carbohydrate substrate for energy requirements, in stress and in the presence of a raised blood amino acid (especially arginine) level. Glucagon, exercise and even physiological sleep likewise have a stimulant effect[343]. Some of these stimuli, including sleep, can be used to diagnose adenohypophysial capacity for GH secretion. The connection between the increase in GH secretion after these non-specific stimuli and the growth effect is not yet clear, but the reaction of secretion to these stimuli is limited in growth disturbances due to GH deficiency.

Some factors inhibit GH secretion. The main one is an increase in the blood sugar level[344], but corticoids[345] and medroxyprogesterone[346] also inhibit it. From the clinical aspect, the finding of an efficacious way of blocking GH secretion by drugs would be a great advance. There are also a number of findings testifying to the existence of an internal feed-back between the GH level and GH-RH secretion (reviewed in Refs. 191, 340).

3.5.3 Mechanism of GH action

The action of GH is one of the most complex hormonal actions in the organism (for review see Ref. 347). It is probably mediated by a plasma protein, originally mentioned as sulphatation factor and recently renamed as somatomedin (for reviews see Refs. 389, 390). The action of GH (or somatomedin) consists primarily in stimulation of growth, stimulation of protein synthesis in the liver, the muscles and other tissues, inhibition of glucose utilisation (in large doses GH has a diabetogenic effect) and stimulation of lipolysis and ketogenesis. The main effect on growth (primarily protein synthesis) of GH is brought about by stimulation of the transport of amino acids[348] and of their incorporation into protein[349]. Since theophylline inhibits the effect of GH on amino-acid uptake and protein synthesis[350], the adenyl cyclase system would also appear to participate in the action of GH, although experiments with dibutyryl cyclic-3′,5′-adenosine monophosphate yielded negative results. The effect of theophylline can be explained otherwise than by a sparing effect on intracellular cyclic-3′,5′-adenosine monophosphate, however[350]. The growth effect of GH is influenced by several other hormones, particularly insulin; in muscle, for example, GH augments the number of cells (cell nuclei), while insulin acts on growth of the cytoplasm[351].

Differentiation between the primary and secondary effects of GH in tissues is a complicated problem[352, 353]. The lipolytic effect seems to be mediated by a reaction requiring protein synthesis as it is blocked by inhibitors of protein synthesis[354]. The question of the effect of GH after body growth is complete is problematical. Summing up the findings to date, Tchobroutsky[355] concluded: (a) that GH is essential for growth during youth; (b) that it is essential for survival during starvation and (c) that it is not essential after completion of growth in adulthood and for the regularly nourished. 'For an adult living in a consumer society, GH is a luxury hormone'.

The prolactin effect of GH[318] is evidently a relic of ancient evolutionary relationships and is due to similarity of the structure of parts of the GH and the prolactin molecule[319, 356].

3.6 PROLACTIN (LUTEOTROPHIC HORMONE, LTH)

Prolactin (for reviews see Refs. 357, 358) is another adenohypophysial hormone with direct tissue action. In addition, in some animals (rat, sheep) it is also a luteotrophic hormone maintaining the function of the corpus luteum. It is a protein and is formed in adenohypophysial acidophils known as lactotrophs or prolactin cells. In some animals these differ sharply from the somatotrophs, but in man the two types of cells are hard to differentiate[143]. Prolactin secretion is regulated by a hypothalamic inhibitory hormone, the prolactin-inhibiting hormone (PIH); various other stimulant and inhibitory factors also participate in the regulation of its secretion.

3.6.1 Biochemistry of prolactin

Prolactin is a protein with a molecular weight of 23 000 to 26 000 (according to the species). It is composed of a single peptide chain and has three disulphide bridges in its molecule (ovine prolactin). The structure of human and primate GH and prolactin is very similar[319], as already mentioned in Section 3.5.1. The structure of ovine prolactin is shown in Table 3.12. Li[356] has compared the physical and chemical characteristics of human GH and ovine prolactin. The ovine prolactin molecule contains 205 amino-acid residues; the N-terminal amino acid is threonine and the C-terminal acid cysteine. Prolactin is one of the most versatile of all the hormones in terms of the variety of its effects[359] and, like GH, its structure also displays considerable variability.

3.6.2 Regulation of prolactin secretion

Prolactin secretion is also regulated by a hypothalamic regulatory hormone, but prolactin holds an exceptional position among the adenohypophysial hormones in that its basic regulation is inhibitory in character and is effected by the prolactin-inhibiting hormone (PIH). There are also a number of reports on the existence of a stimulating hormone, the prolactin-releasing hormone (PRH)[360, 361], especially in birds[362-364]. Like GH secretion, prolactin secretion is influenced by various other factors, e.g. by suckling of young.

3.6.2.1 Prolactin-inhibiting hormone (PIH)

The terminology of this neurohormone is not yet fixed. Hitherto, the term most commonly used was prolactin-inhibiting factor (PIF), but the term prolactin release-inhibiting hormone (PRIH)[361] was also suggested. The author prefers to keep to the term PIH. Pasteels[365, 366] was the first to demonstrate the inhibitory effect of hypothalamic extracts on prolactin secretion. His findings were confirmed in various other laboratories, particularly by Meites team, with *in vitro* (e.g. Ref. 367) and *in vivo* studies (e.g. Ref. 368). At the time of writing this review, the structure of PIH is unknown to the author. The findings on PIH have been summarised in a series of reviews[358, 369, 370].

The mechanism of the inhibitory action of PIH on prolactin secretion is still obscure although experiments with hypothalamic extracts have given rise to new hypotheses. In prolactin secretion, the factors of polarisation of the adenohypophysial lactotrophs and Ca^{2+} ion uptake appear to play a role

Table 3.12 Sequence of amino acids in ovine prolactin. (After Li[388].)

NH$_2$-Thr-Pro-Val-Cys-Pro-Asn-Gly-Pro-Gly-Asp^{10}-
Cys-Gln-Val-Ser-Leu-Arg-Asp-Leu-Phe-Asp^{20}-
Arg-Ala-Val-Met-Val-Ser-His-Tyr-Ile-His^{30}-
Asn-Leu-Ser-Ser-Glu-Met-Phe-Asn-Glu-Phe^{40}-
Asp-Lys-Arg-Tyr-Ala-Gln-Gly-Lys-Gly-Phe^{50}-
Ile-Thr-Met-Ala-Leu-Asn-Ser-Cys-His-Thr^{60}-
Ser-Ser-Leu-Pro-Thr-Pro-Glu-Asp-Lys-Glu^{70}-
Gln-Ala-Gln-Gln-Thr-His-His-Glu-Val-Leu^{80}-
Met-Ser-Leu-Ile-Leu-Gly-Leu-Arg-Ser-Trp^{90}-
Asn-Asp-Pro-Leu-Tyr-His-Leu-Val-Thr-Glu^{100}-
Val-Arg-Gly-Met-Lys-Gly-Val-Pro-Asp-Ala^{110}-
Ile-Leu-Ser-Arg-Ala-Ile-Glu-Ile-Glu-Glu^{120}-
Glu-Asn-Lys-Arg-Leu-Leu-Glu-Gly-Met-Glu^{130}-
Met-Ile-Phe-Gly-Gln-Val-Ile-Pro-Gly-Ala^{140}-
Lys-Glu-Thr-Glu-Pro-Tyr-Pro-Val-Trp-Ser^{150}-
Gly-Leu-Pro-Ser-Leu-Gln-Thr-Lys-Asp-Glu^{160}-
Asp-Ala-Arg-His-Ser-Ala-Phe-Tyr-Asn-Leu^{170}-
Leu-His-Cys-Leu-Arg-Arg-Asp-Ser-Ser-Lys^{180}-
Ile-Asp-Thr-Tyr-Leu-Lys-Leu-Leu-Asn-Cys^{190}-
Arg-Ile-Ile-Tyr-Asn-Asn-Asn-Cys-COOH

* There are three disulphide bridges between the Cys residues 4–11, 58–173 and 190–198

opposite to that in the secretion of the other adenohypophysial hormones. For instance[371], PIH was found to have an inhibitory effect only when sufficient extracellular calcium was present to support spontaneous secretion of prolactin. PIH probably causes stabilisation of the plasma membrane of the lactotrophs, thereby preventing Ca^{2+} ion uptake and the extrusion of secretory granules. High K^+ ion concentrations inhibit the action of PIH, but (unlike the case of the other adenohypophysial hormones) they do not, by themselves, stimulate prolactin secretion[372]. Participation of the adenyl cyclase system is borne out by findings[373] of calcium-dependent stimulation of prolactin release by dibutyryl cyclic-3',5'-adenosine monophosphate.

Although in most other cases the effect of the various hypothalamic RH was shown to be highly specific, TRH was repeatedly found to stimulate not only TSH secretion, but also prolactin secretion *in vitro*[179, 180] and *in vivo*[181]. TRH thus possesses PRH activity.

3.6.2.2 Other factors influencing prolactin secretion

An internal feed-back between the prolactin level and PIH secretion probably participates in the regulation of prolactin secretion[374]. This is borne out by the results obtained with prolactin implants in both the hypothalamus[375] and the adenohypophysis[376]. Some authors suppose PRH to be the factor of this internal feed-back[377]. Motta *et al.*[191] have submitted a survey of further findings supporting the existence of an internal feed-back.

Since, in some species, prolactin is a luteotrophin, one would presume that progesterone or oestrogens would have a feed-back action. Such evidence was actually found (for review see Ref. 378), but it mostly showed that the feed-back action of progesterone and oestrogens was positive. Other factors include the suckling of young, which leads to quick and copious prolactin secretion (for reviews see Refs. 358, 378). Higher parts of the central nervous system, as well as the hypothalamus, undoubtedly participate in this reflex process. A number of non-specific factors, such as stress and various drugs with a central action, also influence prolactin secretion. Special interest is elicited by phenothiazine derivatives with a stimulant effect[379] and by ergocornine derivatives with an inhibitory effect[380]. Since certain mammary carcinomas are evidently prolactin-dependent, the possibility of inhibiting prolactin secretion by pharmacological means would be clinically invaluable. Another remarkable finding is that in the same manner as these substances influence prolactin secretion (and probably in the opposite way to their effect on PIH secretion) they also influence the growth of the adenohypophysis after oestrogen treatment. This adenohypophysial growth is stimulated by perphenazine[381] and inhibited by ergocornine[182].

3.6.3 Mechanism of prolactin action

Prolactin has very diverse effects which are markedly dependent on the phylogenetic position of the test animal. They include (for reviews see Refs. 358, 378) an osmoregulatory effect, a growth effect, an effect on the migration of amphibians to water to mate, nest-building in birds, the development of a maternal instinct and a whole series of other reactions associated with reproduction or the feeding of the young. In higher mammals, the main effect of prolactin is its action on the mammary gland, where it causes proliferation and milk production. Here it interacts with other hormones, in particular insulin, corticoids and thyroxine. A detailed analysis of the effects of prolactin in the mammary gland has been published by Denamur[382]. Armstrong[296] has studied the luteotrophic action of prolactin in various mammals. Its main effects in the corpus luteum are inhibition of 20α-hydroxysteroid dehydrogenase, stimulation of cholesterol synthesis and its conversion to progesterone as well as an increase in fatty acid synthesis and esterification with cholesterol.

Little is known of the mechanism of action of prolactin in the mammary gland. Its proliferative effect *in vitro* is dependent on the presence of steroid hormones, particularly progesterone and cortisol[383]. Prolactin stimulates amino acid incorporation into the proteins and lactose formation in the mammary gland[384, 385], with a delay of 2–3 days after administration. Prolactin raises the incorporation of phosphorus into casein in mammary gland explants *in vitro*[387]; this effect (like the effect on protein synthesis and RNA synthesis) is blocked by actinomycin D.

In males, prolactin appears to play an important role as an additional growth factor for some organs of the genital tract, e.g. the prostate (for review see Ref. 391).

References

1. Gorbman, A. and Bern, H. A. (1962). *A Textbook of Comparative Endocrinology*, 29 (New York and London: J. Wiley and Sons)
2. Leeman, S. E. and Hammerschlag, R. (1967). *Endocrinology*, **81**, 803
3. Salhanick, H. H., Zarrow, I. G. and Zarrow, M. X. (1959). *Endocrinology*, **45**, 314
4. Hodges, J. R. and Vernikos-Danellis, J. (1962). *Acta Endocrinol.*, **39**, 79
5. Schreiber, V. and Kmentová, V. (1958). *Physiol. Bohemoslov.*, **7**, 437
6. De Nicola, A. F., Clayman, M. and Johnstone, R. N. (1968). *Gen. Comp. Endocrinol.*, **11**, 332
7. Kimura, T. (1963). *Endocrinol. Jap.*, **10**, 146
8. Snyder, J. and D'Angelo, S. A. (1963). *Proc. Soc. Exp. Biol. Med.*, **112**, 1
9. Schreiber, V., Přibyl, T. and Roháčová, J. (1971). *Physiol. Bohemoslov.*, **20**, 147
10. Lippert, T. H. and Watson, N. G. (1969). *Med. Exp.*, **19**, 119
11. Kar, A. B., Chowdhury, S. R., Chowdhury, A. R., Kamboj, V. P. and Chandra, H. (1963). *Steroids*, **5**, 519
12. Singh, H. and Carroll, K. K. (1970). *Lipids*, **5**, 121
13. Dixit, P. K. and Lazarow, A. (1962). *Endocrinology*, **71**, 745
14. Hiroshige, T., Itoh, S. and Sakakura, M. (1969). *Endocrinol. Jap., Suppl.*, **1**, 41
15. Rennels, E. G. (1965). *Endocrinology*, **76**, 984
16. Rennels, E. G. and Hood, J. F. (1964). *Science*, **144**, 416
17. Vylcheva, L. V. and Khadzhiolov, A. A. (1966). *Biochemistry*, **31**, 820
18. Deminatti, M. (1961). *Compt. Rend.*, **253**, 321
19. Borell, U. and Westman, A. (1949). *Acta Endocrinol.*, **3**, 111
20. Vértes, Z., Vértes, M., Kovács, J., Lelkes, J. and Nagy, P. (1966). *Acta Physiol. Acad. Sci. Hung.*, **30**, 322
21. Vértes, M., Kovács, S., Lelkes, J. and Vértes, Z. (1967). *Acta. Physiol. Acad. Sci. Hung.*, **32**, 81
22. Aboliņš, L. (1952). *Exp. Cell Res.*, **3**, 1

23. Laschet, U., Hasse, M. and Rieche, K. (1960). *First Int. Congr. Endocrinology, Adv. Abstr. Short Commun.*, 146 (F. Fuchs, editor) (Copenhagen: Periodica)
24. Hess, M., Corrigan, J. J. and Hodak, J. A. (1961). *Proc. Soc. Exp. Biol. Med.*, **106,** 420
25. Dhom, G. and Stöcker, E. (1964). *Experientia*, **20,** 384
26. Jiang, N.-S. (1964). *Biochim. Biophys. Acta*, **87,** 347
27. Kraicer, J. (1964). *Biochim. Biophys. Acta*, **87,** 703
28. Kraicer, J., Herlant, M. and Duclos, P. (1967). *Can. J. Physiol. Pharmacol.*, **45,** 947
29. Lee, K. L., Bowers, C. Y. and Miller, O. N. (1968). *Endocrinology*, **83,** 763
30. Salaman, D. F. and Kirby, K. S. (1968). *Nature (London)*, **217,** 454
31. Kraicer, J. and Cheng, S. C. (1968). *Can. J. Physiol. Pharmacol.*, **46,** 431
32. Valotaire, Y. and Duval, J. (1969). *Bull. Soc. Chim. Biol.*, **51,** 1233
33. Valotaire, Y., Duval, J. and Jouan, P. (1969). *Bull. Soc. Chim. Biol.*, **51,** 211
34. Messier, B. (1969). *Acta Endocrinol.*, **61,** 133
35. Treolar, O. L. and Leathem, J. H. (1969). *Proc. Soc. Exp. Biol. Med.*, **132,** 968
36. Städtler, F., Stöcker, E., Dhom, G. and Tietze, H. U. (1970). *Acta Endocrinol.*, **64,** 324
37. Salaman, D. F. (1970). *J. Endocrinol.*, **48,** 125
38. Hymer, W. C., Mastro, A. and Griswold, E. (1970). *Science*, **167,** 1629
39. Valotaire, Y. and Duval, J. (1970). *Biochim. Biophys. Acta*, **224,** 63
40. Goluboff, L. G., MacRae, M. E., Ezrin, C. and Sellers, E. A. (1970). *Endocrinology*, **87,** 1113
41. Goodner, C. J. and Dowling, J. T. (1963). *Diabetes*, **12,** 368
42. Ducommun, S. (1965). *Ann. Endocrinol. (Paris)*, **26,** 385
43. Tonoue, T. and Yamamoto, K. (1967). *Biochem. Biophys. Res. Commun.*, **26,** 315
44. Tonoue, T. and Yamamoto, K. (1967). *Endocrinology*, **81,** 1029
45. Tonoue, T. and Yamamoto, K. (1968). *Japan J. Physiol.*, **18,** 481
46. Grieshaber, C. K. and Hymer, W. C. (1968). *Proc. Soc. Exp. Biol. Med.*, **128,** 459
47. Lee, K. L., Bowers, C. Y. and Miller, O. N. (1968). *Endocrinology*, **83,** 754
48. Hoshino, S. and Yamamoto, K. (1969). *Gunma Symp. Endocrinol.*, **6,** 249
49. Litonjua, A. D., Guerrero, L. M., Capco, P. and Payabyab, N. (1969). *Acta Med. Philippina*, **5,** 115
50. Tixier-Vidal, A. and Gourdji, D. (1970). *J. Cell. Biol.*, **46,** 130
51. Matsuzaki, S. (1970). *Endocrinol. Jap.*, **17,** 379
52. Samli, M. H. and Barnett, C. A. (1971). *Endocrinology*, **88,** 540
53. Levey, H. A. and Roberts, S. (1957). *Amer. J. Physiol.*, **189,** 86
54. Goodner, C. J. (1964). *Endocrinology*, **75,** 846
55. Goodner, C. J. (1965). *Endocrinology*, **76,** 1022
56. Goodner, C. J. (1966). *Diabetes*, **15,** 115
57. Migliorini, R. H. and Antunes-Rodrigues, J. (1970). *Experientia*, **26,** 191
58. Goodner, C. J. and Freinkel, N. (1960). *J. Clin. Invest.*, **39,** 991
59. Field, J. B., Pastan, I., Herring, B. and Johnson, P. (1960). *Endocrinology*, **67,** 804
60. Dumont, J. E. (1960). *Biochim. Biophys. Acta*, **42,** 157
61. Goodner, C. J. and Freinkel, N. (1961). *J. Clin. Invest.*, **40,** 261
62. Krass, M. E. and LaBella, F. S. (1967). *Biochim. Biophys. Acta*, **148,** 384
63. Schreiber, V. and Lojda, Z. (1967). *Endocrinol. Exper.*, **1,** 237
64. Arvy, L. (1971). *Histoenzymology of the Endocrine Glands*. (Oxford: Pergamon)
65. Balogh, K. and Cohen, R. B. (1962). *Endocrinology*, **70,** 874
66. Matsuzaki, S. (1968). *Endocrinol. Jap.*, **15,** 223
67. Jacobowitz, D. and Marks, B. H. (1964). *Endocrinology*, **75,** 86
68. Lojda, Z. (1960). *Cesk. Morfol.*, **8,** 148
69. Lojda, Z. and Schreiber, V. (1960). *Cesk. Fysiol.*, **9,** 377
70. Lojda, Z. and Schreiber, V. (1962). *Cesk. Fysiol.*, **11,** 314
71. Lojda, Z. and Schreiber, V. (1964). *J. Physiol. (Paris)*, **56,** 459
72. Bleicher, S., Karnovsky, M. and Freinkel, N. (1961). *J. Clin. Invest.*, **40,** 1024
73. Ruegamer, W. R., Newman, G H., Richert, D. A. and Westerfeld, W. W. (1965). *Endocrinology*, **77,** 707
74. Howe, A. and Thody, A. J. (1967). *J. Endocrinol.*, **39,** 351
75. Moguilevsky, J. A., Dahl, V. and Gómez, C. J. (1964). *Acta Physiol. Latinoam.*, **14,** 304
76. Hoch-Ligeti, C., Brown, T. J. and Grantham, H. H. (1967). *Endocrinology*, **80,** 483
77. Sobel, H. J. (1964). *J. Endocrinol.*, **29,** 1
78. McKerns, K. W. (1963). *Biochim. Biophys. Acta*, **73,** 507

79. Fand, S. B. and Wattenberg, L. W. (1963). *Lab. Invest.*, **12**, 454
80. Andersen, H. and Von Bulow, F. A. (1970), *Progr. Histochem. Cytochem.*, **1**, 29P
81. Dumont, J. E. (1960). *Biochim. Biophys. Acta*, **42**, 157
82. Kroon, D. B. (1949). *Acta Endocrinol.*, **2**, 227
83. Romieu, M., Stahl, A. and Seite, R. (1951). *Compt. Rend. Soc. Biol.*, **145**, 415
84. Romieu, M., Stahl, A. and Seite, R. (1952). *Arch. Anat. Microscop. Morphol. Exp.*, **34**, 369
85. Abolinš-Krogis, A. (1953). *Arkiv Zool.*, **4**, 557
86. Melchior, J. and Micuta, B. S. (1956). *Cancer Res.*, **16**, 520
87. Imoto, T. (1957). *Arch. Histol. (Okayama)*, **13**, 491
88. Kobayashi, H. and Kambara, S. (1959). *Endocrinology*, **64**, 615
89. Iizuka, M. (1960). *Folia Endocrinol. Jap.*, **36**, 15
90. Holmes, R. L. (1961). *J. Endocrinol.*, **23**, 63
91. Pearse, A. G. E. and Van Noorden, S. (1963). *Cytologie de l'adénohypophyse*, 63 (J. Benoit and Ch. da Lage, editors) (Paris: Éd. C.N.R.S.)
92. Abolinš, L. and Abolinš, A. (1949). *Nature (London)*, **164**, 455
93. Schreiber, V. and Kmentová, V. (1959). *Acta Biol. Acad. Sci. Hung.*, **9**, 285
94. Sobel, H. J. (1961). *Endocrinology*, **68**, 101
95. Sobel, H. J. (1964). *J. Endocrinol.*, **30**, 323
96. Sobel, H. J. (1962). *Anat. Rec.*, **143**, 389
97. Arvy, L. (1961). *Compt. Rend. Soc. Biol.*, **155**, 2119
98. Novikoff, A. B. (1961). *The Cell*, 423 (J. Brachet and A. E. Mirsky, editors) (New York: Academic Press)
99. Smith, R. E. (1963). *J. Cell. Biol.*, **19**, 66A
100. Smith, R. E. and Farquhar, M. G. (1966). *J. Cell. Biol.*, **31**, 319
101. Wächtler, K. and Pearse, A. G. E. (1966). *Zellforsch. Mikroskop. Anat.*, **69**, 326
102. Hiroshige, T., Nakatsugawa, T., Matsuoka, Y. and Itoh, S. (1966). *Japan J. Physiol.*, **16**, 94
103. Hiroshige, T., Nakatsugawa, T., Imazeki, T. and Itoh, S. (1966). *Japan J. Physiol.*, **16**, 103
104. Pearse, A. G. E. (1952). *J. Pathol. Bacteriol.*, **64**, 791
105. Feustel, G., Piper, K.-S., Hübner, H.-J. and Luppa, H. (1969). *Acta Histochem.*, **34**, 317
106. Surowiak, J. (1969). *Folia Biol. (Warsaw)*, **17**, 205
107. Pelletier, G. and Novikoff, A. B. (1972). *J. Histochem. Cytochem.*, **20**, 1
108. Mietkiewski, K., Malendowicz, L. and Lukaszyk, O. (1969). *Acta Endocrinol.*, **61**, 293
109. Takagi, I. and Yamamoto, K. (1969). *Japan J. Physiol.*, **19**, 465
110. Smith, R. E. and Farquhar, M. G. (1964). *J. Cell. Biol.*, **23**, 87A
111. Smith, R. E. and Farquhar, M. G. (1970). *J. Histochem. Cytochem.*, **18**, 237
112. Pearse, A. G. E. (1956). *J. Pathol. Bacteriol.*, **72**, 471
113. Vanha-Perttula, T. (1966). *Esterases of the Rat Adenohypophysis. Acta Physiol. Scand., Suppl. 69*, 283
114. Vanha-Perttula, T. and Hopsu, V. K. (1965). *Histochemie*, **4**, 372
115. Adams, E. and Smith, E. L. (1951). *J. Biol. Chem.*, **191**, 651
116. Meyer, R. K. and Clifton, K. H. (1956). *Arch. Biochem.*, **62**, 198
117. Ellis, S. (1960). *J. Biol. Chem.*, **235**, 1694
118. Ellis, S. (1963). *Biochem. Biophys. Res. Commun.*, **12**, 452
119. Vanha-Perttula, T. P. J. and Hopsu, V. K. (1965). *Ann. Med. Exp. Biol. Fenniae (Helsinki)*, **43**, 32
120. Ellis, S. and Perry, M. (1966). *J. Biol. Chem.*, **241**, 3679
121. McDonald, J. K., Ellis, S. and Reilly, T. J. (1966). *J. Biol. Chem.*, **241**, 1494
122. Ellis, S. and Nuenke, J. M. (1967). *J. Biol. Chem.*, **242**, 4623
123. Ellis, S., Nuenke, J. M. and Grindeland, R. E. (1968). *Endocrinology*, **83**, 1029
124. McDonald, J. K., Leibach, F. H., Grindeland, R. E. and Ellis, S. (1968). *J. Biol. Chem.*, **243**, 4143
125. McDonald, J. K., Reilly, T. J., Zeitman, B. B. and Ellis, S. (1968). *J. Biol. Chem.*, **243**, 2028
126. Vanha-Perttula, T. (1969). *Endocrinology*, **85**, 1062
127. Perdue, S. F. and McShan, W. H. (1962). *J. Cell. Biol.*, **15**, 159
128. Hymer, W. C. and McShan, W. H. (1963). *J. Cell. Biol.*, **17**, 67

129. Tesar, J. T. (1967). *Fed. Proc.*, **26**, 534
130. Grinberg, R., Volpert, E. M. and Werner, S. C. (1963). *J. Clin. Endocrinol. Metab.*, **23**, 140
131. Reichlin, S., Volpert, E. M. and Werner, S. C. (1966). *Endocrinology*, **78**, 302
132. Axelrod, J. and Daly, J. (1965). *Science*, **150**, 892
133. Jaffe, R. B. (1969). *Abstr. 51st Meeting Endocrine Soc.*, New York
134. Narahara, H. T. and Williams, R. H. (1958). *Amer. J. Physiol.*, **193**, 476
135. Zor, U., Kaneko, T., Schneider, H. P. G., McCann, S. M., Lowe, J. P., Bloom, G., Borland, B. and Field, J. B. (1969). *Proc. Nat. Acad. Sci. U.S.A.*, **63**, 918
136. Chase, L. R., Aurbach, G. D. and Wilber, J. F. (1969). *Abstr. 51st Meeting Endocrine Soc.*, New York
137. Guillemin, R., Burgus, R. and Vale, W. (1969). *Progress in Endocrinology*, 577 (C. Gual, éditor) (Amsterdam: Excerpta Medica)
138. Wilber, J. F., Peake, G. T. and Utiger, R. D. (1969). *Endocrinology*, **84**, 758
139. Benoit, J. and Da Lage, Ch. (éditors) (1963). *Cytologie de l'adénohypophyse*, (Paris: Éd. C.N.R.S.)
140. Pearse, A. G. E. and Van Noorden, S. (1963). *Can. Med. Assoc. J.*, **88**, 462
141. Malamed, S., Portanova, R. and Sayers, G. (1971). *Proc. Soc. Exp. Biol. Med.*, **138**, 920
142. Pooley, A. S. (1971). *Endocrinology*, **88**, 400
143. Leleux, P. and Robyn, C. (1971). *Karolinska Symp. on Res. Methods in Reprod. Endocrinol.*, **3**, 168; *Acta Endocrinol. Suppl. 153*, 168
144. Nakane, P. K. (1971). *Karolinska Symp. on Res. Methods in Reprod. Endocrinol.*, **3**, 190; *Acta Endocrinol. Suppl. 153*, 190
145. Sayers, G., Portanova, R., Beall, R. J. and Malamed, S. (1971). *Karolinska Symposia on Res. Methods in Reprod. Endocrinol.*, **3**, 11; *Acta Endocrinol. Suppl. 153*, 11
146. Green, J. D. and Harris, G. W. (1947). *J. Endocrinol.*, **5**, 136
147. Harris, G. W. (1955). *Neural Control of the Pituitary Gland*, (London: E. Arnold)
148. Schally, A. V., Bowers, C. Y. and Locke, W. (1964). *Amer. J. Med. Sci.*, **248**, 114
149. McCann, S. M., Dhariwal, A. P. S. and Porter, J. C. (1968). *Ann. Rev. Physiol.*, **30**, 589
150. Guillemin, R. (1964). *Rec. Prog. Horm. Res.*, **20**, 89
151. Burgus, R. and Guillemin, R. (1970). *Ann. Rev. Biochem.*, **39**, 499
152. Schally, A. V., Bowers, C. Y., Kuroshima, A., Ishida, Y., Redding, T. W. and Kastin, A. J. (1965). *Proc. XXIII Int. Congr. Physiol. Sci., Tokyo. Excerpta Medica Int. Congr. Series*, **87**, 275
153. Schreiber, V. (1969). *Progress in Endocrinology*, 555 (C. Gual, editor) (Amsterdam: Excerpta Medica)
154. Meites, J. (editor) (1970). *Hypophysiotropic Hormones of the Hypothalamus: Assay and Chemistry*, (Baltimore: Williams and Wilkins)
155. Bates, R. W. and Condliffe, P. G. (1966). *The Pituitary Gland*, Vol. 1, 374 (G. W. Harris and B. T. Donovan, editors) (London: Butterworths)
156. Fontaine, Y. (1969). *Progress in Endocrinology*, 453 (C. Gual, editor) (Amsterdam: Excerpta Medica)
157. Pierce, J. G., Liao, T.-H., Cornell, S. S. and Carlsen, R. B. (1971). *Structure–Activity Relationships of Protein and Polypeptide Hormones*, 91 (M. Margoulies and F. C. Greenwood, editors) (Amsterdam: Excerpta Medica)
158. Pierce, J. G. (1971). *Endocrinology*, **89**, 1331
159. Pierce, J. G., Bahl, O. P., Cornell, J. S. and Swaminathan, N. (1971). *J. Biol. Chem.*, **246**, 2321
160. Schreiber, V., Kočí, J., Eckertová, A., Franc, Z. and Kmentová, V. (1961). *Physiol. Bohemoslov.*, **10**, 417
161. Guillemin, R., Yamazaki, E., Jutisz, M. and Sakiz, E. (1962). *Compt. Rend.*, **25**, 1018
162. Schreiber, V. (1967). *Neuroendocrinology*, **2**, 175
163. Guillemin, R., Burgus, R. and Vale, W. (1969). *Progress in Endocrinology* (C. Gual, editor) (Amsterdam: Excerpta Medica)
164. Schally, A. V. and Bowers, C. Y. (1971). *Proc. 6th Midwest Conf. on Thyroid and Endocrinology*, 25 (University of Missouri Press)
165. Schally, A. V., Bowers, C. Y., Redding, T. W. and Barrett, J. F. (1966). *Biochem. Biophys. Res. Commun.*, **25**, 165
166. Folkers, K., Enzmann, F., Bøler, J., Bowers, C. Y. and Schally, A. V. (1969). *Biochem. Biophys. Res. Commun.*, **37**, 123

167. Burgus, R., Dunn, T. F., Desiderio, D. and Guillemin, R. (1969). *Compt. Rend.*, **269**, 1870
168. Bowers, C. Y., Schally, A. V., Weill, A., Reynolds, G. A. and Folkers, K. (1971). *Further Advance in Thyroid Research*, 129 (K. Fellinger and R. Höfer, editors) (Vienna: Verlag der Wiener Medizinischen Akademie)
169. Schally, A. V., Arimura, A., Bowers, C. Y., Wakabayashi, I., Kastin, A. J., Redding, T. W., Mittler, J. C., Nair, R. M. G., Pizzolato, P. and Segal, A. J. (1970). *J. Clin. Endocrinol. Metab.*, **31**, 291
170. Vale, W., Rivier, J. and Burgus, R. (1971). *Endocrinology*, **89**, 1485
171. Schreiber, V., Eckertová, A., Franc, Z., Rybák, M., Gregorová, I., Kmentová, V. and Jirgl, V. (1963). *Physiol. Bohemoslov.*, **12**, 1
172. Sinha, D. and Meites, J. (1965). *Neuroendocrinology*, **1**, 4
173. Geschwind, I. I. (1970). *Hypophysiotropic Hormones of the Hypothalamus: Assay and Chemistry*, 298 (J. Meites, editor) (Baltimore: Williams and Wilkins)
174. Wilber, J., Peake, G., Mariz, I., Utiger, R. and Daughaday, W. (1968). *Clin. Res.*, **16**, 277
175. Vale, W. and Guillemin, R. (1967). *Experientia*, **23**, 855
176. Vale, W., Burgus, R. and Guillemin, R. (1967). *Experientia*, **23**, 843
177. Schreiber, V., Přibyl, T. and Roháčová, J. (1972). *Sb. Lekar.*, **74**, 26
178. Pittman, J. A., Dubovsky, E. and Beschi, R. S. (1970). *Biochem. Biophys. Res. Commun.*, **40**, 1246
179. Tashjian, A. H., Barowsky, N. J. and Jensen, D. K. (1971). *Biochem. Biophys. Res. Commun.*, **43**, 516
180. La Bella, F. S. and Vivian, S. R. (1971). *Endocrinology*, **88**, 787
181. Jacobs, L. S., Snyder, P. J., Wilber, J. F., Utiger, R. D. and Daughaday, W. H. (1971). *J. Clin. Endocrinol. Metab.*, **33**, 996
182. Schreiber, V., Přibyl, T. and Rocháčová, J. (1973). *Physiol. Bohemoslov.*, **22**, 11
183. Aron, M., Van Caulert, C. and Stahl, J. (1931). *Compt. Rend. Soc. Biol.*, **107**, 64
184. Brown-Grant, K. (1966). *The Pituitary Gland*, Vol. 2, 235 (G. W. Harris and B. T. Donovan, editors) (London: Butterworths)
185. Reichlin, S. (1966). *Neuroendocrinology*, Vol. 1, 445 (L. Martini and W. F. Ganong, editors) (New York and London: Academic Press)
186. Knigge, K. M. (1964). *Major Problems in Neuroendocrinology*, 261 (E. Bajusz, editor) (Basel: Karger)
187. Rose, S., Nelson, J. and Bradley, T. R. (1960). *Ann. N. Y. Acad. Sci.*, **86**, 647
188. Guillemin, R., Yamazaki, E., Gard, D. A., Jutisz, M. and Sakiz, E. (1963). *Endocrinology*, **73**, 564
189. Bowers, C. Y., Lee, K. L. and Schally, A. V. (1968). *Endocrinology*, **82**, 303
190. Vale, W., Burgus, R. and Guillemin, R. (1968). *Neuroendocrinology*, **3**, 34
191. Motta, M., Fraschini, F. and Martini, L. (1969). *Frontiers in Neuroendocrinology*, *1969*, 211 (W. F. Ganong and L. Martini, editors) (New York, London and Toronto: Oxford Univ. Press)
192. D'Angelo, S. A. (1961). *Endocrinology*, **69**, 834
193. Levey, H. A. and Roberts, S. (1962). *Endocrinology*, **71**, 244
194. Goodman, H. M. (1969). *Progress in Endocrinology*, 114 (C. Gual, editor) (Amsterdam: Excerpta Medica)
195. Engel, F. L. and Lebovitz, H. E. (1966). *The Pituitary Gland*, Vol. 2, 563 (G. W. Harris and B. T. Donovan, editors) (London: Butterworths)
196. Pastan, I. (1969). *Progress in Endocrinology*, 98 (C. Gual, editor) (Amsterdam: Excerpta Medica)
197. Pastan, I., Roth, J. and Macchia, V. (1966). *Proc. Nat. Acad. Sci. U.S.A.*, **56**, 1802
198. Pastan, I. and Katzen, R. (1967). *Biochem. Biophys. Res. Commun.*, **29**, 792
199. Yamashita, K. and Field, J. B. (1970). *Biochem. Biophys. Res. Commun.*, **40**, 171
200. Gilman, G. A. and Rall, T. W. (1966). *Fed. Proc.*, **25**, 617
201. Kendall-Taylor, P. (1972). *J. Endocrinol.*, **52**, 533
202. Pastan, I. and Macchia, V. (1967). *J. Biol. Chem.*, **25**, 5757
203. Pastan, I. and Wollman, S. H. (1967). *J. Cell. Biol.*, **35**, 262
204. Williams, C., Rocmans, P. A. and Dumont, J. E. (1970). *Biochim. Biophys. Acta*, **222**, 474
205. Levey, G. S., Roth, J. and Pastan, I. (1969). *Endocrinology*, **84**, 1009

206. Purves, H. D. (1966). *The Pituitary Gland*, Vol. 1, 147 (G. W. Harris and B. T. Donovan editors) (London: Butterworths)
207. Li, C. H., Dixon, J. S. and Chung, D. (1961). *Biochim. Biophys. Acta*, **46**, 324
208. Evans, H. M., Sparks, L. L. and Dixon, J. S. (1966). *The Pituitary Gland*, Vol. 1, 317 (G. W. Harris and B. T. Donovan, editors) (London: Butterworths)
209. Schwyzer, R. (1969). *Protein and Polypeptide Hormones*, 201 (M. Margoulies, editor) (Amsterdam: Excerpta Medica)
210. Doepfner, W. (1969). *Progress in Endocrinology*, 407 (C. Gual, editor) (Amsterdam: Excerpta Medica)
211. Craig, L. C., Kac, H., Chen, H. C. and Printz, M. P. (1971). *Structure–Activity Relationships of Protein and Polypeptide Hormones*, 176 (M. Margoulies and F. C. Greenwood, editors) (Amsterdam: Excerpta Medica)
212. Novales, R. R. (1967). *Neuroendocrinology*, Vol. 2, 241 (L. Martini and W. F. Ganong, editors) (New York and London: Academic Press)
213. Schnabel, E. and Li, C. H. (1960). *J. Biol. Chem.*, **235**, 2010
214. Hechter, O. and Braun, T. (1971). *Structure–Activity Relationships of Protein and Polypeptide Hormones*, 212 (M. Margoulies and F. C. Greenwood, editors) (Amsterdam Excerpta Medica)
215. Fortier, C. (1966). *The Pituitary Gland*, Vol. 2, 195 (G. W. Harris, and B. T. Donovan, editors) (London: Butterworths)
216. Mangili, G., Motta, M. and Martini, L. (1966). *Neuroendocrinology*, Vol. 1, 297 (L. Martini and W. F. Ganong, editors) (New York and London: Academic Press)
217. Schally, A. V. and Bowers, C. Y. (1964). *Metabolism*, **13**, 1190
218. Schally, A. V., Lipscomb, H. and Guillemin, R. (1962). *Endocrinology*, **71**, 164
219. Schally, A. V., Arimura, A., Bowers, C. Y., Kastin, A. J., Sawano, S. and Redding, T. W. (1968). *Rec. Prog. Horm. Res.*, **24**, 497
220. Geshwind, I. I. (1969). *Frontiers in Neuroendocrinology 1969*, 389 (W. F. Ganong and L. Martini, editors) (New York, London and Toronto: Oxford Univ. Press)
221. Vernikos-Danellis, J. (1965). *Endocrinology*, **76**, 240
222. Uemura, T. (1968). *Endocrinol. Jap.*, **15**, 130
223. Kraicer, J., Milligan, J. V., Gosbee, J. L., Conrad, R. G. and Branson, C. M. (1969). *Science*, **164**, 426
224. Zimmerman, G. and Fleischer, N. (1970). *Endocrinol.*, **87**, 426
225. Yates, F. E. (1967). *The Adrenal Cortex* 144 (A. Eisenstein, editor) (Boston: Little, Brown and Co.)
226. Davidson, J. M., Feldman, S., Smith, E. R. and Weick, R. F. (1969). *Progress in Endocrinology*, 542 (C. Gual, editor) (Amsterdam: Excerpta Medica)
227. Ingle, D. J. (1938). *Amer. J. Physiol.*, **124**, 369
228. Fortier, C. (1959). *Proc. Soc. Exp. Biol. Med.*, **100**, 16
229. Cox, G. S., Hodges, J. R. and Vernikos, J. (1958). *J. Endocrinol.*, **17**, 177
230. Davidson, J. M. and Feldman, S. (1963). *Endocrinology*, **72**, 936
231. Davidson, J. M. and Feldman, S. (1967). *Acta Endocrinol.*, **55**, 240
232. Vernikos-Danellis, J. (1964). *Endocrinology*, **75**, 514
233. Vernikos-Danellis, J. and Marks, B. H. (1970). *Hypophysiotropic Hormones of the Hypothalamus: Assay and Chemistry*, 60 (J. Meites, editor) (Baltimore: Williams and Wilkins)
234. Halász, B. and Szentágothai, J. (1960). *Acta Morphol. Acad. Sci. Hung.*, **9**, 251
235. Motta, M., Mangilli, G. and Martini, L. (1965). *Endocrinology*, **77**, 392
236. Motta, M., Fraschini, F., Piva, F. and Martini, L. (1968). *Mem. Soc. Endocrinol.*, **17**, 3
237. Garren, L. D., Davis, W. W., Gill, G. N., Moses, H. L., Ney, R. L. and Crocco, R. M. (1969). *Progress in Endocrinology*, 102 (C. Gual, editor) (Amsterdam: Excerpta Medica)
238. Roth, J., Pastan, I., Lefkowitz, R. J., Pricer, W. E., Freychet, P. and Neville, D. M. (1971). *Structure–Activity Relationships of Protein and Polypeptide Hormones*, 228 (M. Margoulies and F. C. Greenwood, editors) (Amsterdam: Excerpta Medica)
239. Lefkowitz, R. J., Roth, J. and Pastan, I. (1970). *Science*, **150**, 633
240. Schwyzer, R., Schiller, P., Fauchère, J.-L., Karlaganis, G. and Pelican, G.-M. (1971). *Structure–Activity Relationships of Protein and Polypeptide Hormones*, 167 (M. Margoulies and F. C. Greenwood, editors) (Amsterdam: Excerpta Medica)

241. Rasmussen, H. (1971). *Structure–Activity Relationships of Protein and Polypeptide Hormones*, 194 (M. Margoulies and F. C. Greenwood, editors) (Amsterdam: Excerpta Medica)
242. Haynes, R. C. (1958). *J. Biol. Chem.*, **233**, 1220
243. Haynes, R. C., Koritz, S. B. and Peron, F. G. (1969). *J. Biol. Chem.*, **234**, 1421
244. Garren, L. D., Ney, R. L., and Davis, W. W. (1965). *Proc. Nat. Acad. Sci. U.S.A.*, **53**, 443
245. Koritz, S. B. and Hall, P. F. (1964). *Biochemistry (Wash.)*, **3**, 1298
246. Urquhart, J. and Li, C. C. (1969). *Progress in Endocrinology*, 880 (C. Gual, editor) (Amsterdam: Excerpta Medica)
247. Koritz, S. B. (1969). *Protein and Polypeptide Hormones*, 171 (M. Margoulies, editor) (Amsterdam: Excerpta Medica)
248. Butcher, R. W. and Sutherland, E. W. (1969). *Protein and Polypeptide Hormones*, 176 (M. Margoulies, editor) (Amsterdam: Excerpta Medica)
249. Farese, R. V. (1969). *Protein and Polypeptide Hormones*, 181 (M, Margoulies, editor) (Amsterdam: Excerpta Medica)
250. Ferguson, J. J. (1969). *Protein and Polypeptide Hormones*, 185 (M. Margoulies, editor) (Amsterdam: Excerpta Medica)
251. Garren, L. D. (1969). *Protein and Polypeptide Hormones*, 189 (M. Margoulies, editor) (Amsterdam: Excerpta Medica)
252. Staehelin, M. and Maier, R. (1969). *Protein and Polypeptide Hormones*, 193, (M. Margoulies, editor) (Amsterdam: Excerpta Medica)
253. Butcher, R. W. and Sutherland, E. W. (1969). *Protein and Polypeptide Hormones*, 150 (M. Margoulies, editor) (Amsterdam: Excerpta Medica)
254. Bally, P. R. and Tilbury, K. L. (1969). *Protein and Polypeptide Hormones*, 154, (M. Margoulies, editor) (Amsterdam: Excerpta Medica)
255. Stock, K. and Westermann, E. (1969). *Protein and Polypeptide Hormones*, 159 (M. Margoulies, editor) (Amsterdam: Excerpta Medica)
256. Rodbell, M. and Birnbaumer, L. (1969). *Protein and Polypeptide Hormones*, 878 (M. Margoulies, editor) (Amsterdam: Excerpta Medica)
257. Gemzell, C. and Roos, P. (1966). *The Pituitary Gland*, Vol. 1, 492 (G. W. Harris and B. T. Donovan, editors) (London: Butterworths)
258. Segaloff, A. (1966). *The Pituitary Gland*, Vol. 1, 518 (G. W. Harris and B. T. Donovan, editors) (London: Butterworths)
259. Canfield, R. E. and Bell, J. J. (1969). *Progress in Endocrinology*, 402 (C. Gual, editor) (Amsterdam: Excerpta Medica)
260. Amoroso, E. C. and Porter, D. G. (1966). *The Pituitary Gland*, Vol. 2, 364 (G. W. Harris and B. T. Donovan, editors) (London: Butterworths)
261. Maghuin-Rogister, G. and Hennen, G. (1971). *Structure–Activity Relationships of Protein and Polypeptide Hormones*, 112 (M. Margoulies and F. C. Greenwood, editors) (Amsterdam: Excerpta Medica)
262. Butt, W. R. and Kennedy, J. F. (1971). *Structure–Activity Relationships of Protein and Polypeptide Hormones*, 115 (M. Margoulies and F. C. Greenwood, editors) (Amsterdam: Excerpta Medica)
263. Saxena, B. B. and Rathnam, P. (1971). *Structure–Activity Relationships of Protein and Polypeptide Hormones*, 122 (M. Margoulies and F. C. Greenwood, editors) (Amsterdam: Excerpta Medica)
264. Sayers, G., Beall, R. J. and Seelig, S. (1972). *Science*, **175**, 1131
265. Papkoff, H. (1971). *Structure–Activity Relationships of Protein and Polypeptide Hormones*, 73 (M. Margoulies and F. C. Greenwood, editors) (Amsterdam: Excerpta Medica)
266. Ward, D. N. and Liu, W.-K. (1971). *Structure–Activity Relationships of Protein and Polypeptide Hormones*, 80 (M. Margoulies and F. C. Greenwood, editors) (Amsterdam: Excerpta Medica)
267. Bahl, O. P. (1971). *Structure–Activity Relationships of Protein and Polypeptide Hormones*, 99 (M. Margoulies and F. C. Greenwood, editors) (Amsterdam: Excerpta Medica)
268. De la Llosa, P. and Jutisz, M. (1969). *Protein and Polypeptide Hormones*, 229 (M. Margoulies, editor) (Amsterdam: Excerpta Medica)

269. Roos, P. (1969). *Progress in Endocrinology*, 377 (C. Gual, editor) (Amsterdam: Excerpta Medica)
270. Crooke, A. C. (1969). *Progress in Endocrinology*, 1218 (C. Gual, editor) (Amsterdam: Excerpta Medica)
271. McCann, S. M., Taleisnik, S. and Friedman, H. M. (1960). *Proc. Soc. Exp. Biol. Med.*, **104**, 432
272. Igarashi, M. and McCann, S. M. (1964). *Endocrinology*, **74**, 446
273. Matsuo, H., Baba, Y., Nair, R. M. G., Arimura, A. and Schally, A. V. (1971). *Biochem. Biophys. Res. Commun.*, **43**, 1334
274. White, W. F. (1970). *Hypophysiotropic Hormones of the Hypothalamus: Assay and Chemistry*, 249 (J. Meites, editor) (Baltimore: Williams and Wilkins)
275. Schally, A. V., Baba, Y., Arimura, A., Redding, T. W. and White, W. F. (1971). *Biochem. Biophys. Res. Commun.*, **42**, 50
276. Schally, A. V., Arimura, A., Baba, Y., Nair, R. M. G., Matsuo, H., Redding, T. W. and Debeljuk, L. (1971). *Biochem. Biophys. Res. Commun.*, **43**, 393
277. Arimura, A., Matsuo, H., Baba, Y., Debeljuk, L., Sandow, J. and Schally, A. V. (1972). *Endocrinology*, **90**, 163
278. Schally, A. V., Arimura, A., Kastin, A. J., Matsuo, H., Baba, Y., Redding, T. W., Nair, R. M. G., Debeljuk, L. and White, S. F. (1971). *Science*, **173**, 1036
279. Kobayashi, T., Kigawa, T., Mizuno, M. and Watanabe, T. (1971). *Karolinska Symp. on Res. Methods in Reproductive Endocrinol.*, **3**, 27; *Acta Endocrinol. Suppl.*, **153**, 27
280. Samli, M. H. and Geschwind, I. I. (1968). *Endocrinology*, **82**, 225
281. Wakabayashi, K., Schneider, H. P. G., Watanabe, D. B., Crighton, D. B. and McCann S. M. (1968). *Fed. Proc.*, **27**, 269
282. Jutisz, M. and de la Llosa, M. P. (1968). *Excerpta Medica Int. Congress Series 157*, 137
283. Jutisz, M. and de la Llosa, M. P. (1967). *Endocrinology*, **81**, 1193
284. Watanabe, S., Dhariwal, A. P. S. and McCann, S. M. (1968). *Endocrinology*, **82**, 674
285. McCann, S. M., Wakabayashi, K., Ashworth, R., Schneider, H. P. G., Kamberi, I. and Coates, P. (1969). *Progress in Endocrinology*, 571 (C. Gual, editor) (Amsterdam: Excerpta Medica)
286. Yang, W. H., Sairam, M. R., Papkoff, H. and Li, C. H. (1972). *Science*, **175**, 637
287. Papkoff, H., Solis-Wallckerman, J., Martin, M. and Li, C. H. (1971). *Arch. Biochem. Biophys.*, **143**, 226
288. Harris, G. W. and Campbell, H. J. (1966). *The Pituitary Gland*, Vol. 2, 99 (G. W. Harris and B. T. Donovan, editors) (London: Butterworths)
289. Donovan, B. T. (1966). *The Pituitary Gland*, Vol. 2, 49 (G. W. Harris and B. T. Donovan, editors) (London: Butterworths)
290. Davidson, J. M. (1969). *Frontiers in Neuroendocrinology 1969*, 343 (W. F. Ganong and L. Martini, editors) (New York, London and Toronto: Oxford Univ. Press)
291. Davidson, J. M. (1966). *Neuroendocrinology*, Vol. 1, 565 (L. Martini and W. F. Ganong, editors) (New York and London: Academic Press)
292. Flerkó, B. (1966). *Neuroendocrinology*, Vol. 1, 613 (L. Martini and W. F. Ganong, editors) (New York and London: Academic Press)
293. Hohlweg, W. (1934). *Klin. Wochschr.*, **13**, 92
294. Yen, S. S. C. and Tsai, C. C. (1971). *J. Clin. Endocrinol. Metab.*, **33**, 882
295. Harris, G. W. and Naftolin, F. (1970). *Brit. Med. Bull.*, **26**, 6
296. Armstrong, D. T. (1969). *Progress in Endocrinology*, 89 (C. Gual, editor) (Amsterdam: Excerpta Medica)
297. Ahrén, K. and Hamberger, L. (1969). *Progress in Endocrinology*, 75 (C. Gual, editor) (Amsterdam: Excerpta Medica)
298. Means, A. R. and Vaitukaitis, J. (1972). *Endocrinology*, **90**, 39
299. Lee, C. Y. and Ryan, R. J. (1971). *Endocrinology*, **89**, 1515
300. Rajaniemi, H. and Vanha-Perttula, T. (1972). *Endocrinology*, **90**, 1
301. Marsh, J. M., Butcher, R. W., Savard, K. and Sutherland, E. W. (1966). *J. Biol. Chem.* **241**, 5436
302. Marsh, J. M. (1969). *Progress in Endocrinology*, 83 (C. Gual, editor) (Amsterdam: Excerpta Medica)
303. Hall, P. F. and Eik-Nes, K. B. (1962). *Biochim. Biophys. Acta*, **63**, 411
304. Means, A. R. and Hall, P. F. (1967). *Endocrinology*, **81**, 1151
305. Irby, D. C. and Hall, P. F. (1971). *Endocrinology*, **89**, 1367

306. Yokoe, Y., Irby, D.C. and Hall, P.F. (1971). *Endocrinology*, **88**, 195
307. Li, C. H. and Starman, B. (1964). *Biochim. Biophys. Acta*, **96**, 175
308. Li, C. H., Liu, W.-K. and Dixon, J. S. (1966). *J. Amer. Chem. Soc.*, **88**, 2010
309. Li, C. H. and Yamashiro, D. (1970). *J. Amer. Chem. Soc.*, **92**, 26
310. Li, C. H. and Yamashiro, D. (1970). *J. Amer. Chem. Soc.*, **92**, 7608
311. Li, C. H., Papkoff, H., Fonss-Bech, P. and Condliffe, P. G. (1956). *J. Biol. Chem.*, **218**, 41
312. Harris, J. I., Li, C. H., Condliffe, P. G. and Pon, N. G. (1954). *J. Biol. Chem.*, **209**, 133
313. Li, C. H., Papkoff, H. and Hayashida, T. (1959). *Arch. Biochem. Biophys.*, **85**, 97
314. Li, C. H. (1957). *Fed. Proc.*, **16**, 775
315. Papkoff, H., Li, C. H. and Liu, W.-K. (1962). *Arch. Biochem. Biophys.*, **96**, 367
316. Squire, P. G. and Pedersen, K. O. (1961). *J. Amer. Chem. Soc.*, **83**, 476
317. Tashjian, A. H. and Levine, L. (1969). *Progress in Endocrinology*, 440 (C. Gual, editor) (Amsterdam: Excerpta Medica)
318. Forsyth, I. A. and Folley, S. J. (1970). *Ovo-Implantation, Human Gonadotropins and Prolactin*, 266 (P. O. Hubinont, F. Leroy, C. Robyn and P. Leleux, editors) (Basel: Karger)
319. Bewley, T. A. and Li, C. H. (1970). *Science*, **168**, 1361
320. Franz, J., Haselbach, C. H. and Liberto, O. (1962). *Acta Endocrinol.*, **41**, 336
321. Deuben, R. R. and Meites, J. (1964). *Endocrinology*, **74**, 408
322. Meites, J. (1964). *Proc. Second Int. Congr. Endocrinol. Excerpta Medica Int. Congress Series 83*, 522 (Amsterdam: Excerpta Medica)
323. Reichlin, S. and Schalch, D. S. (1969). *Progress in Endocrinology*, 584 (C. Gual, editor) (Amsterdam: Excerpta Medica)
324. Schally, A. V., Arimura, A., Wakabayashi, I., Sawano, S., Barrett, J. F., Bowers, C. Y., Redding, T. W., Mittler, J. C. and Schally, A. V. (1970). *Hypophysiotropic Hormones of the Hypothalamus: Assay and Chemistry*, 208 (J. Meites, editor) (Baltimore: Williams and Wilkins)
325. Krulich, L. and McCann, S. M. (1969). *Endocrinology*, **85**, 319
326. Schally, A. V., Baba, Y., Nair, R. M. G. and Bennett, C. D. (1971). *J. Biol. Chem.*, **246**, 6647
327. Veber, D. F., Bennet, C. D., Milkowksi, J. D., Gal, G., Denkewalter, R. G. and Hirschman, R. (1971). *Biochem. Biophys. Res. Commun.*, **45**, 235
328. Schally, A. V., Arimura, A., Wakabayashi, I., Redding, T. W., Dickerman, E. and Meites, J. (1972). *Experientia*, **28**, 205
329. Schally, A. V., Muller, E. E. and Sawano, S. (1968). *Endocrinology*, **82**, 271
330. Coates, P. W., Ashby, E. A., Krulich, L., Dhariwal, A. P. S. and McCann, S. M. (1968). *Proc. XXVI Meeting of the Electron-Microscopy Society of America*, 232
331. Mittler, J. C., Sawano, S., Wakabayashi, I., Redding, T. W. and Schally, A. V. (1970). *Proc. Soc. Exp. Biol. Med.*, **133**, 890
332. De Virgiliis, G., Meldolesi, J. and Clementi, F. (1968). *Endocrinology*, **83**, 1278
333. Couch, E. F., Arimura, A., Schally, A. V., Saito, M. and Sawano, S. (1969). *Endocrinology*, **85**, 1084
334. Coates, P. W., Ashby, E. A., Krulich, L., Dhariwal, A. P. S. and McCann, S. M. (1970). *Amer. J. Anat.*, **128**, 389
335. Müller, E. E., Pecile, A., Naimzada, M. K. and Ferrario, G. (1969). *Experientia*, **25**, 750
336. Hertelendy, F., Tood, H., Peake, G. T., Machlin, L. J., Johnston, G. and Pounds, G. (1971). *Endocrinology*, **89**, 1256
337. Peake, G. T., Steiner, A. L. and Daughaday, W. H. (1972). *Endocrinology*, **90**, 212
338. Parsons, J. A. (1970). *J. Physiol (London)*, **210**, 973
339. Schoffield, J. G. and Stead, M. (1971). *FEBS Lett.*, **13**, 149
340. Reichlin, S. (1966). *The Pituitary Gland*, Vol. 2, 270 (G. W. Harris and B. T. Donovan, editors) (London: Butterworths)
341. Pecile, A. and Müller, E. E. (1966). *Neuroendocrinology*, Vol. 1, 537 (L. Martini and W. F. Ganong, editors) (New York and London: Academic Press)
342. Glick, S. M. (1969). *Frontiers in Neuroendocrinology 1969*, 141 (W. F. Ganong and L. Martini, editors) (New York, London and Toronto: Oxford Univ. Press)
343. Mace, J. W., Gotlin, R. W. and Beck, P. (1972). *J. Clin. Endocrinol. Metab.*, **34**, 339
344. Hunter, W. M., Willoughby, J. M. T. and Strong, J. A. (1968). *J. Endocrinol.*, **40**, 297

345. Frantz, A. G. and Rabkin, M. T. (1964). *New Engl. J. Med.*, **271**, 1375
346. Simon, S., Schiffer, M. S., Glick, S. M. and Schwartz, E. (1967). *J. Clin. Endocrinol. Metab.*, **27**, 1633
347. Evans, H. M., Briggs, J. H. and Dixon, J. S. (1966). *The Pituitary Gland*, Vol. 1, 439 (G. W. Harris and B. T. Donovan, editors) (London: Butterworths)
348. Kostyo, J. L. and Engel, F. L. (1960). *Endocrinology*, **67**, 708
349. Liberti, J. P., Wood, D. M. and Du Vall, C. H. (1972). *Endocrinology*, **90**, 311
350. Payne, S. G. and Kostyo, J. L. (1970). *Endocrinology*, **87**, 1186
351. Cheek, D. B. and Hill, D. E. (1970). *Fed. Proc.*, **29**, 1503
352. Rilema, J. A. and Kostyo, J. L. (1971). *Endocrinology*, **88**, 240
353. Jackson, C. D., Irving, C. C. and Sells, B. N. (1970). *Biochim. Biophys. Acta*, **217**, 64
354. Caldwell, A. B. and Fain, J. N. (1970). *Hormones Metab. Res.*, **2**, 3
355. Tchobroutsky, G. (1970). *Ann. Endocrinol. (Paris)*, **31**, 111
356. Li, C. H. (1968). *Perspectives Biol. Med.*, **11**, 498
357. Lyons, W. R. and Dixon, J. S.. (1966). *The Pituitary Gland*, Vol. 1, 527 (G. W. Harris and B. T. Donovan, editors) (London: Butterworths)
358. Meites, J. and Nicoll, C. S. (1966). *Ann. Rev. Physiol.*, **28** 57
359. Bern, H. A. and Nicoll, C. S. (1969). *Progress in Endocrinology*, 433 (C. Gual, editor) (Amsterdam: Excerpta Medica)
360. Mishkinsky, J., Khazen, K. and Sulman, F. G. (1968). *Endocrinology*, **82**, 611
361. Arimura, A., Dunn, J. D. and Schally, A. V. (1972). *Endocrinology*, **90**, 378
362. Nicoll, C. (1965). *J. Exp. Zool.*, **158**, 203
363. Kragt, C. L. and Meites, J. (1965). *Endocrinology*, **76**, 1169
364. Gourdji, D. and Tixier-Vidal, A. (1966). *Compt. Rend.*, **263**, 162
365. Pasteels, J.-L. (1961). *Compt. Rend.*, **253**, 3074
366. Pasteels, J.-L. (1962). *Compt. Rend.*, **254**, 2664
367. Kragt, C. L. and Meites, J. (1967). *Endocrinology*, **80**, 1170
368. Kuroshima, A., Arimura, A., Bowers, C. Y. and Schally, A. V. (1966). *Endocrinology*, **78**, 216
369. Nicoll, C. S., Fiorindo, R. P., McKennee, C. T. and Parsons, J. A. (1970). *Hypophysiotropic Hormones of the Hypothalamus: Assay and Chemistry*, 115 (J. Meites, editor) (Baltimore: Williams and Wilkins)
370. Pasteels, J.-L. (1970). *Ovo-Implantation, Human Gonadotrophins and Prolactin*, 279 (P. O. Hubinont, F. Leroy, C. Robyn and P. Leleux, editors) (Basel, München and New York: Karger)
371. Parsons, J. A. and Nicoll, C. S. (1971). *Neuroendocrinology*, **8**, 213
372. Parsons, J. A. (1970). *J. Physiol. (London)*, **210**, 973
373. Lemay, A. and Labrie, F. (1972). *FEBS Lett.*, **20**, 7
374. Welsh, C. W., Negro-Villar, A. and Meites, J. (1968). *Neuroendocrinology*, **3**, 238
375. Clemens, J. A., Sar, M. and Meites, J. (1969). *Endocrinology*, **84**, 868
376. Spies, H. G. and Clegg, M. T. (1971). *Neuroendocrinology*, **8**, 205
377. Mishkinsky, J., Nir, I. and Sulman, F. G. (1969). *Neuroendocrinology*, **5**, 48
378. Everett, J. W. (1966). *The Pituitary Gland*, Vol. 2, 166 (G. W. Harris and B. T. Donovan editors) (London: Butterworths)
379. Ben-David, M., Danon, A. and Sulman, F. G. (1970). *Neuroendocrinology*, **6**, 336
380. Wuttke, W., Cassell, E. and Meites, J. (1971). *Endocrinology*, **88**, 737
381. Schreiber, V., Přibyl, T. and Roháčová, J. (1972). *Physiol. Bohemoslov.*, **21**, 639
382. Denamur, R. (1969). *Progress in Endocrinology*, 959 (C. Gual, editor) (Amsterdam: Excerpta Medica)
383. Prop, F. J. A. (1969). *Protein and Polypeptide Hormones*, 508 (M. Margoulies, editor) (Amsterdam: Excerpta Medica)
384. Chadwick, A. (1962). *Biochem. J.*, **85**, 554
385. Falconer, I. R. (1967). *Abstr. IV. Meeting FEBS Oslo*, 81
386. Fiddler, T. J. and Falconer, I. R. (1969). *Protein and Polypeptide Hormones*, 320 (M. Margoulies, editor) (Amsterdam: Excerpta Medica)
387. Wang, D. Y., Hallowes, R. C., Smith, R. H. and Amor, V. (1971). *J. Endocrinol.*, **49**, IV
388. Ganong, W. F. (1971). *Review of Medical Physiology*, 317 (Los Altos: Lange)
389. Hall, K. (1972). *Human Somatomedin* (Copenhagen: Periodica)

390. Daughaday, W. H., Hall, K., Raben, M. S., Salmon, W. D., Van den Brande, J. L. and Van Wyk, J. J. (1972). *Nature (London)*, **236**, 107
391. Dorfman, R. I. (1972). *Biochemical Actions of Hormones*, 295 (G. Litwack, editor) (New York and London: Academic Press)

4
Hormonal Regulation of Calcium Metabolism

J. F. HABENER and H. D. NIALL
Massachusetts General Hospital and Harvard Medical School, Boston

4.1	INTRODUCTION		102
4.2	PARATHYROID HORMONE		103
	4.2.1	Chemistry	103
		4.2.1.1 Isolation and structural analysis	103
		4.2.1.2 Chemical synthesis and structure–activity relationships	106
	4.2.2	Biosynthesis	108
		4.2.2.1 Anatomy	108
		4.2.2.2 Biosynthetic precursor (proparathyroid hormone)	108
	4.2.3	Control of secretion	111
		4.2.3.1 Proportionality of control	111
		4.2.3.2 Factors controlling secretion	112
		4.2.3.3 Biochemical mechanisms of secretion	113
		4.2.3.4 Relationship of hormone synthesis to secretion	114
		4.2.3.5 Adaptive increases in parathyroid hormone secretion	114
	4.2.4	Metabolism	114
		4.2.4.1 Nature of the circulating hormone	114
		4.2.4.2 Rate of disappearance, distribution and organ specific destruction	119
	4.2.5	Assay of parathyroid hormone	120
		4.2.5.1 Bioassays	121
		4.2.5.2 Radioimmunoassay	121
	4.2.6	Mode of action of parathyroid hormone	122
		4.2.6.1 Effect of parathyroid hormone on bone	122
		(a) Kinetic responses of bone	123
		(b) Changes in bone cells	124
		(c) Effect of parathyroid hormone on bone matrix	124
		(d) Role 3′,5′-cyclic-AMP in the action of parathyroid hormone on bone	125
		(e) Effect of other variables on bone	125
		4.2.6.2 Action of parathyroid hormone on the kidney	125

	4.2.6.3	Action of the parathyroid hormone on small intestine	127
	4.2.6.4	Other actions of parathyroid hormone	127

4.3 CALCITONIN 128
    4.3.1 Chemistry 128
        4.3.1.1 Isolation and sequence determination 128
        4.3.1.2 Chemical synthesis and structure–activity relations 129
    4.3.2 Biosynthesis and control of secretion of calcitonin 131
    4.3.3 Assay of calcitonin 133
        4.3.3.1 Bioassay 133
        4.3.3.2 Radioimmunoassay 133
    4.3.4 Metabolism 134
        4.3.4.1 Distribution and disappearance 134
        4.3.4.2 Organ-specific destruction 135
    4.3.5 Mode of action of calcitonin 136
        4.3.5.1 Action on bone 136
        4.3.5.2 Action on kidney 137
        4.3.5.3 Action on intestine 137
        4.3.5.4 Biochemical mechanism of action 138
    4.3.6 Therapeutic uses 138

4.4 VITAMIN D 139
    4.4.1 Assay of vitamin D 139
    4.4.2 Chemistry and metabolism 139
    4.4.3 Physiology and mode of action 142

ACKNOWLEDGEMENTS 144

4.1 INTRODUCTION

In 1901, McCallum and Voegtlin[250] first recognised that a humoral substance, parathyroid hormone, was required for the maintenance of a normal concentration of blood calcium. Since that time great progress has been made in our understanding of the functions that parathyroid hormone, calcitonin and vitamin D have in the regulation of calcium metabolism.

    The purpose of this chapter is to review the recent information pertaining to the hormonal influences on calcium metabolism. The major emphasis has been placed on developments in research on parathyroid hormone but, in addition, certain aspects of the chemistry and mode of action of calcitonin and vitamin D have been reviewed. The reader should appreciate that due to the existence of a vast body of literature relating to the subject of calcium metabolism, it has not been possible to present a detailed account of all the achievements in the field. Therefore, we have arbitrarily selected certain topics of particular interest for more detailed discussion.

    A number of excellent reviews have been written in the past few years that cover extensively many areas of calcium metabolism. In particular the reader is referred to the exhaustive chapter on this entire subject by Potts and Deftos in Duncan's *Textbook of Metabolism*[1].

The chemistry of parathyroid hormone and calcitonin has recently been reviewed by Potts et al.[2] and many aspects of the physiology of parathyroid hormone by Parsons and Potts[3]. Comprehensive reviews covering developments in calcitonin research have been written by Hirsh and Munson[4], Copp[5,6], Potts[7], and most recently by Potts et al.[8]. Many accounts of current research in the field are presented in the Proceedings of the *4th Int. Conf. on Parathyroid Hormone and Calcitonin* held at Chapel Hill, North Carolina, in the spring of 1971.

4.2 PARATHYROID HORMONE

4.2.1 Chemistry

4.2.1.1 Isolation and structural analysis

Although the first biologically-active extracts of parathyroid hormone from bovine glands were made by Collip in 1925 using hot 5% hydrochloric acid[9], further purification of the hormone was not achieved until 35 years later when Aurbach[10] and Rasmussen et al.[11] recognised that the hot acid extraction caused extensive cleavage of the polypeptide hormone. Aurbach developed a more suitable extraction method using phenol, which led to extractions of a stable hormone preparation free from cleavages within the peptide chain[11]. Subsequently, in 1964, Rasmussen et al. developed another efficient extraction method using 8 M urea and cysteine in cold hydrochloric acid[12]. The extracted hormone preparation was further purified by chromatography on Sephadex[13] and Carboxymethylcellulose[14]. These procedures made available sufficiently large quantities of relatively pure parathyroid hormone to permit the development of a radioimmunoassay[15] and of a number of studies on the physiological action of the hormone. However, early attempts at structural sequence analysis of even the most highly purified hormone prepared by chromatography on carboxymethylcellulose were impeded by low yields and by contamination of the hormone by as much as 10–20% with non-hormonal peptides.

A significant improvement in purification of the hormone came when Keutmann et al. employed 8 M urea in the buffer used to elute the hormone from carboxymethylcellulose and a highly sensitive renal-cortical adenyl cyclase assay[16] to monitor the extraction procedure. Keutmann was able to separate cleanly the hormonal polypeptide from the contaminants[17]. In addition, the carboxymethylcellulose–urea chromatography separated out two additional variants of the hormone which later proved to be isohormones, closely related in amino acid composition to the major hormonal species[17]. Compositional analysis of all three forms of the hormone showed that each contained 84 amino acids and differed only in the content of threonine and valine[18].

During the past few years the complete amino acid sequences of the major forms of bovine[18,19] and porcine[20] parathyroid hormones have been determined. These hormones have been shown to be single polypeptides of 84 amino acids (Figure 4.1). The complete analysis of the major bovine hormone was accomplished in 1970 by Niall et al.[18] and independently by Brewer and Ronan[19] using the automated degradation method of Edman and Beggs[21,22]. This method utilises the reagent phenylisothiocyanate which couples with the terminal alpha amino group of a peptide chain[23]. Under acidic conditions, the

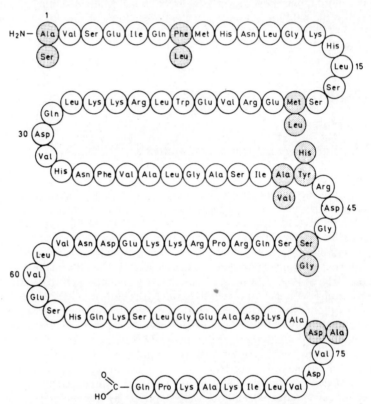

Figure 4.1 Structure of the bovine and porcine parathyroid hormones. Backbone structure is that of the major form of bovine parathyroid hormone. Shaded positions indicate where the porcine hormone differs from the bovine; the residues found in porcine parathyroid hormone are shown in apposition to the residues in the backbone structure. (From Keutmann et al.[17], by courtesy of American Chemical Society.)

terminal amino acid can be cleaved selectively as a heterocyclic derivative (an anilinothiazolinone) thus shortening the chain by one residue. The cleaved amino acid is converted to a more stable isomeric form, a phenylthiohydantoin, and can be identified by gas chromatography.

The amino acid sequence of the first 54 residues was completed using the automated sequenator (Beckman Spinco Model 890)[18,24]. A progressive fall in the yield at each successive cleavage step prohibited further sequencing beyond residue 54. The sequence of the remaining 30 amino acids of the hormone was determined on a fragment comprised of residues 53–84 obtained by tryptic digestion of the hormone after the lysine side chains had been protected specifically against tryptic attack by reaction with maleic anhydride. It had been recognised that all arginine residues in the molecule were accounted for in the first 52 positions. Therefore, after blockade by maleylation of the lysine residues, there were no trypsin-susceptible peptide bonds in the peptide fragment 53–84[18, 24]. Sequence analysis of the two isohormones of bovine parathyroid hormone has not yet been accomplished due to lack of sufficient amounts of these minor hormone variants.

Using techniques similar to those employed for the purification of bovine

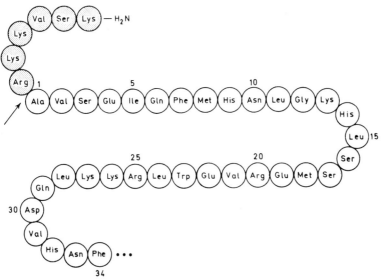

Figure 4.2 Model depicting partial structure of bovine proparathyroid hormone. Sequence is that which has been chemically synthesised and includes the prohormone specific hexapeptide (shaded residues) and the first 34 residue sequence of parathyroid hormone which retains the full biological activity of the complete 84 amino acid peptide shown in Figure 4.1. Arrow indicates peptide bond between prohormone and hormone sequence which is sensitive to cleavage by dilute trypsin. (From Potts *et al.*[2], by courtesy of Academic Press.)

parathyroid hormone, Woodhead *et al.* have succeeded in isolating a completely homogenous preparation of porcine parathyroid hormone[20]. The complete amino acid sequence has now been determined and found to differ from the bovine hormone in seven positions[24] (Figure 4.1). The most notable changes in the sequence of the porcine hormone are the lack of tyrosine and the elimination of one methionine residue, which is replaced by leucine. The latter difference is of particular interest from the standpoint of structure–activity relationships since both the bovine and porcine hormones lose biological activity when their methionines are oxidised to the sulphoxide form.

Human parathyroid hormone has been isolated and purified on a small scale by O'Riordan *et al.*[25] and by Brewer *et al.*[26]. Preliminary sequence analyses of the first amino-terminal 34 or 37 residues have been reported independently by two groups of investigators[26, 27]. The structure found for this segment of the human hormone closely resembles that of the bovine and porcine hormones. However, at the present time the two groups of investigators do not agree on the exact structure of the human parathyroid hormone. One group has found differences in sequence from the bovine hormone at positions 1, 7 and 16[27], whereas the other group has found additional differences at positions 22, 28 and 30[26]. A co-operative effort in the collection of additional human parathyroid tissue will be required before further structural analysis can be accomplished and the correct structure ascertained.

More recently, a small amount of proparathyroid hormone, a biosynthetic precursor to parathyroid hormone (see section on Biosynthesis of Parathyroid Hormone) has been isolated from bovine glands in relatively homogeneous form[28]. Preliminary sequence analysis reveals that it contains six additional amino acids at the amino-terminus of parathyroid hormone, four of which are the basic residues arginine and lysine[29] (Figure 4.2). A prohormone fragment consisting of the prohormone hexapeptide sequence added to the first 34 residues of parathyroid hormone has been synthesised chemically[30] (Figure 4.2). This synthetic proparathyroid hormone should prove valuable in assessment of the biological and immunological properties of the prohormone.

4.2.1.2 Chemical synthesis and structure–activity relationships

Once the complete amino acid sequence of the bovine parathyroid hormone was known, it became possible to undertake a detailed analysis of the structural requirements necessary for biological activity. Earlier work had indicated that the full 84 amino acid sequence of the hormone was not required for biological activity[31,32]; cleavage of the native hormone with dilute hydrochloric acid had shown that a fragment from the amino terminal region comprised of residues 1–29 was biologically active[31].

It was possible to confirm this observation through biological testing of a chemically-synthesised peptide fragment of the hormone consisting of the amino terminal 34 residues. Using a modification of the procedure of Merrifield[33], Tregear et al. have since synthesised a number of shorter amino terminal fragments as well as several structural variants of the fragments in an attempt to define the minimum structure required for biological activity (see later)[35]. The synthetic tetratriacontapeptide was found to possess all the specific biological properties associated with native parathyroid hormone[34]. In addition to stimulating adenyl cyclase in both bone and kidney cells, the synthetic peptide elevated blood calcium in rats and dogs and caused an increase in renal excretion of both cyclic AMP and phosphate in the rat[34]. Although the biological potency of the initial preparations of synthetic 1–34 was only ca. 1/6 that of native parathyroid hormone on a molar basis (400–600 USP units mg^{-1} compared to 1600 USP units mg^{-1} for native parathyroid hormone), subsequent syntheses using methods to minimise exposure to harsh chemical reagents such as hydrogen fluoride and piperidine, have resulted in increased biological activity (ca. 3000 USP units mg^{-1}) which on a molar basis is nearly equivalent to the activity of native parathyroid hormone[35].

Recent studies by Tregear appear to have defined the minimum structural requirement for biological activity on receptors in the renal cortex (Figure 4.3)[35]. A continuous sequence extending from valine at position 2 and including lysine at position 27 is required for activity. An interesting finding was that deletion of as little of the sequence as the single amino-terminal amino acid alanine (synthetic fragment 2–34) led to a sharp decrease in biological activity. Substitution of tyrosine for alanine at position 1 resulted in a biologically-active product.

In addition to the information obtained from the above studies, using hormone analogues containing amino acid deletions, it is now known that chemical additions of amino acids onto the amino terminal alanine of the

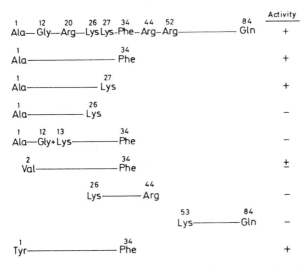

Figure 4.3 The relative biological activities of natural and synthetic fragments of bovine parathyroid hormone determined in renal adenyl cyclase assay. A continuous sequence that extends from alanine, residue 1, to lysine, residue 27, is required for significant biological activity. Deletion of the amino-terminal alanine drastically lowers activity. Activity is retained when tyrosine is substituted for the amino-terminal alanine

peptide 1–34 results in a lowering of biological activity. This observation is of particular interest since it is known that the prohormone of parathyroid hormone contains amino acids extending from the amino terminus[36] (see above discussion and section on Biosynthesis of Parathyroid Hormone) and suggests that the prohormone may not be inherently biologically active.

The finding that there is only one methionine (in position 8) in the active region of the porcine hormone, compared with two in the bovine sequence, has stimulated interest in the synthesis of a peptide fragment with an isosteric substitution of norleucine for methionine. The availability of this synthetic fragment should permit a more complete definition of biological activity. Results recently obtained indicate that methionine is not required for activity at least on renal receptors[37]. Since it is known that oxidation of methionine residues in the hormone to the sulphoxide form destroys all biological activity[20, 32, 38], this observation implies that methionine when present in the more polar oxidised form in the molecule blocks activity, but that the presence of methionine itself is not required for activity.

The availability of methods for the chemical synthesis of specific sequences of a hormone molecule whose structure is known, has proved to be a valuable aid to the biologist. The synthetic fragments of parathyroid hormone have already been used in preliminary studies to characterise antisera to the hormone and to develop radioimmunoassays which are sensitive to discrete regions of the hormone sequence[95]. These assays have been applied in preliminary studies to the investigation of the hormone circulating in blood and have provided an interesting insight into the complex nature of the metabolism of the hormone (see section on Metabolism of Parathyroid Hormone).

4.2.2 Biosynthesis

4.2.2.1 Anatomy

Parathyroid hormone is the only known humoral substance biosynthesised by the parathyroid glands, and under normal conditions the parathyroid glands are the only source of the hormone[39, 40]. Secretion by a number of malignant neoplasms of a substance immunologically related to parathyroid hormone has been demonstrated[41-43]. Presumably ectopic secretion of parathyroid hormone under these conditions results from depression of latent genes by the process of malignant transformation[46]. In the evolutionary scale the parathyroids are first seen in amphibians; they are absent in fishes and lower vertebrates[40]. The parathyroid glands are small reddish-brown bodies derived from the third and fourth branchial pouches and are usually located in or adjacent to the thyroid gland. Their number and size vary considerably among animal species. In mammals they are usually four in number but a fifth or sixth gland may occasionally be present, usually in an aberrant location such as in the mediastinum.

4.2.2.2 Biosynthetic precursor (prohormone)

Until very recently little information has been available concerning the mechanisms involved in the biosynthesis of parathyroid hormone. One of the most recent developments in the field of hormone research has been the discovery that parathyroid hormone is biosynthesised as a higher molecular weight precursor[28, 47], somewhat similar to the biosynthesis of insulin as proinsulin[48] (see Chemistry of Parathyroid Hormone and Figure 4.2).

Hamilton and his co-workers[28] were the first to identify a basic polypeptide that, in addition to parathyroid hormone, was labelled with radioactive amino acids during *in vitro* incubation of bovine parathyroid slices. Due to its greater positive charge the basic peptide eluted separately from parathyroid hormone on ion exchange chromatography; this provided a means for its isolation. It was subsequently shown to produce hypercalcaemia in rats, bone resorption in culture and to cross-react with antisera to parathyroid hormone[49]. In addition, gel filtration studies showed that the molecular weight of this peptide was considerably higher than that of parathyroid hormone.

Although the peptide had been isolated in nearly pure form for over 3 years[49], its precursor nature was not established until Kemper et al.[47] formally demonstrated a precursor relationship to parathyroid hormone by kinetic pulse-labelling experiments (Figure 4.4). The peptide was thus identified as a true biosynthetic precursor of parathyroid hormone. Similar *in vitro* studies

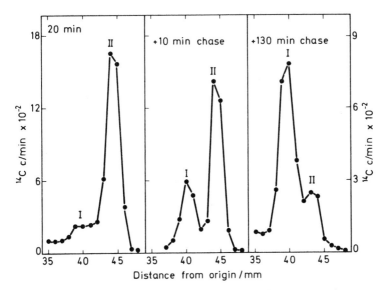

Figure 4.4 *In vitro* demonstration of a biosynthetic precursor (parathyroid hormone) to bovine parathyroid hormone. Polyacrylamide gel electrophoresis was used to analyse urea-hydrochloric acid extracts of bovine parathyroid glands after incubation of gland slices for 20 min with ^{14}C-labelled amino acids (left) followed by a 'chase' incubation with unlabelled amino acids for 10 min (middle) and 130 min (right). Radioactivity is first incorporated into proparathyroid hormone (peak II). In the absence of further incorporation of radioactive amino acids into protein, proparathyroid hormone is converted sequentially to parathyroid hormone (peak I) thus indicating that the proparathyroid hormone is a biosynthetic precursor to the parathyroid hormone. (From Kemper et al.[47], by courtesy of the National Academy of Sciences, U.S.A.)

were subsequently carried out by Habener et al.[50] in tissue from human parathyroid adenomas; these studies promptly led to the identification of a prohormone for human parathyroid hormone as well.

In view of the identification of proparathyroid hormone analogous to the biosynthetic precursor of insulin, proinsulin[48], it now appears that the synthesis of many of the polypeptide hormones may occur via higher molecular weight precursors. Indeed evidence has been reported recently in support of the existence of higher molecular weight forms of gastrin[51], ACTH[52] and glucagon[53]. The function of proparathyroid hormone, or for that matter of prohormones in general, is not known. However, several lines of evidence suggest that proparathyroid hormone may serve as a biologically inactive form of the hormone. Recent studies on the structural requirements for biological activity of chemically synthesised parathyroid hormones indicate that extension of the hormone at the amino terminus lowers biological activity[3] (see Section 4.2.12). In addition, preliminary tests of biological activity of limited quantities of proparathyroid hormone performed directly *in vivo* and *in vitro* have shown that the activity is no more than one-third that of parathyroid hormone[28]. It is not known whether this observed activity of the prohormone is inherent in the molecular or may be due to enzymatic cleavage of prohormone to hormone following injection into test animals or exposure to receptor tissue in the *in vitro* assays. Recent studies have shown that exposure of proparathyroid hormone to dilute solutions of trypsin *in vitro* results in the formation of immunologically-active intact parathyroid hormone[28].

Thus, it is now possible to postulate that a specific intracellular cleavage step may be required for the activation of the hormone in preparation for its storage in the secretory granule and for subsequent secretion. Such a cleavage process could provide a means for regulating the amount of active hormone available for secretion, particularly if subsequent studies should show that the activity of the proparathyroid hormone cleaving enzyme is sensitive to changes in extracellular concentrations of calcium. That the extra peptide sequence of proparathyroid hormone may play a role in the intracellular transport and packaging of the hormone in the secretory granule also has been suggested[47]. At present any proposed functions for proparathyroid hormone must be considered speculative and will require much further work for final elucidation and proof.

The discovery of proparathyroid hormone may have interesting implications in clinical studies concerned with the evaluation of circulating levels of hormone. Recent studies indicate that under certain conditions proparathyroid hormone may be secreted into the circulation. Immunoreactive parathyroid hormone secreted *in vivo* from a small number of parathyroid adenomas appears to be larger than the hormone of molecular weight 9500[54], since it has been observed to elute from gel filtration columns in a position corresponding to a molecular weight of *ca.* 11 000. Recent studies by Martin *et al.*[55] indicate that certain monolayer cultures of human parathyroid cells release into the culture media an immunoreactive basic polypeptide resembling proparathyroid hormone in its electrophoretic behaviour. In addition, the rate of intracellular cleavage of the prohormone to hormone has been reported to vary among different tumours of the parathyroids incubated *in vitro* and the rate is inversely related to the amount of immunoreactive prohormone in the tumour tissue[50]. This suggests that the parathyroid tumours secreting 'large' parathyroid hormone *in vivo* may convert proparathyroid hormone to parathyroid hormone slowly and consequently may accumulate and secrete proparathyroid hormone, analogous to the observed[56] secretion of proinsulin by certain islet cell tumours of the pancreas.

In this context it is interesting to note evidence reported by Riggs *et al.*[42] and Roof *et al.*[43] which shows that the immunochemical characteristics of parathyroid hormone found in the circulation of patients with non-parathyroid neoplasms, presumably secreting parathyroid hormone-like peptides, differ from immunoreactive parathyroid hormone found in the circulation of most patients with parathyroid adenomas. One explanation for this observation may be that proparathyroid hormone is the form of the hormone secreted by these non-parathyroid neoplasms. In addition to the recognised hormonal fragments of parathyroid hormone (see later), proparathyroid hormone and its metabolic degradation products may contribute to the heterogeneity of the hormone in the circulation.

4.2.3 Control of secretion

4.2.3.1 *Proportionality of control*

Classic experiments done in both *in vivo*[57, 58] and *in vitro*[59] model systems have shown that the secretion of parathyroid hormone is inversely proportional to the concentration of ionised calcium perfusing the parathyroid glands. Early

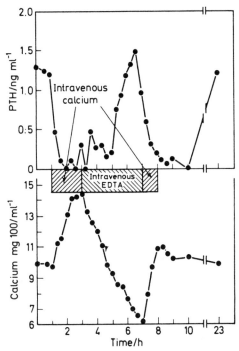

Figure 4.5 Concentrations of parathyroid hormone and calcium in the plasma of a cow in response to sequential infusions of calcium and EDTA. The concentration of parathyroid hormone rises and falls inversely with changes in the calcium concentration. (From Sherwood *et al.*[61], by courtesy of Macmillan.)

studies demonstrated that perfusion of the parathyroids with calcium-deficient fluids resulted in systemic hypercalcaemia, presumably due to increased secretion of parathyroid hormone into the circulation[60]. The availability of a specific radioimmunoassay sufficiently sensitive to measure parathyroid hormone in the circulation of the cow and other mammals has permitted direct confirmation that circulating hormone is increased with hypocalcaemia and decreased with hypercalcaemia (Figure 4.5)[61-63]. The finding in these studies, that the intact 84 amino acid hormone disappears rapidly from blood whenever secretion is abolished but that the hormone is always found circulating in normal animals, showed that the secretion of the hormone is continuous, not intermittent.

Detailed analyses of the changes in parathyroid hormone and calcium concentrations in the peripheral plasma have shown a simple linear and inverse relationship between hormone concentration and calcium over the range of 4 to 12 mg (100 ml)$^{-1}$ [62]. This has led to the conclusion that calcium regulates hormone secretion predominantly through a proportional control mechanism, although there was some indication that when the rate of plasma calcium concentration was changed rapidly, there was a marked rise in hormone concentration early in the course of the infusion greater than that predicted by the proportional mechanism[62]. This rise was sometimes followed by parathyroid hormone concentrations lower than predicted (see Figure 4.5). This has suggested that a derivative control mechanism might operate under conditions of rapid change in calcium concentration[62].

The development of methods for the direct perfusion of isolated goat parathyroids *in situ* with blood containing high concentrations of calcium or magnesium have shown that the changes in concentrations of hormone in peripheral blood occur rapidly (within minutes) and reflect actual changes in rate of hormone secretion from the gland rather than alterations in distribution or rate of destruction[63].

4.2.3.2 *Factors controlling secretion*

Studies done *in vitro* with explants of bovine parathyroid glands have also shown a first-order relationship between hormone release and the total divalent cation (calcium plus magnesium)[64-66]. These studies have shown that calcium and magnesium ions are equivalent in blocking the release of hormone and have suggested the possibility of a cation–ligand interaction in or at the surface of the parathyroid cell. It should be noted that it has not yet been determined whether physiological variations in magnesium concentration affect parathyroid secretion or whether hypomagnesaemia *per se* may influence hormone secretion. Studies in man have suggested that infusions of strontium may also suppress hormone secretion[67].

The divalent cations may not be the sole determinants regulating parathyroid hormone secretion. It has been suggested that the peptide hormones glucagon[68], ACTH[68] and calcitonin[69, 70] may stimulate secretion of parathyroid hormone but the evidence presented is preliminary and not conclusive. As a result of studies in cows, Reitz *et al.*[71] have concluded that calcitonin does not have a direct effect on parathyroid secretory activity and only influences secretory rate secondarily through effects on plasma calcium concentration. Phosphate has been shown to have no direct effect on hormone secretion but stimulates the parathyroid gland through a lowering of the concentration of calcium[62]. There is no convincing evidence of a requirement for involvement of innervation or of a special blood supply, since parathyroid glands in rats transplanted to muscle or other sites continue to secrete hormone[72, 73].

4.2.3.3 Biochemical mechanisms of secretion

Very little information is available at present concerning the biochemical mechanisms by which calcium regulates the synthesis and secretion of parathyroid hormone. It is not known at what level in protein biosynthesis calcium exerts its effects. Earlier evidence[74] suggested that calcium may regulate the flow of amino acids into the parathyroid cell but subsequent studies[75] have not borne out this contention. Nothing is known about the possible effects of calcium on regulation of messenger RNA synthesis or possible action of calcium on translational events, i.e. peptide initiation, elongation or termination. There is no information about whether calcium inhibits biosynthesis directly or whether changes in rates of hormone biosynthesis occur only in response to calcium-mediated changes in glandular secretion.

Calcium has been demonstrated to directly affect biosynthesis of parathyroid hormone *in vitro*. Hamilton *et al.* have found clearcut differences in the amount of radioactive hormone appearing in bovine parathyroid slices during *in vitro* incubations and have found that these changes are inversely related to the calcium concentrations in the media[75, 76]. In these studies synthesis of non-hormonal proteins was found to be independent of the calcium concentration. These same workers found no effect of magnesium on biosynthesis of parathyroid hormone and suggested that whereas calcium affects, directly or indirectly, both synthesis and secretion of hormone, magnesium may affect only secretion[75]. These observations would suggest that there may be two independent receptors sensitive to divalent cations, one which controls secretion and the other, synthesis of hormone.

Recent studies have implicated cyclic AMP as a possible intermediate in the action of calcium on parathyroid hormone secretion[77-80]. Perfusion of parathyroid glands *in vivo* or incubation of glandular tissue *in vitro*[78-80] with solutions containing dibutyryl cyclic AMP, an analogue of cyclic AMP that can freely enter cells, stimulates secretion of bioassayable and immunoassayable parathyroid hormone even in the presence of high concentrations of calcium which ordinarily suppress hormone secretion. In addition, studies on the activity of adenyl cyclase in canine parathyroid homogenates have shown an unusual sensitivity to calcium[79]. Concentrations of ionic calcium as low as 1×10^{-7} molar inhibit activity of the enzyme[77, 79]. Since these ionic calcium concentrations are at least 100-fold less than the concentrations of calcium required to produce significant inhibition of adenyl cyclase in other tissues, the data suggest that adenyl cyclase may play a specific role in mediating the known effects of calcium on parathyroid function. In view of the important role of adenyl cyclase in the regulation of hormone secretion in a variety of endocrine tissues[81], it would not be surprising if further studies ultimately proved that cyclic AMP serves as the 'second messenger' in the action of calcium on the secretory activity of the parathyroids.

4.2.3.4 Relationship of hormone synthesis to secretion

Whatever may be the mechanism of control exerted by calcium it can be argued that hormone secretion in the parathyroid gland must be closely coupled to new hormone biosynthesis The parathyroid is unusual among endocrine glands in that the content of stored hormone is very low, compared, for example, to pituitary or islet cell tissue[1]. Microscopic studies reveal sparse numbers of secretory granules, an observation that is particularly evident in human parathyroid tissue[82]. Glandular content of hormone measured by radioimmunoassay and bioassay has been estimated between 40 and 150 µg g^{-1} of wet weight[41, 73, 75, 83]. At a normal secretory rate of 2–10 µg min^{-1} in the cow[61, 62], hormone stores are only sufficient to provide needs for 70 min in the basal state (total parathyroid weight in a cow is *ca.* 1g[84]). When stimulated by a fall in plasma calcium, glandular secretion may increase as much as five-fold, even further increasing the requirement for rapid biosynthesis in order to provide hormone for secretion[62, 85]. With maximal stimulation of hormone secretion, increased rates of biosynthesis sufficient to replenish the entire gland content of hormone every few minutes are required. Such considerations lead one to the conclusion that synthesis of hormone may be a rate limiting step in hormone secretion.

4.2.3.5 Adaptive increases in parathyroid hormone secretion

Under conditions of chronic hypocalcaemic stimulation the parathyroids develop hormone secretory rates many-fold higher than the maximum rates of hormone production seen under normal conditions[86-88]. An example of adaptation of gland activity is seen in the parturient cow. Remarkable adaptation of parathyroid gland activity occurs in the pregnant cow apparently due to an as yet unexplained resistance to the action of parathyroid hormone, which develops in the late stages of pregnancy[87, 88]. When the animals give birth and begin to lactate, a severe spontaneous hypocalcaemia occurs. Concomitant with the marked fall in blood calcium, rates of hormone secretion increase to a level 50–100-fold or greater than that of the normal animal[86-88]. In spite of these greatly exaggerated secretory responses in the animals, hormone secretion has been shown to fall acutely when infusions of calcium are administered which raise the blood calcium to normal or above. The rate of secretion in any individual animal is still linearly and inversely controlled by the serum calcium (Figure 4.6). The results suggest that in these animals with compensatory 'secondary hyperplasia' of the parathyroids, the amount of glandular tissue capable of hormone biosynthesis is greatly increased.

4.2.4 Metabolism

4.2.4.1 Nature of the circulating hormone

A significant advance in our understanding of the secretion and metabolism of parathyroid hormone has been the recognition that immunoreactive parathyroid hormone in the circulation is heterogeneous; it consists not only

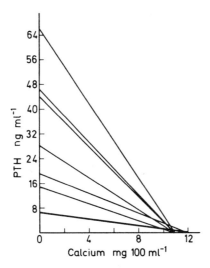

Figure 4.6 Comparison of the relationship between plasma calcium and plasma parathyroid hormone concentrations in a number of parturient cows that have developed adaptive increases in glandular secretion during pregnancy. The concentration of parathyroid hormone in the parturient cows, determined during the development of spontaneous hypocalcaemia which occurs at the time of parturition, is linearly and inversely proportional to the calcium, as is seen in normal cows (heavy line). However, the slope of response relating parathyroid hormone to calcium is 2–8 times greater than normal, reflecting a greatly increased secretory capacity of the hyperplastic glands. Despite increased rates of hormone production, control of hormone secretion by calcium is retained since hormone secretion falls to zero in all animals at *ca.* 12 mg % in plasma calcium. (From Buckle *et al.*[86], by courtesy of Excerpta Medica.)

of intact hormone but also of a number of metabolically-formed peptide fragments[54, 89, 90]. Historically, Berson and Yalow were the first to provide evidence that hormone in plasma differed immunochemically from hormone extracted from the parathyroids[91]. They showed that plasma samples from certain patients with hyperparathyroidism frequently reacted differently from standard extracted hormone in the radioimmunoassay and further that the degree of difference varied with the particular antiserum used. In addition, the rate of disappearance of hormone after parathyroidectomy appeared more rapid when measured with one antiserum compared to another. Although these studies clearly demonstrated heterogeneity of hormone in the circulation, the explanation of the heterogeneity was not then apparent. Studies of hormone secretion *in vitro* first produced evidence that heterogeneity may be due to the presence of hormonal fragments. Sherwood et al.[89] and Arnaud and his colleagues[92] studied hormone secretion *in vitro* from parathyroid explants maintained as surviving organ cultures and showed that most of the immunoreactive hormone found in the culture media was smaller in size than the 1–84 polypeptide extractable from bovine parathyroids, whose structure had been determined. These observations led to the interpretation that most of the hormone is cleaved prior to secretion *in vivo* as well as *in vitro*[89].

However, subsequent studies of hormone secreted by monolayer cultures of parathyroid cells[93] and also hormone secreted directly from the parathyroid

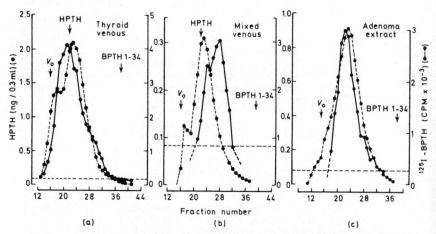

Figure 4.7 Bio-gel p-10 filtration patterns of immunoreactive parathyroid hormone in blood samples obtained simultaneously by venous catheterisation of (A) inferior thyroid vein and (B) superior vena cava in a patient with a parathyroid adenoma. This is compared to (C) filtration of a partially purified hormone extract prepared from human parathyroid adenomas. Hormone concentrations indicated in mixed venous samples are only approximate because of non-parallel response in assay. Iodine-125 labelled bovine parathyroid hormone ([^{123}I]BPTH) was co-chromatographed in each filtration as a marker of intact hormone (dotted lines). Arrows mark void volume of column (V_o), elution position of intact human parathyroid hormone (HPTH) and the synthetic bovine 1–34 amino-terminal peptide (BPTH 1–34). Dashed horizontal line indicates sensitivity limit of the radioimmunoassay. Note that thyroid venous sample resembles adenoma extract in eluting slightly before markers of HPTH and [^{123}I]BPTH. In contrast, activity in the mixed venous sample elutes later than the markers. (From Habener et al.[54], by courtesy of the National Academy of Science, USA.)

glands *in vivo*[54], have indicated that this concept is probably incorrect. Gel filtration and immunoassay studies have shown that the form of the hormone released from the gland *in vivo* is at least as large as the extracted peptide and that the smaller circulating form or forms of parathyroid hormone probably arise from degradation of the secreted hormone during peripheral circulation (Figure 4.7)[54]. A large circulating fragment with a molecular weight of *ca.* 7000, resulting from cleavage of the hormone, was identified in the course of the gel filtration studies[94].

Recently it has become possible to evaluate the chemical nature and biological significance of the hormone in the circulation more completely by characterising the antisera used for the analysis of plasma samples and fractions from gel-filtration studies[94, 95]. Quite limited regions of the parathyroid hormone appear to satisfy the binding requirements of the various antisera[95]. For example, the antibodies in one particular antiserum (GP-1) can be shown to react as well with an equimolar mixture of peptides 1–29 or 1–34 and 53–84 from the bovine sequence as with intact 1–84 bovine hormone. Therefore it has been possible to prepare antisera by absorption methods that recognise exclusively a limited region of the sequence at either the amino or carboxy terminal end of the hormonal molecule[95]. Studies with radioimmunoassays using the absorbed antisera indicate that the large hormonal fragment is missing a portion of the amino terminal sequence (Figure 4.8)[94].

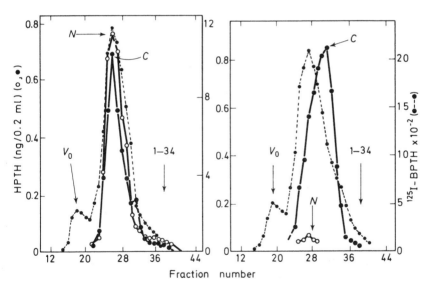

Figure 4.8 Gel filtration pattern (Bio-gel P-10) of immunoreactive parathyroid hormone obtained from the thyroid vein by venous catheterisation (left) and from the general circulation by venipuncture (right). Plasma (0.5–0.8 ml) and 50 pg (20 000 c/min) of iodine-125 labelled bovine parathyroid hormone consisting of native hormone purified from gland extracts [(^{125}I]BPTH) were co-chromatographed. Immunoreactive parathyroid hormone was measured in aliquots of each gel fraction against a human parathyroid hormone standard using radioimmunoassays that specifically measured an amino-terminal sequence (*N*-assay) and a carboxy-terminal sequence (*C*-assay). Arrows indicate void volume of column (V_o) and elution position of synthetic bovine peptide 1–34 (1–34). (From Habener *et al.*[94], by courtesy of Macmillan.)

As indicated in Section 4.2.1.2 existing information on the structural requirements for biological activity makes it probable that any fragment which does not contain the amino-terminal 27 residues in intact form will be biologically inert. Since the only immunoreactivity detected by the amino-terminal assay corresponds to a small amount of intact hormone and the fragment is present in much higher concentrations than native, uncleaved hormone, it is likely that much of the immunoreactive hormone detected in the circulation represents hormone whose biological activity has been destroyed. Further, since the concentration of intact hormone has been shown to constitute less than 10–15% of the total immunoreactive hormone in the

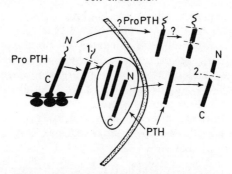

Figure 4.9 Diagram schematically illustrating suggested sequential cleavages of pro-parathryoid hormone (proPTH) and parathyroid hormone (PTH) during course of biosynthesis, secretion and metabolism. (1) Cleavage of prohormone to 84 amino acid hormone which is transported and stored in secretory granule. (2) Cleavage of secreted hormone during passage in the circulation as an event in metabolism of the hormone. Proposed secretion of proparathyroid hormone is problematic at present. If secreted, it would presumably be cleaved in the circulation similar to the cleavage of PTH

circulation (see Figure 4.8), the rate of cleavage of intact hormone to hormonal fragments must be considerably faster than the rate of disappearance from the circulation of the predominant hormonal fragment.

Recent work by Canterbury and Reiss[96,97] and by Goldsmith et al.[98] have confirmed the presence of hormonal fragments in the circulation of man. In addition to finding the large C-terminal fragment with their antisera, these workers have identified in the peripheral circulation a smaller fragment of the hormone of ca. 4000 mol. wt. This smaller fragment was found to be biologically active in an in vitro renal adenyl cyclase assay (the larger fragment was inactive), which indicates that it probably consists of the amino-terminal sequence of the hormone and that cleavage of the hormone is not necessarily a process of hormone inactivation.

The above considerations may ultimately help to clarify disagreements in reports about the diagnostic significance of hormone levels measured in plasma. Reiss et al.[99] and more recently Berson et al.[100] have reported finding that nearly all patients with hyperparathyroidism have peripheral hormone concentrations exceeding those found in normal individuals. In contrast, Potts et al.[101] and Arnaud et al.[102] found considerable overlap in hormone concentrations in hyperparathyroid patients and normal individuals. It seems likely that antisera, used in radioimmunoassays by these different workers, are sensitive to different limited regions of the hormonal molecule. That is, different antisera may vary in sensitivity for detection of active, intact hormonal fragments.

At present the heterogeneity of circulating parathyroid hormone precludes valid measurement of absolute concentrations of hormone in plasma. Further studies of the peripheral metabolism of the hormone will be required to determine the clinical significance of the fragments which might be measured. In addition, provision of appropriate hormone standards and thorough characterisation of antisera will be necessary to avoid misleading cross-reactivity. A summary of the current concepts concerning the alterations in the hormone that occur from the time of initial biosynthesis in the cell and the secretion and metabolism of the hormone in the circulation are shown schematically in Figure 4.9.

It should be noted that although present evidence[54,97] favours the view that the second cleavage occurs peripherally after secretion of the hormone into the circulation, an alternate possibility exists, i.e. that fragments[89,92] of the intact hormone may be secreted from the gland. Since most of the hormone secreted *in vivo* consists of intact hormone[54], the amount of fragment secreted would have to be very small, and further, in order to constitute the predominant form of the hormone in the circulation, the rate of disappearance of the fragments from the circulation would have to be many-fold slower than the disappearance of intact hormone ($T_{1/2}$ of intact parathyroid hormone injected into animals is 8–20 min (see Section 4.2.4). The possibility of such a prolonged disappearance time for hormonal fragments has been suggested by Berson and Yalow[91].

4.2.4.2 Rate of disappearance, distribution and organ-specific destruction

A number of studies have been done in man and animals to determine the rate of disappearance of parathyroid hormone from the circulation either following injection of exogenous hormone[62,91,103,104] or following acute suppression of endogenously secreted hormone by parathyroidectomy or administration of calcium infusions[62]. The data indicate that parathyroid hormone is short-lived and is rapidly cleared from the circulation with a half-time of disappearance of from 8 to 20 min. However, in view of the recently recognised complex nature of the state of the hormone in the circulation, it is probably fair to say that at least based on immunoassay studies, the actual disappearance of biologically-active hormone from the circulation is uncertain. It has not been possible to measure the disappearance of hormone by bioassay techniques because of the insensitivity of current bioassays and the extremely low levels of the hormone present in the circulation even under conditions of maximum stimulation of endogenous hormone or injection of exogenous hormone.

The volume of distribution of parathyroid hormone in the body has been determined by injection of the hormone into experimental animals. The distribution volume corresponds to ca. 30% of the body weight which is equivalent to the extracellular fluid volume[62]. Although it is difficult to relate any apparent volume of distribution to an actual physiological compartment such as extracellular fluid, these results indicate that specificity of hormone interaction with target organs is probably dependent upon receptor binding rather than on restricted distribution.

Present evidence indicates that the liver and kidney are the organs primarily involved in the metabolism of parathyroid hormone[103, 105]. Simultaneous sampling of arterial and venous blood across various organs in the dog has been done during a continuous infusion of bovine parathyroid hormone until steady-state concentrations of the hormone were obtained. Radioimmunoassay of the hormone concentration in these cross-organ blood samples has demonstrated a 20–30% fall in hormone concentration across the kidney and liver while no fall was detected across lung or hind-limb[105]. Renal excretion does not appear to play a role in hormone disappearance since less than 1% of the administered dose of hormone could be detected in the urine.

The rate of disappearance of injected parathyroid hormone from blood has been measured by radioimmunoassay in intact nephrectomised and partially hepatectomised rats[103]. Both bilateral nephrectomy and partial hepatectomy, but not adrenalectomy or splenectomy, resulted in a significant decrease in the rate of disappearance of hormone, indicating that kidney and liver contribute to hormone metabolism[103].

Injection of tritiated acetamidino-parathyroid hormone has shown that the hormone localises rapidly and specifically in the kidneys of thyroparathyroidectomised rats within 30 min after subcutaneous injection[106]. The hormone was rapidly metabolised to trichloroacetic acid soluble components with a resultant rapid loss of radioactivity from the kidney. In these studies localisation of hormone in liver and bone was not clearly demonstrable[106].

4.2.5 Assay of parathyroid hormone

Three general assay methods have been developed for the measurement of parathyroid hormone: (1) *in vivo* bioassays, (2) *in vitro* bioassays and (3) radioimmunoassay. Each method has its own characteristic advantages and disadvantages, and, as a result, no ideal assay system is currently available for the specific measurement of biologically active hormone at the level of sensitivity required for the detection of concentrations of hormone normally found in the circulation of man and animals. Although the bioassays are specific for the measurement of biologically-active hormone, their inherent sensitivity is low. Conversely, the radioimmunoassays are sufficiently sensitive to measure normal circulating levels of hormone in plasma but they cross-react with biologically inactive metabolic fragments of hormone. Because of the very low concentrations of hormone found in tissue and blood, direct chemical methods for assay of polypeptide hormones are not available. Ultimately, when it is developed, a competitive binding radioligand assay may prove to be the most sensitive and specific assay method.

4.2.5.1 Bioassays

A number of bioassays are in use. They are based on the hypercalcaemic response produced by solutions containing hormone after injection into a variety of test animals such as dog,[107] rat[108-110], mouse[111] and chick[112]. Although these assays directly assess biological activity they are relatively insensitive, requiring from 0.05 to 500 μg of hormone per dose. Since parathyroid hormone circulates at concentrations of ca. 0.001 μg ml^{-1} it is not possible to use in vivo bioassays for the measurement of circulating hormone.

More recently in vitro bioassays have been developed which measure various biochemical responses induced in isolated target tissues (bone and kidney) by parathyroid hormone. Assays based on the hormone mediated resorption of calcium-45 labelled bone in tissue culture[113] or on the activation of adenyl cyclase and consequent generation of cyclic AMP in isolated bone cells[114] or renal cortical cell membranes[16] have been described. The sensitivity of these assays is approximately ten times greater than the in vivo assays, but the inhibitory effect of serum or plasma limits the volume of the test dose to ca. 0.05 ml (up to 0.5 ml can be administered to an animal in the in vivo bioassays). Therefore, the overall sensitivity for the measurement of hormone in serum or plasma is no greater than it is for the in vivo assay.

However, an in vitro bioassay for parathyroid hormone based on the hormonal inhibition of CO_2 production from citrate in mouse calvaria has recently been reported and may prove to be useful in measurements of physiological concentrations of hormone[115]. The log(dose) response plot is linear over the range 0.0025–0.15 μg ml^{-1} and the response is independent of serum concentration.

4.2.5.2 Radioimmunoassay

The radioimmunoassay for parathyroid hormone first developed by Berson et al.[15] is sufficiently sensitive to provide a direct measure of the concentration of hormone in peripheral blood. Radioimmunoassays have subsequently been developed by a number of workers and have proven useful in a wide variety of physiological and clinical studies of hormone secretion and metabolism[99, 101, 102, 116]. Most immunoassays can detect as little as 10–30 pg of hormone, making it possible to measure normal circulating levels of hormone.

Although the radioimmunoassay is highly sensitive for the detection of hormone, it is now known that many, if not all, antisera cross-react to varying degrees with biologically-inactive hormonal fragments in the circulation (see Section 4.2.4). In spite of the differences in cross-reactivity of antisera, it is apparent from the results of many clinical centres that the radioimmunoassay is generally useful in the differentiation of patients with hyperparathyroidism from normal individuals and patients with hypercalcaemia due to causes other than hyperparathyroidism[99-101, 117]. Thus, it must be appreciated that the radioimmunoassay results do validly reflect the overall secretory activity of the parathyroids; immunoreactive metabolic fragments of the hormone, as well as intact hormone are increased in states of parathyroid overactivity. Difficulty in the interpretation of assay results comes when attempts are made to assess acute changes in hormone secretion[99, 118] or when absolute concentrations of hormone are compared in different laboratories using different antisera[94].

4.2.6 Mode of action of parathyroid hormone

Parathyroid hormone has been shown to exert its principal physiological and biochemical effects on three major organs, the bone, kidney and small intestine. The result of the action of the hormone on each of these organs is to raise the calcium concentration of the extracellular fluid. The hormone exerts its effect on bone by mobilising calcium through increasing resorption of bone (mineral and matrix). In the kidney the hormone acts to increase tubular resorption of calcium and to decrease resorption of phosphate leading to a decreased renal excretion of calcium and to phosphaturia. The hormone increases the transport of calcium across the intestinal mucosa resulting in an increased absorption of dietary calcium.

The biochemical mechanisms through which parathyroid hormone exerts its effects on these target organs are not completely understood. However, the recent availability of purified parathyroid hormone has now permitted detailed investigations to be made of the effects of the hormone in intact animals and in isolated tissue preparations. It appears certain that at least one of the actions of parathyroid hormone on bone and kidney, and possibly small intestine, like those of many other of the polypeptide hormones, is mediated through the activation of adenal cyclase in membrane receptors located at the cell surface and by the consequent intracellular formation of 3'5'-cyclic AMP.

4.2.6.1 Effect of parathyroid hormone on bone

The role that parathyroid hormone plays in the metabolic destruction of bone was first conclusively demonstrated by Barnicot[119] and by Chang[120], who showed that parathyroid tissue grafted adjacent to bone caused osseous erosions. Later Gaillard[121] and Raisz[122] showed that extracts of parathyroid glands directly caused resorption of bone in tissue culture. Talmage et al.[123] have demonstrated that parathyroid extracts cause hypercalcaemia in nephrectomised, fasting rats, thereby implicating bone as the source of newly-mobilised calcium.

The importance of the effects of parathyroid hormone on bone can be appreciated when it is realised that 99% of the total body calcium resides in the skeleton[1]. The skeleton represents one of the largest masses of target tissue directly affected by any hormone and contains ca. 1.5 kg of elemental calcium in the average adult man[1]. Despite the enormous reservoirs of calcium in the skeleton, very little ($>$ 1%) is freely miscible with the extracellular fluid[124]. Active resorption of bone mineral is required to mobilise significant amounts of calcium.

(a) *Kinetic responses of bone*—Recent experimental studies have confirmed a long-standing observation that the earliest response of an animal to an injection of parathyroid hormone is a transient hypocalcaemia rather than hypercalcaemia which occurs later (Figure 4.10)[125, 247, 248]. Kinetic analysis of the distribution of radiocalcium has shown that the initial hypocalcaemic response is due to a sudden shift of calcium into the skeleton. The characteristic physiological response of bone to parathyroid hormone can be produced *in vitro* by raising the calcium concentration of the culture media[126]. Parsons *et al.* have recently shown that the administration of a small dose of calcium given with parathyroid hormone to an animal enhances calcium mobilisation from the skeleton an hour or two later[3, 127]. However, in order to obtain the enhancement in the hypercalcaemic response, the calcium must be given within a short time of the administration of the parathyroid hormone; there

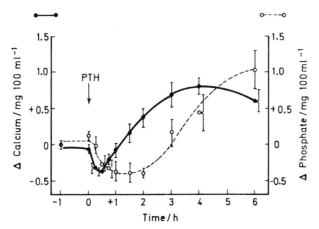

Figure 4.10 Mean changes in plasma calcium (●) and inorganic phosphate (○) at selected time intervals after parathyroid hormone injection in a dog. Vertical bars show standard errors of the means. (From Parsons *et al.*[125], by courtesy of Lippincott.)

is no enhancement if the interval between the hormone and the calcium injections > 8 min[127].

This enhancement of the skeletal response to parathyroid hormone by calcium may also manifest itself on a chronic basis. Vitamin D deficient and/or thyroparathyroidectomised rats with hypocalcaemia fail to respond to injections of parathyroid hormone; the response can be restored by bringing the plasma calcium to normal by either injecting calcium or increasing the dietary calcium absorption[128]. It should be pointed out that parathyroid hormone has been shown to effectively produce a hypercalcaemic response in rats immediately following thyroparathyroidectomy in spite of severe hypocalcaemia[129, 130]. To resolve this conflict, further investigation is required where both the duration of hypocalcaemia and the rate of calcium mobilisation are taken into account.

The influx of calcium into bone is apparently due to a transient, specific increase in permeability of the bone cell membrane to calcium and may serve to activate components of the parathyroid hormone response in mobilising bone calcium. The response probably requires increases in intracellular concentrations of both calcium and 3'5'-cyclic AMP[131, 132].

The time-course of the hypercalcaemia induced by the action of parathyroid hormone on bone is complex and consists of identifiable fast and slow components[133]. Present evidence suggests that the earliest hypercalcaemic response (following the immediate hypocalcaemic response described above) represents an outflow of calcium from bone cells that does not require protein synthesis and probably represents activation of pre-existing enzyme and transport systems involved in the mobilisation of bone mineral[134, 135]. Support for this supposition is provided by experiments employing an inhibitor of protein synthesis, such as Actinomycin D. Pre-treatment of rats with Actinomycin D does not prevent the initial rise in plasma calcium caused by parathyroid hormone but does eliminate the later response[136]. An additional observation to support the existence of two components in mobilisation of calcium from bone, has recently been reported by Milhaud et al.[137]. They obtained evidence from radiocalcium studies which suggests that one of the early effects of parathyroid hormone is to block bone formation. This effect may be on the osteoblastic cells of the bone.

(b) *Changes in bone cells*—The known physiological and biochemical responses of bone to parathyroid hormone appear to correlate in many ways with histological and histochemical changes that have been observed in the bone cells. There is evidence from numerous studies that the hormone causes changes in all three types of bone cells, osteoclasts, osteoblasts, and osteocytes. In general it is believed that the osteoclasts are responsible for the destruction of bone, the osteoblasts are involved in new bone formation, and osteocytes may maintain the metabolic function of 'stable' bone[1, 3]. The activity of all three cell types is inter-related in a complex manner.

Several lines of evidence have shown that parathyroid hormone stimulates both growth and metabolic activity of osteoclasts[138, 139]. The effect of parathyroid hormone on osteoblasts appears to be biphasic. The initial response is a rapid depression of metabolic activity[140-142], but after several days of hormone administration the growth of osteoblastic cells has been shown to increase, leading to the formation of large amounts of new bone tissue[143]. Interestingly, it has been observed that the skeleton in patients with hyperparathyroidism due to excessive secretion of parathyroid hormone often shows areas of marked bone proliferation as well as areas of destruction[144]. Clearly the cellular changes produced by parathyroid hormone require further study.

There is some evidence to suggest that parathyroid hormone also mobilises calcium from bone by an action of osteocytes[145-148]. The appearance of osteocytes and the bone around them has been noted to change after exposure to parathyroid hormone. There is no indication of how such an effect might be produced and no convincing evidence has been reported of parathyroid hormone-induced biochemical changes in osteocytes.

(c) *Effect of parathyroid hormone on bone matrix*—In addition to effects on bone mineral, parathyroid hormone also has important effects on bone matrix. Mineral accounts for only half the fresh weight of the skeleton and parathyroid hormone exerts a potent destructive effect on the collagen and ground substance that makes up the bone matrix. Mineral and matrix are intimately linked together in bone and it has not been possible to clearly separate the chemical processes needed to dissolve the two components. While changes in plasma calcium serve as an indicator of bone mineral destruction, plasma and urinary hydroxyproline measurements appear to correlate with destruction of matrix[149, 150]. Pinnel and Krane have observed that the ratios of galactosyl-hydroxylysine to glucosyl-galactosyl-hydroxylysine differ in collagen from skin and bone and can be used as a highly specific indicator of bone collagen destruction[151].

(d) *Role of 3′5′-cyclic AMP in the action of parathyroid hormone on bone* — A large body of evidence has accumulated over the past 2 or 3 years indicating that most, if not all, of the known actions of parathyroid hormone on bone are expressed through the activation of the enzyme adenyl cyclase and the formation of 3′5′-cyclic AMP[152, 253]. The activation of adenyl cyclase in bone cells by parathyroid hormone has recently been used to develop a sensitive *in vitro* bioassay for the hormone that can measure as little as 5 ng of hormone[114] (see Section 4.2.5).

The studies described above imply that a rise in the concentration of cyclic AMP in bone cells leads to the resorption of bone material and transfer of calcium from bone to extracellular fluid. The mechanism through which cyclic AMP mediates these effects in bone is not known. However, important actions of cyclic AMP on intracellular reactions have been described in a number of other systems, whose hormonal responses are mediated by increases in the intracellular content of cyclic AMP; by analogy one may propose that a similar mechanism exists in bone tissue[81]. These tissues have been found to contain cyclic AMP-dependent protein kinases[155]. Cyclic AMP has been shown to activate two of these kinases by dissociating them from an inhibitor protein[156]. The kinase enzymes appear to be responsible for catalysing the phosphorylation of a number of proteins such as histones, ribosomal proteins and microtubular proteins. The latter is of some interest since secretory processes are thought to involve microtubules. A microtubular system has been identified in bone and might be involved in transfer of calcium across cells and membranes from bone to the extracellular fluid[157].

(e) *Effect of other variables on bone* — A cautionary note should be raised in the interpretation of *in vivo* studies of the action of parathyroid hormone on bone. A number of variables can modify the mobilisation of calcium induced by parathyroid hormone. Even in the same animal the responses to repeated injections of parathyroid hormone vary unpredictably[125]. Factors which may influence the response include the action of calcitonin on the bone, the activity of Vitamin D in the animal, the acid–base balance[158], levels of circulating phosphates[159] and changes in blood flow which can alter the partial pressure of oxygen in the bone cell[160] as well as change the absorption and rate of destruction of the administered hormone.

4.2.6.2 Action of parathyroid hormone on the kidney

Parathyroid hormone has two known separate actions on the kidney. The first, the earliest to be discovered, is to promote renal excretion of phosphate[161]. For a great many years this was thought to be the sole effect of the hormone on the kidney and it was not until 1955 that Talmage and co-workers demonstrated a renally-mediated conservation of calcium by observing that parathyroidectomy in animals resulted in increased calcium excretion and that this effect could be prevented by the administration of parathyroid extract[162]. The work of Bernstein and co-workers[251] further demonstrated that in addition to phosphaturia, parathyroid hormone specifically decreased the renal clearance of calcium.

The effect of the hormone in preventing proximal tubular reabsorption of calcium and distal tubular re-absorption of phosphate has been elegantly and conclusively demonstrated by Widrow et al.[164] and Goldberg et al.[165] through studies using stop–flow micropuncture techniques. The latter workers have demonstrated that parathyroid hormone very effectively inhibits prioxmal tubular resorption of sodium as well as phosphate—the difference being

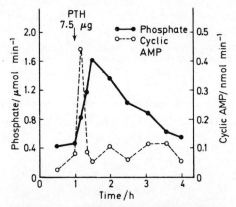

Figure 4.11 The effect of parathyroid hormone on the excretion into the urine of phosphate and cyclic-AMP by a parathyroidectomised rat. Parathyroid hormone (7.5 μg) was injected intravenously over a 2 min period at the time shown by the arrow. (From Chase and Aurbach[166], by courtesy of AAAS.)

that nearly all sodium is re-absorbed in the distal tubule, whereas phosphate is not, resulting in the excretion of increased amounts of phosphate but not of sodium. These same workers also made the important observation that these effects of parathyroid hormone on phosphate resorption could be mimicked by infusing dibutyryl-cyclic AMP into the kidney. It is now known that similar to its action on bone, parathyroid hormone acts on the cells of the renal cortex to stimulate adenyl cyclase activity and the formation of 3′5′-cyclic AMP[166, 167]. The concentration of cyclic AMP in the urine rises within minutes of the administration of the hormone and precedes the rise in urinary phosphate (Figure 4.11)[166].

Administration of calcium infusions to animals and man which suppress endogenous secretion of parathyroid hormone have also been shown to decrease excretion of cyclic AMP into the urine[168]. Although current studies strongly suggest that the increase in intracellular concentrations of cyclic AMP produced in response to parathyroid hormone accounts for the phosphaturic action of the hormone on the kidney, it is not yet certain whether the action of the hormone on calcium transport by the kidney also is attributable to an increase in intracellular concentration of cyclic AMP[152]. It is of interest to note that patients with the clinical disorder pseudohypoparathyroidism secrete much less cyclic AMP into the urine in response to injection of parathyroid hormone than do normal individuals[169]. Such patients have been shown to lack the renal receptor for parathyroid hormone action. This disorder represents another example of clinical manifestations of hormone deficiency not actually due to lack of hormone but rather to a failure of the end organ to respond. Marcus and Aurbach have developed a sensitive *in vitro* bioassay for parathyroid hormone based on activation of adenyl cyclase activity in an isolated preparation of renal cortical cell membranes[16].

4.2.6.3 Action of parathyroid hormone on the small intestine

The effect of parathyroid hormone on the small intestine was the last major action of the hormone to be discovered and has not been well studied. This is in large part due to the unavailability of suitable *in vitro* systems in which to evaluate the biochemical actions of hormone on intestinal cells. However, a number of experimental studies done in man and in animals have now established that parathyroid hormone increases intestinal absorption of calcium by increasing transport of the ion from the gut lumen[170, 171]. Many of the early workers were unable to reproduce the effects of parathyroid hormone on the intestine and it was subsequently found that dietary calcium has a profound influence on parathyroid-hormone-mediated changes in intestinal calcium transport[172]. Animals on adequate or high intakes of dietary calcium are refractory to the effects of parathyroidectomy or administration of parathyroid hormone. The nature of this adaptive response to dietary calcium is not completely understood but may involve changes in the levels of intracellular Vitamin D-dependent proteins involved in calcium transport.

The biochemical mechanisms involved in the action of parathyroid hormone on the intestine are not well understood. Recent studies using isolated vascularly perfused small intestine of the rat have shown a stimulation of Vitamin D induced calcium absorption within 30 min after the administration of parathyroid hormone[173]. This relatively short time required for a response indicates that activation of existing transport mechanisms probably occurs rather than induction of new protein synthesis. Limited evidence indicates that the cyclic AMP mechanism may also be involved in the action of parathyroid hormone on the intestine. However, the biochemical aspects of cyclic AMP interactions have not been worked out as well in the intestine as they have in bone and kidney. Further work is required to assess more fully the physiological significance of the action of parathyroid hormone on the intestinal transport of calcium and the mechanism of the stimulating effect.

4.2.6.4 Other actions of parathyroid hormone

Parathyroid hormone may exert an effect on organs other than bone, kidney and liver. Parathyroidectomy of lactating rats has been shown to increase calcium content of milk in spite of hypocalcaemia[174]. The hormone appears to greatly increase calcium in saliva[175]. Parathyroid hormone given acutely has been reported to produce a transient systemic hypotension. However, the purity of the hormone preparations used was not known and it is possible that the hypotensive effect was due to a contaminating substance. Charbon and workers have presented evidence that this hypotensive effect is due to a marked increase in blood flow to liver and kidney[176]. There is also evidence that parathyroid hormone may be important in tooth development and in maintaining normal metabolism of the lens. Teeth which erupt during parathyroid deficiency remain permanently acalcified[177]. Lenticular cataracts are a frequent manifestation of chronic hypoparathyroidism[1].

4.3 CALCITONIN

In 1962 Copp and co-workers discovered a potent hypocalcaemic, hypophosphatemic peptide hormone during regional perfusion of the thyroid and parathyroid glands of experimental animals with calcium[178]. They called this newly-discovered hormone calcitonin. The hormone was subsequently shown to originate from a very specific type of cell of neuroectodermal origin known as the parafollicular or C-cell[179] located in the thyroid in mammals and in a separate gland, the ultimobranchial body, in lower non-mammalian vertebrates[180]. In the intervening 10 years since its discovery, calcitonin has been isolated on a large scale from a number of mammalian species (hog, cow, sheep, human) by several pharmaceutical companies[2] and from salmon and avian ultimobranchial bodies by Copp et al.[181]. Further, chemical purification of the hormone extracts by a number of laboratories has led to the final characterisation and structural analysis of seven different calcitonins from five animal species including three isohormonal variants from the salmon[154].

Several of the calcitonins have been synthesised chemically in amounts sufficient to allow widespread usage in a large number of physiological, biochemical and therapeutic studies. It is known that elevation of blood calcium stimulates secretion of calcitonin and that the hormone in turn causes a prompt lowering of the blood calcium primarily through its action on the skeleton to block mineral resorption, and on the kidney, to promote calcium excretion[4-8].

Thus in many ways calcitonin acts to regulate calcium in a manner opposite to that of parathyroid hormone and it has been suggested that calcitonin may serve as a physiological antagonist of parathyroid hormone in maintaining calcium homeostasis[1]. However, recent evidence casts some doubt on this hypothesis since, at least in normal adult man, calcitonin circulates at extremely low concentrations and does not appear to play a significant physiological role in calcium metabolism.

4.3.1 Chemistry of calcitonins

4.3.1.1 Isolation and sequence determination

The minute amounts of calcitonin contained in mammalian thyroid glands made isolation and purification of the hormone difficult. Thus, large scale extraction of mammalian thyroid glands and salmon ultimobranchial tissue was required and was successfully undertaken using acetic acid, urea and cysteine[182-184]. Calcitonin from the pig was purified 44 000-fold from the extracts by gel filtration on Sephadex G-50, G-25 and finally by ion exchange chromatography on carboxymethylcellulose[182].

It was not possible to isolate calcitonin from normal human thyroid because of the extremely low content of hormone in the normal glands and the difficulty of obtaining sufficient quantities of the tissue. Thus, human calcitonin was obtained from a medullary carcinoma of the thyroid, a tumour composed of calcitonin-rich C-cells, by extraction with butanol–acetic acid, gel filtration and counter-current distribution[185].

The amino acid sequences of seven calcitonins from five different species have been determined[154]. The sequence analyses were accomplished using the sequential degradation method of Edman where amino acids are successively removed and identified from the amino terminal end of the hormonal polypeptide[154].

Examination of the structures of the calcitonins reveals both common features and differences (cf. legend to Figure 4.12). The amino acid sequences of the calcitonins are similar in that they each consist of a 32 amino acid chain with a 1–7 disulphide bridge at the amino terminus and a carboxyl terminal prolinamide. Only nine positions in the sequence are common in all the hormone structures and seven of these nine are in the region of the disulphide loop. The remaining two residues in common are the carboxyl terminal prolinamide and the glycine at position 28. Between positions 10 and 27 there is considerable similarity between ovine, bovine and porcine calcitonins but both the human and salmon molecules differ markedly from these three and from one another. The amino acid substitutions in this region, however, are conservative and in general preserve the general overall chemical properties of the molecule; the content and location of hydrophobic and hydrophilic as well as acidic and basic residues remain relatively constant.

Methionine is found at two positions in the calcitonins, 8 and 25. In human calcitonin, oxidation of the methionine at position 8 to the sulphoxide form causes complete loss of biological activity[186]. Oxidation or alkylation of the single methionine at position 25 of porcine calcitonin, however, is not associated with any loss in activity[182]. It should be noted, however, that methionine is not required for biological activity since salmon calcitonin lacks methionine.

The successful synthesis of porcine calcitonin followed shortly after elucidation of the amino acid sequence and was accomplished independently by Rittel et al.[187] and Guttman et al.[188]. Both groups used the classical or solution-phase method of peptide synthesis. Similar methods have been used for the synthesis of human calcitonin by Seiber et al.[189] and of salmon calcitonin by Guttman et al.[190].

4.3.1.2 Chemical synthesis and structure–activity relations

Evaluation of the biological activity of a number of synthetic fragments and derivatives of porcine calcitonin has provided interesting information about

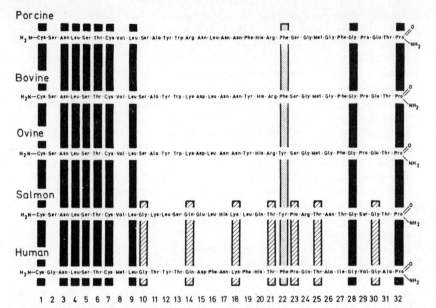

Figure 4.12 The amino acid sequences of calcitonins from different species. The salmon sequence shown is that of the predominant or major form. Residues identical in all species are indicated by the solid bars. Additional residues common to salmon and human calcitonins are indicated by cross-hatched bars. The stippled bar (residue 22) shows a sequence position invariably accompanied by an aromatic residue. (From Potts et al.[2], by courtesy of Academic Press.)

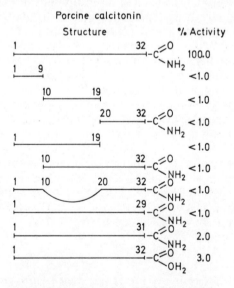

Figure 4.13 The relative biological activities of synthetic fragments of the porcine calcitonin molecule. (Adapted from Sieber et al.[191] and Guttmann et al.[192].)

structural requirements for activity (Figure 4.13)[191, 192]. The most striking finding is that the complete sequence of 32 amino acids is required for activity. Fragments of the molecule whether derived from the amino terminal, middle or carboxy terminal region of the molecule are completely devoid of activity. Shortening the molecule by even one amino acid or removal of the *N*-terminal amino group or *C*-terminal amide function resulted in a dramatic decrease in hypocalcaemic activity. Thus, in contrast to parathyroid hormone, calcitonin does not contain a specific active region of the sequence.

The observations on synthetic analogues of calcitonin so far have been limited. Further syntheses of additional analogues should provide even more information about structure–activity relations in the calcitonin molecule.

4.3.2 Biosynthesis and control of secretion of calcitonin

Calcitonin is synthesised and secreted by a highly specialised cell of neuroectodermal origin known as the parafollicular or the 'C'-cell[179]. The cells arise from the last or sixth branchial pouch during embryonic development and in lower vertebrates, aggregates of cells form a separate body, the ultimobranchial body[180]. In mammals the cells migrate to the thyroid as parafollicular cells and are located in the basement membrane of the thyroid follicle. Pearse and co-workers have shown that these cells selectively take up and store amine precursors such as 5-hydroxytryptophan and dihydroxyphenylalanine or DOPA, in a manner similar to cells of adrenal medulla and ganglionic tissue[193].

No work has yet been done to define the mechanisms of calcitonin biosynthesis and very little detailed information is available concerning the factors that regulate its synthesis. However, it has become clear that secretion of calcitonin is controlled principally by the changes in the calcium concentration of blood perfusing the glands. Numerous studies have demonstrated that secretion of calcitonin increases when the calcium concentration of the blood flowing through the thyroid is increased and decreases when the calcium concentration is lowered[4-8]. This response of calcitonin to calcium is just the opposite of that observed with parathyroid hormone.

The development of radioimmunoassays for calcitonin has made possible detailed studies of the factors that control secretion of calcitonin in a wide variety of animal species including man. Immunoassay measurements have clearly shown that in the rabbit[194], pig[195] and cow[196] calcitonin is continuously secreted under normal physiological concentrations of blood calcium. For example, increases of a hundred-fold in hormone secretion have been readily detected in the effluent thyroid blood from the pig when stimulated by calcium infusion (Figure 4.14)[195]. Analysis of the radioimmunoassay results have indicated that blood calcium controls calcitonin secretion by a directly proportional mechanism[8]. As pointed out earlier, little parathyroid hormone is stored; whereas the amount of calcitonin stored in the parafollicular cells is very large in relation to the rates of secretion of hormone normally found[5]. In fact, stores of preformed hormone are sufficient to support secretion of calcitonin for many hours without requiring biosynthesis of new hormone[1].

Recent work has shown that in addition to calcium other substances may stimulate the secretion of calcitonin. Studies have shown release of calcitonin in response to infusions of glucagon[195, 197] gastrin[195], (and the synthetic gastrin fragment, pentagastrin) and cholecystakinin. Calcitonin secretion has been evoked in the pig by distillation of calcium salts into the intestine in amounts

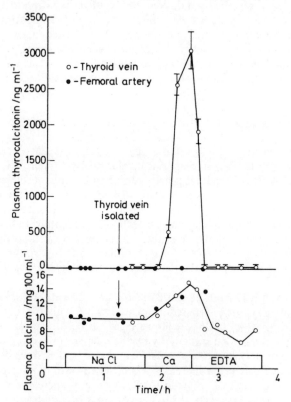

Figure 4.14 Measurement of endogenous porcine calcitonin by radioimmunoassay in the thyroid effluent blood (O) and femoral arterial blood (●) of a pig during calcium and EDTA infusions. Calcium perfusion provokes over a thousand-fold increase in calcitonin secretion. The absence of a detectable increase in concentration of calcitonin in the femoral arterial blood (no thyroid venous blood enters the circulation) serves as a control for the specificity of the techniques. (From Cooper et al.[195], by courtesy of Lippincott.)

insufficient to produce significant changes in blood calcium[198]. Further, Munson et al. have proposed that secretion of calcitonin occurs in response to acute intake of dietary calcium and serves as a physiological control mechanism to prevent transient hypercalcaemia following ingestion of meals containing calcium[199]. Munson has shown that thyroidectomised rats develop periodic hypercalcaemia following feeding which is not seen in rats with intact thyroids[199].

A number of studies have been reported describing the secretion of calcitonin in humans by medullary carcinoma of the thyroid, a malignant tumour composed of parafollicular 'C' cells[8]. Excessive amounts of calcitonin are invariably found by radioimmunoassay in the blood and in the tumours of patients.

However, in spite of the clear demonstration of calcitonin secretion in normal animals and in patients with medullary thyroid carcinoma, the importance of calcitonin in normal human physiology has not yet been established. It appears that the stores of calcitonin in the normal human thyroid are extremely low and the circulating hormone has not been readily detectable by any immunoassay methods that are currently available[200]. Much additional work must be done in order to define the physiological role, if any, that calcitonin plays in maintaining mineral homeostasis in man.

4.3.3 Assay of calcitonin

4.3.3.1 Bioassays

Most bioassays for calcitonin measure the hypocalcaemic effect produced by the hormone following administration to young rats or mice. The sensitivity of the response is dependent upon both age and diet of the test animal. Young rats (less than 6 weeks old) fed low calcium diets are much more sensitive to the hypocalcaemic effects of the hormone than are older rats on regular or phosphate-deficient diets[5, 6]. The rat assay as modified by Sturtridge and Kumar can detect as little as 0.2 MRC milliunits (2 ng) of hormone[201]. An *in vitro* bioassay of similar sensitivity has been developed by Raisz *et al.* which is based on the inhibition of parathyroid stimulated release of 45Ca from embryonic rat bone when calcitonin is added to the culture media[202].

4.3.3.2 Radioimmunoassay

The most sensitive assay for measurement of calcitonin is the radioimmunoassay. Deftos and co-workers have developed specific, sensitive assays for a number of different species of calcitonin including human[200], salmon[8], bovine[196] and porcine[203], in which there is little or no cross-reactivity. The assays can detect 5–10 MRC μunits (50–100 pg) of hormone ml^{-1} of plasma; this has enabled these workers to readily measure circulating levels of calcitonin in animals under basal and stimulated conditions. (See Section 4.3.2). The results of immunoassay measurements of calcitonin in the circulation would appear to be much simpler to interpret than results obtained by assay of parathyroid hormone (see Section 4.3.5). This is so because (a) no biologically active fragments of calcitonin are known (see Sections 4.3.1.1 and 4.3.1.2) and (b) no immunoreactive fragments of calcitonin have been detected in the circulation using a wide variety of different antisera[163].

4.3.4 Metabolism

4.3.4.1 Distribution and disappearance

A number of investigators have determined the volume of distribution, rate of disappearance and metabolic clearance rates of calcitonin following injection or infusion of the hormone into a number of animal species. The results have shown that, as with parathyroid hormone, calcitonin is cleared extremely rapidly from the circulation with a half-life of 2–20 min depending upon the species of origin of the calcitonin studied[163, 194]. The volume of distribution has

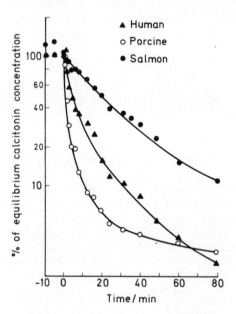

Figure 4.15 Simultaneous disappearance of human ▲ porcine, ○ and salmon ● calcitonins from the circulation of a dog as measured by specific radioimmunoassays following the termination of an infusion of all three calcitonins. Note slower disappearance of salmon calcitonin compared to porcine and human calcitonin which correlates with the greater relative biological potency of the salmon hormone. (From F. R. Singer, unpublished data.)

been found to correspond to the extra-cellular fluid space[1]. The findings are consistent with a closely regulated system where, immediately after secretion, the hormone is widely distributed throughout the body and thereby made available for interaction with receptors; the hormone in the circulation is short-lived; its biological effects depend upon continued secretion.

It has been known for several years that the five species of calcitonins thus far isolated and purified vary markedly in their effectiveness in producing hypocalcaemia in man and test animals[4-8]. Recent evidence suggests that the difference in biological activity between the different calcitonins may be due to differences in their rates of metabolism. Habener and Singer have compared the disappearance and metabolic clearance rates of salmon, human and porcine calcitonins following simultaneous infusion of all three calcitonins into dogs (Figure 4.15)[163, 204]. The decreases in hormone concentrations were measured by separate radioimmunoassays which were completely specific for each calcitonin. Using these techniques, it was possible to make a closely-controlled interspecies comparison of the rates of hormone disappearance in a single experiment[163].

The relative rates of disappearance from the circulation of these three species of calcitonins were found to reflect their relative hypocalcaemic potencies. Salmon calcitonin, the most potent of all calcitonins known, has the slowest rate of disappearance indicating that it is more resistant to metabolic destruction than the less potent human and porcine species of calcitonins.

4.3.4.2 Organ-specific destruction

The inactivation of the calcitonins probably occurs in one or more specific organ sites. Early studies of the inactivation of calcitonin produced by incubation with tissue homogenates suggested that liver and kidney may be organs responsible for the metabolism of the hormone[205]. Singer et al. have recently obtained direct information on the organs involved in calcitonin metabolism in studies where, during infusion of the calcitonins into dogs, blood was drawn simultaneously from in-dwelling arterial and venous catheters placed across the kidney, liver, lung and hind limb and the concentrations of calcitonin were measured by radioimmunoassay[206]. The results of these studies have clearly demonstrated that among the different species of calcitonins, differences exist in the sites of destruction. The sole site of metabolism of the more stable salmon calcitonin was shown to be the kidney, whereas porcine calcitonin with the more rapid rate of disappearance is destroyed in kidney, liver and muscle alike. The hormone is presumably destroyed in the kidney since $<1\%$ of the infused hormone could be detected in the urine[206].

It has been proposed that the explanation for the observed differences in the rates of metabolism among the different species of calcitonins must be related to primary structural differences in the molecules[163, 206]. Salmon calcitonin differs from porcine calcitonin in 19 of the 32 amino acid residues, and in eleven positions it contains amino acids not found in any other calcitonin (see Section 4.3.1). These structural differences are thought to result in a greater stability of the salmon calcitonin molecule and to protect it against rapid metabolic destruction. The resistance of the salmon hormone to destruction might in fact account for its greater biological potency. Marx et al.[207] have reported that salmon calcitonin may have greater affinity than other calcitonins for binding to isolated receptor membranes from kidney. This also may be a reflection of a greater structural stability of the calcitonin from the salmon.

4.3.5 Mode of action of calcitonin

A large body of evidence has accumulated indicating that the principal, if not the sole, physiologically important action of calcitonin in the regulation of calcium metabolism is on the bone[1, 4-6, 208]. In addition, studies done thus far have suggested that the hypocalcaemic effect of the hormone is probably not due to changes in renal excretion, intestinal absorption or soft tissue distribution of calcium[4-6].

4.3.5.1 Action on bone

It has been shown by both *in vivo* and *in vitro* studies that the hormone acts on the skeleton through an enhanced mineral accretion by bone[1, 4-6]. Studies by O'Riordan and Aurbach[208] have demonstrated the inhibition of skeletal calcium absorption in rats given calcitonin intravenously a short time after

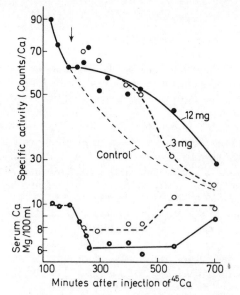

Figure 4.16 Rate of decline in specific activity of ^{45}Ca after intravenous injection in untreated control rats and in animals given two doses of a relatively crude calcitonin preparation. 3 mg dose (○), 12 mg dose (●). Lower portion of figure shows changes in total serum calcium in rats given calcitonin at time indicated by vertical arrow. Results are average values with a minimum of four rats per group. (From O'Riordan and Aurbach[208], by courtesy of Lippincott.)

the administration of radioactive calcium (Figure 4.16). No acceleration in the rate of disappearance of radiocalcium from blood was observed, suggesting that the hormone had not increased mineral deposition in bone. Rather, there was a dramatic flattening in the rate of fall in the specific activity in the animals given hormone, reflecting a sudden reduction in the rate of entry of non-radioactive calcium into the blood. Additional studies by Parsons *et al.* have shown a pronounced reduction in the artero-venous difference in calcium when calcitonin is added to the fluid during perfusion of the isolated tibia in the cat[209]. They calculated that the magnitude of the effect would be sufficient to account for the overall hypocalcaemic effect observed in the whole animal. A direct inhibition of calcitonin on bone resorption *in vitro* has been demonstrated by Aliapoulios *et al.*[210] and by Friedman and Raisz[211] using either histological methods or by showing a reduction in the rate of release of radioactive calcium from bone.

It has become apparent that calcitonin may act as a physiological antagonist of parathyroid hormone. When administered at the same time as parathyroid hormone, calcitonin completely blocks the bone resorption induced by parathyroid hormone and prevents any rise in blood calcium. However, direct antagonism of the action of parathyroid hormone is not the mechanism by which calcitonin acts, since calcitonin has been shown to be effective in animals that have undergone parathyroidectomy[4, 212].

Further, inhibition of bone resorption has been demonstrated *in vitro* when calcitonin is either added alone or is used to block parathyroid hormone-induced resorption: reduced excretion of hydroxyproline following administration of calcitonin has provided further evidence for suppression of bone resorption[213, 214]. The fact that the hypocalcaemic effect of calcitonin depends on suppression of bone resorption is further reflected in the observation that young actively growing animals with rapid bone turnover are far more sensitive to the hormone than adult animals[5]. Administration of even extremely large amounts of calcitonin to normal adult man does not significantly alter the calcium concentration of the plasma[4, 5]. In contrast, patients with Paget's disease of bone, a skeletal disorder characterised by markedly elevated rates of bone turnover, are exquisitely sensitive to the effects of administered calcitonin and show a prompt hypocalcaemic response to the hormone[154].

4.3.5.2 Action on kidney

Calcitonin has been shown to exert an action on the kidney in promoting urinary excretion of phosphate, calcium, sodium, potassium and magnesium[1]. It is felt that calcitonin promotes excretion of these ions through a direct inhibition of tubular re-absorption rather than through changes in renal blood flow or tubular secretion. However, the physiological importance of these renal effects of calcitonin has not been established since the doses of hormone used are higher than the reported levels of endogenous hormone in the blood.

4.3.5.3 Action on intestine

Whether calcitonin exerts a significant effect on ion transport in the intestine remains controversial. Several authors have reported experimental evidence which supports the conclusion that the intestine is not necessary for the action of calcitonin and that calcitonin has no effect on normal intestinal calcium absorption[1]. On the other hand, Olson et al.[173] have recently reported that calcitonin produces an immediate drop in the calcium absorption in isolated vascularly perfused intestines of rats given vitamin D.

Important information about the physiological effects of calcitonin, other than the already known effect of inhibition of bone calcium resorption, may come from studies on the action of calcitonin in lower, non-mammalian vertebrates. It is now known that calcitonin from a number of submammalian vertebrates is chemically similar to calcitonin of mammals and is extremely potent in mammals[5]. Studies of the effect of calcitonin in fish, amphibians and birds may provide important insight into other aspects of its physiological action. For example, high concentrations of calcitonin have been found in the blood and ultimobranchial tissue of the dog fish shark which, unlike mammals, has a cartilaginous skeleton[5]. The concentrations of calcitonin in the blood of salmon have been shown to vary several-fold during migration of the fish from salt to fresh water and to vary significantly between females and males[216].

4.3.5.4 Biochemical mechanism of action

Investigations from a number of laboratories have focused on the mechanism of action of calcitonin at the cellular and subcellular levels. Studies by Aurbach et al.[153] and Murad et al.[217] have shown that, as has been found for parathyroid hormone, calcitonin acts through the stimulation of adenyl cyclase, leading to increased formation of cyclic AMP, in both renal and skeletal tissue. Current evidence indicates that calcitonin and parathyroid hormone act on bone and kidney through separate receptors[153]. Little is known about the particular type of bone cell or cells that serves as the target for the action of calcitonin. Some ultrastructural evidence suggests that it may act on the osteoclast[218]. It seems unlikely, however, that parathyroid hormone and calcitonin, recognised as physiological antagonists, would act on the same cell through increase in intracellular cyclic AMP.

4.3.6 Therapeutic uses

Soon after calcitonin was isolated and synthesised in sufficient quantities to allow its use of clinical studies, the hormone proved to be useful in the therapy of patients with a variety of skeletal disorders. The hormone has been particularly useful in the treatment of patients with Paget's disease of bone, a disorder of bone metabolism characterised by greatly accelerated bone turnover and often accompanied by painful deformity of the skeleton[219-222,252]. As was anticipated from preceding animal studies, salmon calcitonin compared to the mammalian calcitonins was found to be unusually effective in reducing pain and bone resorption in a large number of patients with extensive Paget's disease[252]. Calcitonin has also been effective in the treatment of acute hypercalcaemia stemming from malignant osteolytic metastases[223].

The hormone holds promise as a possible beneficial agent in the treatment of post-menopausal osteoporosis.

4.4 VITAMIN D

The need for Vitamin D for normal skeletal growth and metabolism has been recognised since the beginning of the century[1]. The cure of rickets by ultra-violet irradiation of affected subjects, or by administration of cod liver oil, was documented nearly 50 years ago. The potent anti-rachitic factor in liver and irradiated skin was structurally identified as cholecalciferol (D_3) by Askew and Bourdillon in 1931[39] and was assumed to be the functional form of the vitamin until 1966, when Lund and DeLuca[224] demonstrated the existence of a metabolite fraction of Vitamin D_3 that possessed biological activity equivalent to that of the parent vitamin. Numerous studies undertaken during the past 5 years have shown that Vitamin D itself is inactive[243]. Enzymatically-catalysed hydroxylations of the vitamin are specifically required for calcium mobilising activity in intestine, bone and possibly kidney, and may in turn be regulated by calcium, parathyroid hormone[244] and possibly by calcitonin[245]. In many ways, Vitamin D acts as a hormone, since the formation of the active metabolites are regulated by a series of controlled enzymatic processes.

4.4.1 Assay of vitamin D

The standard classical method for measuring Vitamin D has been the tibia line test, a bioassay which measures the width of the calcium deposit stimulated by Vitamin D in the endochondral area of tibia from rachitic rats[226]. The assay is much more sensitive than chemical tests and gives a response in the range between 0.25 and 2.0 International Units (one I.U. = 0.025 µg of Vitamin D_3) with a good linear log dose response curve.

Even more sensitive assays for the measurement of Vitamin D_3 and 25-hydroxy D_3 have recently been developed utilising the techniques of competitive protein binding radioassay. Belsey has partially purified specific binding proteins from serum of Vitamin D deficient rats [227] and Haddad has partially purified a similar protein from rat kidney[228]. The assays are sufficiently sensitive to measure as little as 0.001 µg of Vitamin D ml^{-1} of serum. Preliminary application of these assays has shown that they can detect circulating levels of the vitamin in normal individuals; the assays should open up new avenues of investigation in the area of mineral homeostasis.

4.4.2 Chemistry and metabolism

The vitamins D are a group of sterols which occur in nature. All of the compounds are structurally-related derivatives of cholesterol and all possess the physiological property of curing or preventing rickets, a disease characterised by skeletal abnormalities, including failure of calcification[1]. Although numerous sterols exist which have anti-rachitic activity there are three compounds of biological importance to man. Cholecalciferol, Vitamin D_3, is the structural form that occurs in nature in the fish liver oils and is synthesised from 7-dehydrocholesterol in the skin in response to ultraviolet irradiation. Ergocalciferol, Vitamin D_2, occurs in the plant kingdom and is produced from ergosterol by irradiation. Dehydrotachysterol (DHT_2) is a synthetically-produced analogue of the vitamin, formed by extensive irradiation of ergosterol; DHT_2 has proven to be more potent on a weight basis than either D_3

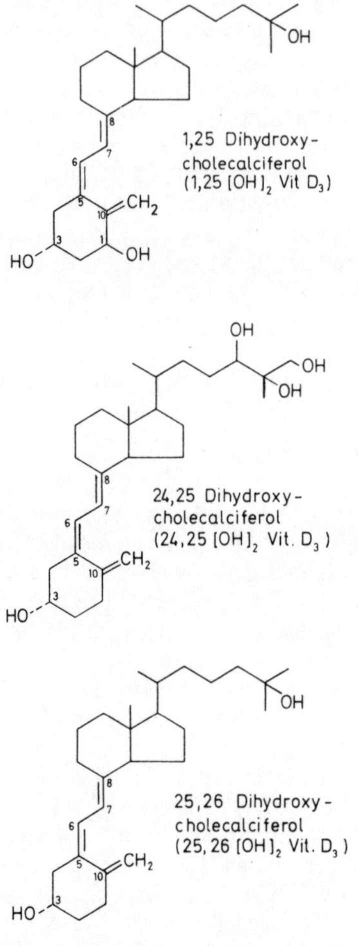

Figure 4.17 Structural formulae of the three dihydroxylated forms of cholecalciferol (vitamin D_3) identified in rat intestine

or D_2; this is probably due to its unique property of being fully active in promoting intestinal transport of calcium without requiring further hydroxylation in the kidney.

After formation in the skin or absorption from the diet, Vitamin D is stored in the body. The vitamin is then converted to the 25-hydroxylated derivative, 25-hydroxycholecalciferol (or 25-hydroxyergocalciferol) by a specific hydroxylase in the liver[225]. Although 25-hydroxy D_3 is more active than D_3 on isolated target tissues, it has recently been discovered that further hydroxylation of the vitamin must occur before it can express full activity on target tissues *in vivo*. 25-hydroxy-D is metabolised in the kidney to 1,25-dihydroxy Vitamin D and possibly to additional dihydroxy metabolites including 24,25-dihydroxy-D and 25,26-dihydroxy-D (Figure 4.17)[229]. The latter two compounds have not yet been fully characterised and their biological actions are not well understood. It should be noted, however, that the metabolite referred to in the literature as 21,25-dihydroxy-D_3 has recently been identified as 24,25-dihydroxy-D_3.

The characteristic time lag that occurs before increased calcium transport is manifest in the small intestine following the administration of either Vitamin D or 25-hydroxy D_3 is markedly reduced following D repletion with 1,25-dihydroxy-D_3 (Figure 4.18)[231]. This dihydroxy metabolite is also more potent on a weight basis than either the Vitamin D_3 or hydroxy-D_3[231].

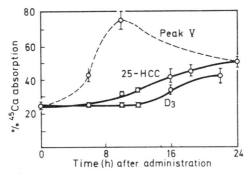

Figure 4.18 Response of the intestinal absorption of calcium by vitamin D-deficient chicks given five international units of vitamin D_3, (○); 25-hydroxy-D_3, (●) or 1,25-dihydroxy-D_3 (peak V), (○). The assay involved the measurement of disappearance of ^{45}Ca from ligated loops of duodenum *in vivo*. (From DeLuca *et al.*[243], by courtesy of Excerpta Medica.)

Furthermore, while pretreatment of animals with actinomycin D, the antibiotic which inhibits protein synthesis by blocking messenger RNA transcription, abolishes the intestinal action of both vitamin D_3 and 25-hydroxy-D_3, it does not interfere with the intestinal effect of 1,25-dihydroxy-D_3[232]. The probable explanation for the actinomycin D effect is that the antibiotic interferes with the metabolism of 25-hydroxy-D_3 to the dihydroxy metabolites in the kidney[233].

1,25-dihydroxycholecalciferol can also act to mobilise calcium from bone with a total effect greater than that of 25-hydroxy-D_3[234]. The dihydroxy metabolite is 100-times more active than 25-hydroxy-D_3 in mobilising calcium from foetal rat calvaria *in vitro* and it alone is effective in mobilising bone calcium in anephric rats (Figure 4.19)[235]. The skeletal actions of vitamin D_3, 25-dihydroxy-D_3 and 1,25-dihydroxy-D_3 are all blocked by pre-treatment of animals with actinomycin D, suggesting the possibility that some further metabolite of 1,25-dihydroxy-D_3 may be required for final expression of Vitamin D action on bone[234].

Work with synthetic analogues of Vitamin D has re-inforced the concept that small structural changes in the vitamin can produce marked changes in the specificity and effectiveness of its actions. One analogue, 5,6-*trans*-25-hydroxycholecalciferol, has been shown to stimulate active transport of calcium by the intestine, but has little, if any, activity in stimulating mobilisation of calcium from bone[236]. The steric configuration of the hydroxyl group of the

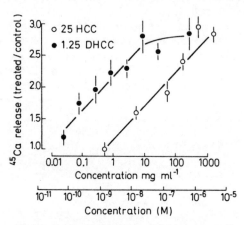

Figure 4.19 Comparison of response of calcium-45-labelled foetal rat bones to 25-hydroxycholecalciferol (25-HCC) and 1,25-dihydroxycholecalciferol (1,25-DHCC). Paired bone shafts, pre-labelled *in vitro* for 24 h in media containing ^{45}Ca, were incubated for 48 h with and without vitamin D. Points indicate the means and vertical lines and standard error for the ratio of treated to control cultures. (From Raisz *et al.*[235], by courtesy of AAAS.)

A-ring in this analogue corresponds to that of the 1-hydroxyl group in 1,25-dihydroxy-D_3, providing evidence that a hydroxyl function must be present on C-1 of Vitamin D compounds for the stimulation of calcium transport in the intestine. The unusual properties of this stereoisomer make it a promising drug for the treatment of calcium abnormalities associated with chronic renal failure where it is desirable to increase dietary absorption without further mobilisation of calcium from the skeleton.

4.4.3 Physiology and mode of action

The activation of Vitamin D through its hydroxylation reactions appears to be a regulated process in both liver and kidney. Experiments done *in vitro* have suggested that the 1-hydroxylation is product-inhibited[229]. The extent of formation *in vitro* of 1,25-dihydroxy-D_3 and 24,25-dihydroxy-D_3 from vitamin D_3 has been shown to be inversely related to the concentration of serum calcium and strontium (but not phosphorus)[237]. Formation of 1,25-dihydroxy-D_3 occurs when the serum calcium falls below 9.2 mg % while formation of 24,25-dihydroxy-D_3 occurs when the serum calcium exceeds 9.5 mg %. The formation of one metabolite seems to preclude formation of the other, indicating a close metabolic regulation based on the serum calcium levels, or some other metabolic determinant that varies with the serum calcium.

Recent evidence has suggested that parathyroid hormone may be required for the hypocalcaemic stimulation of 1,25-dihydroxy-D_3 synthesis to occur[244]. Increased formation of 1,25-dihydroxy-D_3 does not occur in animals made hypocalcaemic by parathyroidectomy, but is seen after parathyroid hormone is given by injection even before calcium levels increase in response to the hormone. In addition, there is indirect evidence implicating calcium as

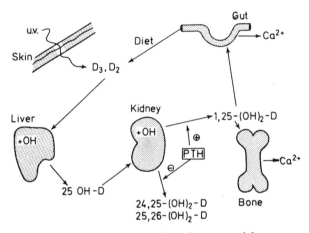

Figure 4.20 Schema depicting the sequential organ-specific hydroxylations of vitamin D that are required for conversion to the biologically active dihydroxylated metabolite (1,25-$(OH)_2$-D). Parathyroid hormone (PTH) is thought to preferentially stimulate conversion of the mono-hydroxylated metabolite (25-OH-D) to 1,25-$(OH)_2$-D rather than to the less active 24,25- and 25,26-dihydroxy metabolites (24, 25-$(OH)_2$-D and 25,26-$(OH)_2$-D)

a regulatory factor in the activation of Vitamin D. Experimental work by Fairbanks and Mitchell[239], Rottensten[240], and Nicolaysen et al.[249]. has established that young rats absorb calcium more efficiently after a period on low calcium diet than after a similar period on a diet high in calcium content. Nicolaysen et al.[249] suggested that the adaptation mechanism is regulated by some endogenous factor which was shown to be dependent on Vitamin D. Boyle et al. have suggested that 1,25-dihydroxy-D_3 is the endogenous factor since even slight hypocalcaemia stimulates its synthesis[238]. Further evidence in support of a regulation of Vitamin D activation in man is the rarity of hypervitaminosis D (toxicity resulting from excessive amounts of Vitamin D), despite wide variations in exposure to ultraviolet light and dietary intake of Vitamin D. In fact, induction of hypervitaminosis D requires an intake of 50–100 times the minimum daily dose of the vitamin needed to prevent rickets.[1] It is possible to demonstrate in animals that a similar margin exists between the dose of vitamin necessary to evoke the maximum physiological response and the dose necessary to produce the manisfestations of hypervitaminosis D on either the skeleton or the intestine[1].

Recent evidence has suggested that Vitamin D may exert its effect on intestinal cells through the formation of a specific calcium binding protein from a pre-existing precursor protein[215]. A calcium binding protein has been isolated from rat intestinal mucosa. It has a molecular weight of 8200 as determined by ultracentrifugation and an apparent molecular weight of 13 000 as determined by gel filtration[241]. In carefully controlled studies the appearance of calcium binding protein has been demonstrated in intestinal mucosa following administration of Vitamin D to rats; the appearance of the calcium binding protein is paralleled by a stoichiometric decrease in a presumed protein precursor which has a molecular weight 1000 daltons higher than the calcium binding protein[242].

A schematic summary of the currently proposed steps in the metabolic activation of Vitamin D is shown in Figure 9.20. In summary, several lines of evidence indicate that Vitamin D is a hormone vitally involved in calcium transport and in the maintenance of calcium homeostasis. These are: (1) large stores of biologically-inactive precursors exist; (2) activation of precursors occurs at specific sites, i.e. liver and kidney and the metabolically active products are secreted into the circulation; and (3) the activations steps are under metabolic control.

Acknowledgements

The authors thank Ms. Jody Jenkins for her excellent assistance in the preparation of this manuscript.

Dr. Habener is a recipient of a Research Career Development Award from the United States Public Health Service.

References

1. Potts, J. T., Jr. and Deftos, L. J. (1969). *Duncan's Textbook of Metabolism*, 904 (P. K. Bondy, editor) (Philadelphia: W. B. Saunders Co.)
2. Potts, J. T., Jr., Keutmann, H. T., Niall, H. D. and Tregear, G. W. (1971). *Vitamins and Hormones*, **29**, 41
3. Parsons, J. A. and Potts, J. T., Jr. (1972). *Clin. Endocrinol. Metab.*, **1**, 33
4. Hirsh, P. F. and Munson, P. L. (1969). *Physiol. Rev.*, **48**, 548
5. Copp, D. H. (1969). *Ann. Rev. Pharmacol.*, **9**, 327
6. Copp, D. H. (1970). *Ann. Rev. Physiol.*, **32**, 61
7. Potts, J. T., Jr. (1970). *Fed. Proc. (Fed. Amer. Soc. Exp. Biol.)*, **29**, 1200
8. Potts, J. T., Jr., Niall, H. D. and Deftos, L. J. (1971). *Current Topics in Experimental Endocrinology*, **1**, 151
9. Collip, J. B. (1925). *J. Biol. Chem.*, **63**, 395
10. Aurbach, G. D. (1959) *J. Biol. Chem.*, **234**, 3179
11. Rasmussen, H. and Craig, L. C. (1961). *J. Biol. Chem.*, **236**, 759
12. Rasmussen, H., Sze, Y. L. and Young, R. (1964). *J. Biol. Chem.*, **239**, 2852
13. Aurbach, G. D. and Potts, J. T., Jr. (1964). *Endocrinology*, **75**, 290
14. Potts, J. T., Jr., Aurbach, G. D., Sherwood, L. M. (1966). *Recent Prog. Horm. Res.*, **22**, 101

15. Berson, S. A., Yalow, R. S., Aurbach, G. D. and Potts, J. T., Jr. (1963). *Proc. Nat. Acad. Sci. USA*, **49,** 613
16. Marcus, R. and Aurbach, G. D. (1969). *Endocrinology*, **85,** 801
17. Keutmann, H. T., Aurbach, G. D., Dawson, B. F., Niall, H. D., Deftos, L. J. and Potts, J. T., Jr. (1971). *Biochemistry*, **10,** 2779
18. Niall, H. D., Keutmann, H. T., Sauer, R., Hogan, M., Dawson, B. F., Aurbach, G. D. and Potts, J. T., Jr. (1970). *Hoppe-Seyler's Physiologische Chemie*, **351,** 1586
19. Brewer, H. B. and Ronan, R. (1970). *Proc. Nat. Acad. Sci. USA*, **67,** 1962
20. Woodhead, J. S., O'Riordan, J. L. H., Keutmann, H. T., Stoltz, M. C., Dawson, B. F., Niall, H. D., Robinson, C. J. and Potts, J. T., Jr. (1971). *Biochemistry*, **10,** 2787
21. Edman, P. and Begg, G. (1967). *Europ. J. Biochem.*, **1,** 80
22. Niall, H. D., Penhasi, H., Gilbert, P., Myers, R. C., Williams, F. S. and Potts, J. T., Jr. (1969). *Fed. Proc. (Fed. Amer. Soc. Exp. Biol.)*, **28,** 661
23. Edman, P. (1950). *Acta Chem. Scand.*, **4,** 283
24. Potts, J. T., Jr., Keutmann, H. T., Niall, H. D., Habener, J. F., Tregear, G. W., Deftos, L. J., O'Riordan, J. L. H. and Aurbach, G. D. (1971). *Calcium, Parathyroid Hormone and the Calcitonins* (R. V. Talmage and P. L. Munson, editors), 159 (Amsterdam: Excerpta Medica)
25. O'Riordan, J. L., Potts, J. T., Jr. and Aurbach, G. D. (1971). *Endocrinology*, **89,** 234
26. Brewer, H. B., Jr., Fairwell, T., Ronan, R., Sizemore, G. W. and Arnaud, C. D. (1972). *Proc. Nat. Acad. USA*, **69,** 3585
27. Niall, H. D., Sauer, R., Jacobs, J., Keutmann. H. T., Segre, G. V., O'Riordan, J. L. H., Aurbach, G. D. and Potts, J. T., Jr. (1973). *Proc. Nat. Acad. Sci. USA* (in the press)
28. Cohn, D. V., MacGregor, R. R., Chu, L. L. H., Kimmel, J. R. and Hamilton, J. W. (1972). *Proc. Nat. Acad. Sci. USA*, **69,** 1521
29. Hamilton, J. W., Niall, H. D., Jacobs, J. W., Keutmann, H. T., Cohn, D. V. and Potts, J. T., Jr. (1973). Amino-terminal sequence of a biosynthetic precursor to bovine parathyroid hormone (in preparation)
30. van Rietschoten, J., Tregear, G. W., Niall, H. D. and Potts, J. T., Jr. (1973). Solid-phase synthesis of the amino-terminal region of bovine proparathyroid hormone (in preparation)
31. Potts, J. T., Jr, Keutmann, H. T., Niall, H. D., Deftos, L. J., Brewer, A. B. and Aurbach, G. D. (1968). *Parathyroid Hormone and Thyrocalcitonin (calcitonin)* (R. V. Talmage and L. F. Belanger, editors) (Amsterdam: Excerpta Medica)
32. Rasmussen, H. and Craig, L. C. (1962). *Biochem. Biophys. Acta*, **56,** 332
33. Merrifield, R. B. (1969). *Advan. Enzymol.*, **32,** 221
34. Potts, J. T., Jr., Tregear, G. W., Keutmann, H. T., Niall, H. D., Sauer, R., Deftos, L. J., Dawson, B. F., Hogan, M. L. and Aurbach, G. D. (1971). *Proc. Nat. Acad. Sci. USA*, **68,** 63
35. Tregear, G. W., van Rietschoten, J., Keutmann, H. T., Niall, H. D., Gear, S. G., Parsons, J. A. and Potts, J. T., Jr. (1972). Bovine Parathyroid Hormone: Structural Requirements for Biological Activity. Submitted to *Endocrinology*
36. Habener, J. F., Kemper, B., Potts, J. T., Jr. and Rich, A. (1972). *Endocrinology*, **92,** 219
37. Potts, J. T., Keutmann, H. T., Niall, H. D., Tregear, G. W., Habener, J. F., O'Riordan, J. L. H., Murray, T. M., Powell, D. and Aurbach, G. D. (1971). *Endocrinology*, 333 (S. Taylor, editor) (London: Heinemann Med. Books, Ltd.)
38. Tashjian, A. H., Ontjis, D. A. and Munson, P. L. (1964). *Biochemistry*, **3,** 1175
39. Fourman, P., Royer, P., Levell, M. J. and Morgan, D. B. (1968). *Calcium Metabolism and the Bone* (Philadelphia: F. A. Davis Co.)
40. Rasmussen, H. (1968). *Endocrinology*, 4th edn., 847 (R. H. Williams, editor) (Philadelphia: W. B. Saunders Co.)
41. Sherwood, L. M., O'Riordan, J. L. H., Aurbach, G. D. and Potts, J. T., Jr. (1967). *J. Clin. Endocrinology. Metab.*, **27,** 140
42. Riggs, L. B., Arnaud, C. D., Reynolds, J. C. and Smith, L. H. (1971). *J. Clin. Invest.*, **50,** 2079
43. Roof, B. S., Carpenter, B., Fink, D. J. and Gordan, G. S. (1971). *Amer. J. Med.*, **50,** 686
44. Mavligit, G. M., Cohen, J. L. and Sherwood, L. M. (1971). *New Engl. J. Med.*, **285,** 154

45. Knill-Jones, R. P., Buckle, R. M., Parsons, V., Calne, R. Y., and Williams, R. (1970). *New Engl. J. Med.*, **282,** 704
46. Omenn, G. S., Roth, S. I. and Baker, W. H. (1969). *Cancer,* **24,** 1004
47. Kemper, B., Habener, J. F., Potts, J. T., Jr. and Rich, A. (1972). *Proc. Nat. Acad. Sci. USA,* **69,** 643
48. Steiner, D. F. and Oyer, P. E. (1967). *Proc. Nat. Acad. Sci. USA,* **57,** 473
49. Hamilton, J. W., MacGregor, R. R., Chu, L. L. H. and Cohn, D. V. (1971). *Endocrinology,* **89,** 1440
50. Habener, J. F., Kemper, B., Potts, J. T.. Jr. and Rich, A. (1972). *Science,* **178,** 630
51. Berson, S. A. and Yalow, R. S. (1971). *Gastroenterology,* **60,** 215
52. Yalow, R. S. and Berson, S. A. (1971). *Biochem. Biophys. Res. Commun.,* **44,** 439
53. Noe, B. D. and Bauer, G. E. (1971). *Endocrinology,* **89,** 642
54. Habener, J. F., Powell, D., Murray, T. M., Mayer, G. P. and Potts, J. T., Jr. (1971). *Proc. Nat. Acad. Sci. USA,* **68,** 2986
55. Martin, T. J., Greenberg, P. B. and Michelangeli, V. (1973). *Clin. Sci.,* **44,** 1
56. Melani, F., Ryan, W. G., Rubenstein, A. H. and Steiner, D. F. (1970). *New Engl. J. Med.,* **283,** 713
57. Patt, H. M. and Luckhardt. A. B. (1942). *Endocrinology,* **31,** 384
58. Copp, D. H. and Davidson, A. G. F. (1961). *Proc. Soc. Exp. Biol. Med.,* **107,** 342
59. Raisz, L. G. (1963). *Nature (London),* **197,** 1115
60. Copp, D. H. and Henze, K. G. (1964). *Endocrinology,* **75,** 49
61. Sherwood, L. M., Potts, J. T., Jr., Care, A. D, Mayer, G. P. and Aurbach, G. D. (1966). *Nature (London),* **209,** 52
62. Sherwood, L. M., Mayer, G. P., Ramberg, C. F., Kronfeld, D. S., Aurbach, G. D. and Potts, J. T., Jr. (1968). *Endocrinology,* **83,** 1043
63. Care, A. D., Sherwood, L. M., Potts, J. T., Jr. and Aurbach, G. D. (1966). *Nature (London),* **209,** 55
64. Targovnik, J. H., Rodman, J. H. and Sherwood, L. M. (1971). *Endocrinology,* **88,** 1477
65. Sherwood, L. M., Herrman, I. and Bassett, C. A. (1970). *Nature (London),* **225,** 1056
66. Au, W. Y. W., Poland, A. P., Stern, P. H. and Raisz, L. G. (1970). *J. Clin. Invest.,* **49,** 1639
67. Massry, S. G., Stern, L., Targoff, C. and Kleeman, C. R. (1970). *Clin. Res ,* **18,** 123
68. Cushard, W. S., Jr., Bercovitz, M., Canterbury, J. M. and Reiss, E. (1971). *J. Clin. Invest.,* **50,** 23a
69. Deftos, L. J., Swenson, V., Bode, H., Hyak, A., Neer, R. and Potts, J. T., Jr. (1972). *Clin. Res.,* **20,** 217
70. Oldham, S. B., Fisher, J. A. and Arnaud, C. D. (1971). *Clin. Res.,* **19,** 128
71. Reitz, R. E., Mayer, G. P., Deftos, L. J. and Potts, J. T., Jr. (1971). *Endocrinology,* **89,** 932
72. Gittes, R. F. and Radde, I. C. (1966). *Endocrinology,* **78,** 1015
73. Aurbach, G. D. and Potts, J. T., Jr. (1964). *Advan. Metab. Dis.,* **1,** 45
74. Raisz, L. G. and O'Brien, J. E. (1963). *Amer. J. Physiol.,* **205,** 816
75. Hamilton, J. W., Spierto, F. W., MacGregor, R. R. and Cohn, D. V. (1971). *J. Biol. Chem.,* **246,** 3224
76. Hamilton, J. W. and Cohn, D. V. (1969). *J. Biol. Chem.,* **244,** 5421
77. Dufresne, L. R. and Gitelman, H. J. (1972). *Calcium, Parathyroid Hormones and the Calcitonins,* 202 (R. V. Talmage and P. L. Munson, editors) (Amsterdam: Excerpta Medica)
78. Sherwood, L. M., Abe, M., Rodman, J. S., Lundberg, W. B. and Targovnik, J. H. (1972). *Calcium, Parathyroid Hormone and the Calcitonins,* 183 (R. V. Talmage and P. L. Munson, editors) (Amsterdam: Excerpta Medica)
79. Dufresne, L. R., Andersen, R. and Gitelman, H. J. (1971). *Clin. Res.,* **19,** 529
80. Hargis, C. K., Bowser, E. N., Henderson, W. J., Martinez, N. J. and Williams, G. A. (1972). *Clin. Res.,* **20,** 429
81. A. Robinson, G. G. Nahas and T. Lubos, editors. *Ann. New York Acad. Sci.,* **185**
82. Roth, S. I. (1962). *Arch. Path.,* **73,** 495
83. Kenny, A. D., Roth, S. I. and Castleman, B. J. (1964). *J. Clin. Endocrinol.,* **24,** 375
84. Mayer, G. P., Ramberg, C. F., Jr. and Kronfeld, D. S. (1968). *Dairy Science,* **10,** 1288
85. Ramberg, C. F., Jr., Mayer, G. P., Kronfeld, D. S., Aurbach, G. D., Sherwood, L. M. and Potts, J. T., Jr. (1967). *Amer. J. Physiol.,* **213,** 878

86. Buckle, R. M., Aurbach, G. D. and Potts, J. T., Jr. (1968). In *Protein and Polypeptide Hormones*, 389 (Amsterdam: Excerpta Medica)
87. Mayer, G. P., Ramberg, C. F., Jr, Kronfeld, D. S., Buckle, R. M., Sherwood, L. M., Aurbach, G. D. and Potts, J. T., Jr. (1969). *Amer. J. Vet. Res.*, **30**, 1587
88. Potts, J. T., Jr., Reitz, R. E., Kaye, M., Buckle, R. M., Richardson, J., Deftos, L. J. and Aurbach, G. D. (1969). *Arch. Int. Med.*, **124**, 408
89. Sherwood, L. M., Rodman, J. S. and Lundberg, W. B. (1970). *Proc. Nat. Acad. Sci. USA*, **67**, 1631
90. Canterbury, J. M. and Reiss, E. (1971). *J. Lab. Clin. Med.*, **78**, 814
91. Berson, S. A. and Yalow, R. S. (1968). *J. Clin. Endocrinol.*, **28**, 1037
92. Arnaud, C. D., Tsao, H. S. and Oldham, S. B. (1970). *Proc. Nat. Acad. Sci. USA*, **67**, 415
93. Martin, T. J , Greenberg, P. B. and Melick, R. A., (1972). *J. Clin. Endocrinol.*, **34**, 437
94. Habener, J. F., Segre, G. V., Powell, D., Murray, T. M. and Potts, J. T., Jr. (1972). *Nature New Biol.*, **238**, 152
95. Segre, G. V., Habener, J. F., Powell, D., Tregear, G. W. and Potts, J. T., Jr. (1972). *J. Clin. Invest.*, **51**, 3163
96. Canterbury, J. M. and Reiss, E. (1972). *Proc. Soc. Exp. Biol. Med.*, **140**, 1393
97. Canterbury, J. M. and Reiss, E. (1973). *J. Clin. Invest.*, **52**, 524
98. Goldsmith, R. S., Furszyfer, J., Johnson, W. J., Fournier, A. E., Sizemore, G. W. and Arnaud, C. D. (1973). *J. Clin. Invest.*, **52**, 173
99. Reiss, E. and Canterbury, J. M. (1969). *N. Engl. J. Med.*, **280**, 1381
100. Berson, S. A. and Yalow, R. S. (1971). *Amer. J. Med.*, **50**, 623
101. Potts, J. T., Jr. *et al.* (1971). *Amer. J. Med.*, **50**, 639
102. Arnaud, C. D., Tsao, H. S. and Littledike, E. T. (1971). *J. Clin. Invest.*, **50**, 21
103. Fang, V. S. and Tashjian, A. H., Jr. (1972). *Endocrinology*, **90**, 1177
104. O'Riordan, J. L. H., Page, J., Kerr, D. N. S., Walls, J., Moorhead, J., Crockett, R. E., Franz, H. and Ritz, E. (1970). *Quart. J. Med.*, **39**, 359
105. Singer, F. R., Habener, J. F., Green, E. and Potts, J. T., Jr. (1972). (Unpublished observations)
106. Zull, J. E. and Repke, D. W. (1972). *J. Biol. Chem.*, **247**, 2195
107. Collip, J. B. and Clark, E. P. (1925). *J. Biol. Chem.*, **66**, 133
108. Biering, A. (1950). *Acta Pharmacologica Toxicologica*, **6**, 40
109. Amer, M. S. (1968). *Endocrinology*, **82**, 166
110. Munson, P. L. (1961). *The Parathyroids* (R. O. Greep and R. V. Talmage, editors), 94 (Springfield, Ill.: Thomas)
111. Bethune, J. E., Inoue, H. and Turpin, R. A. (1967). *Endocrinology*, **81**, 67
112. Parson, J. A., Dashwood, M. R., Bangham, D. R. and Robinson, C. J. (1972). *J. Endocrinol.*, **53**, 21
113. Raisz, L. G. and Niemann, I. (1969). *Endocrinol*, **85**, 446
114. Peck, W. A., Messinger, K. and Carpenter, J. (1972). *Abstracts Fourth Internat. Conf. Endocrinol.*, **234**, 93 (Amsterdam: Excerpta Medica)
115. Chu, L. L. H., MacGregor, R. R., Hamilton, J. W. and Cohn, D. V. (1971). *Endocrinology*, **89**, 1425
116. Reiss, E. and Canterbury, J. M. (1968). *Proc. Soc. Exp. Biol. Med.*, **128**, 501
117. Buckle, R. M. (1970). *Lancet*, **2**, 234
118. Murray, T. M., Peacock, M., Powell, D., Monchik, J. M. and Potts, J. T., Jr. (1972). *Clin. Endocrinol.*, **1**, 235
119. Barnicot, N. A. (1948). *J. Anatomy*, **82**, 233
120. Chang, H. (1951). *Anatomical Record*, **111**, 23
121. Gaillard, P. J. (1955). *Exp. Cell. Res.*, Supp. **3**, 154
122. Raisz, L. G., *et al.* (1968). *Parathyroid Hormone and Thyrocalcitonin* (Calcitonin), 370 (R. V. Talmage and L. F. Belanger, editors) (Amsterdam: Excerpta Medica)
123. Talmage, R. V. and Elliot, J. R. (1956). *Endocrinology*, **59**, 27
124. Neer, R. M., Berman, M., Fisher, L. and Rosenberg, L. E. (1967). *J. Clin. Invest.*, **46**, 1364
125. Parsons, J. A., Neer, R. M. and Potts, J. T., Jr. (1971). *Endocrinology*, **89**, 735
126. Talmage, R. V., Cooper, C. W. and Park, H. Z. (1970). *Vitamins and Hormones*, **28**, 103
127. Parsons, J. A., Reit, B. and Robinson, C. J. (1972). *J. Endocrinol.*, **52** (in the press)
128. Au, W. and Raisz, L. G. (1967). *J. Clin. Invest.*, **46**, 1572

129. Munson, P. L. (1955). *Ann. N.Y. Acad. Sci.*, **60**, 776
130. Causton, A., Charlton, B. and Rose, G. A. (1965). *J. Endocrinol.*, **33**, 1
131. Rasmussen, H. (1970). *Science*, **170**, 404
132. Chase, L. R., Fedak, S. A. and Aurbach, G. D. (1969). *Endocrinology*, **84**, 761
133. Talmage, R. V. (1967). *Clin. Orthopoed.*, **54**, 163
134. Candlish, J. K. and Taylor, C. F. (1970). *J. Endocrinol.*, **48**, 143
135. Parsons, J. A. and Robinson, C. J. (1968). *Parathyroid Hormone and Calictonin*, 329 (R. V. Talmage and L. F. Belanger, editors) (Amsterdam: Excerpta Medica)
136. Rasmussen, H. Arnaud, C. D. and Hawker, C. (1964). *Science*, **144**, 1019
137. Milhaud, G., Le Du and Perault-Staub, A. M. (1971). *Revue Europ. D. Etudes Clin. Biol.*, **16**, 451
138. Vaughn, J. M. (1970). *The Physiology of Bone* (Oxford: Oxford University Press)
139. Bingham, P., Brazell, I. and Owen, M. (1969). *J. Endocrinol.*, **45**, 387
140. Gaillard, P. J. (1965). *Texas Reports on Biology and Medicine*, **23**, 259
141. Flanagan, B. and Nichols, G. (1964). *Endocrinology*, **74**, 180
142. Owen, M. and Bingham, P. (1968). *Parathyroid Hormone and Thyrocalcitonin* (Calcitonin), 216 (R. V. Talmage and R. F. Belanger, editors) (Amsterdam: Excerpta Medica)
143. Pugsley, L. I. and Selye, H. (1933). *J. Physiol.*, **79**, 113
144. Kalu, D. N. (1970). *Lancet.*, **1**, 1363
145. Belanger, L. F., et al. *Mechanisms of Hard Tissue Destruction* (1963) 531 (R. F. Sognnaes, editor) (Washington, D.C.: American Assoc. Advan. Sci.)
146. Talmage, R. V., et al. (1965). *The Parathyroid Glands: Ultrastructure, Secretion and Function*, 107 (P. J. Gaillord, R. V. Talmage and A. M. Budy, editors) (Chicago: University of Chicago Press)
147. Baud, C. A. (1966). *IV European Symposium on Calcified Tissues*, **4** (P. Gaillard and A. Van den Hoof, editors) (Amsterdam: Excerpta Medica)
148. Belanger, L. F. (1969). *Calcified Tiss. Res.*, **4**, 1
149. Bates, W. K., McGowen, J. and Talmage, R. V. (1962). *Endocrinology*, **71**, 189
150. Keiser, H. R. et al. (1964). *J. Clin. Invest.*, **43**, 1073
151. Pinell, S. R., and Krane, S. M. (1972). *Parathyroid Hormone and the Calcitonins* (R. V. Talmage, editor) (Amsterdam: Excerpta Medica)
152. Aurbach, G. D., Marcus, R., Heersche, J and Marx, S. (1971). *Ann. N.Y. Acad. Sci.*, **185**, 386
153. Aurbach, G. D., Marcus, R., Heersche, J. N. M., Winickoff, R. N. S. and Marx, S. J. (1972). *Proc. Fourth Parathyroid Conf., Calcium, Parathyroid Hormone and the Calcitonins*, 502 (R. V. Talmage and P. L. Munson, editors) (Excerpta Medica: Amsterdam)
154. Potts, J. T., Jr., Niall, H. D., Keutmann, H. T. and Lequin, R. M. (1972). *Proc. Fourth Parathyroid Conf., Calcium, Parathyroid Hormone and the Calcitonins*, 121 (R. V. Talmage and P. L. Munson, editors) (Amsterdam: Excerpta Medica)
155. Kuo, J. F. and Greengard, P. (1969). *Proc. Nat. Acad. Sci. USA*, **64**, 1349
156. Tao, M., Salas, M. L. and Lipman, F. (1970). *Proc. Nat. Acad. Sci. USA*, **67**, 408
157. Holtrop, M. E. and Weinger, J. M. (1972). *Calcium, Parathyroid Hormone and the Calcitonins*, 365 (Amsterdam: Excerpta Medica)
158. Cuisinier-Gleizes, P., Mathieu, H. and Royez, P. (1967). *Rev. Francaise d. Clin. Biol.*, **12**, 907
159. Neufield, A. H. and Collip, J. B., (1942). *Endocrinology*, **30**, 135
160. Goldhaber, P. (1963). *Mechanisms of Hard Tissue Destruction*, 609 (R. F. Sognnoes, editor) (Washington, D.C.: Amer. Assoc. Advan. Sci.)
161. Greenwald, I. (1911). *Amer. J. Physiol.*, **28**, 103
162. Talmage, R. V., Kraintz, F. W. and Buchanan, G. D. (1955). *Proc. Soc. Exp. Biol. Med.*, **88**, 600
163. Habener, J. F., Singer, F. R., Deftos, L. J., Neer, R. M. and Potts, J. T., Jr. (1971). *Nature New Biol.*, **232**, 91
164. Widrow, S. H. and Levinsky, N. G. (1962). *J. Clin. Invest.*, **41**, 2151
165. Goldberg, M. (1972). *Parathyroid Hormone and the Calcitonins*, 273 (R. V. Talmage, editor) (Amsterdam: Excerpta Medica)
166. Chase, L. R. and Aurbach, G. D. (1968). *Science*, **159**, 545
167. Melson, G. L., Chase, L. R. and Aurbach, G. D. (1970). *Endocrinology*, **86**, 511
168. Chase, L. R. and Aurbach, G. D. (1967). *Proc. Nat. Acad. Sci. USA*, **58**, 518

169. Chase, L. R., Melson, G. L. and Aurbach, G. D. (1969). *J. Clin. Invest.*, **48,** 1832
170. Talmage, R. V. and Elliott, J. R. (1958). *Endocrinology*, **61,** 256
171. Rasmussen, H. (1959). *Endocrinology*, **65,** 517
172. Shah, B. G. and Draper, H. H. (1966). *Amer. J. Physiol.*, **211,** 963
173. Olson, E. B., Jr., DeLuca, H. F. and Potts, J. T., Jr. (1972). *Calcium, Parathyroid Hormone and the Calcitonins*, 240 (Amsterdam: Excerpta Medica)
174. Toverud, S. U. and Munson, P. C. (1956). *Ann. N.Y. Acad. Sci.*, **64,** 336
175. Kraintz, L., Kraintz, F. W. and Talmage, R. V. (1965). *Proc. Soc. Exp. Biol. Med.*, **120,** 118
176. Charbon, G. A. (1968). *Europ. J. Pharmacol.*, **3,** 275
177. Albright, F. and Reifenstein, E. C. (1948). *The Parathyroid Glands and Metabolic Bone Disease* (Baltimore: Williams and Wilkins)
178. Copp, D. H., Cameron, E. C., Cheney, B. A., Davidson, A. G. F. and Henze, K. G. (1962). *Endocrinology*, **70,** 638
179. Foster, G. V., MacIntyre, I. and Pearse, A. G. E. (1964). *Nature (London)*, **202,** 1029
180. Copp, D. H., Cockroft, D. W. and Kueh, Y. (1967). *Science*, **158,** 924
181. Copp, D. H. and Parkes, C. O. (1968). *Parathyroid and Thyrocalcitonin (Calcitonin)*, 74 (R. V. Talmage and L. F. Belanger, editors) (Amsterdam: Excerpta Medica)
182. Brewer, H. B., Jr., Keutmann, H. T., Potts, J. T., Jr., Reisfeld, R. A., Schlueter, R. and Munson, P. L. (1968). *J. Biol. Chem.*, **243,** 5739
183. Bell, P. M. (1968). *Calcitonin: Symposium on Thyrocalcitonin and the C Cells*, 77, (S. F. Taylor, editor) (London: Heinemann)
184. Gudmundsson, T. V., Byfield, P. G. H., Galante, L. and MacIntyre, I. (1968). *Calcitonin: Symposium on Thyrocalcitonin and the C Cells*, 51 (S. F. Taylor, editor) (London: Heinemann)
185. Riniker, B., Neher, R. *et al.* (1968). *Helv. Chim. Acta*, **51,** 1738
186. Neher, R., Riniker, B., Maier, R., Byfield, P. G. H., Gudmundsson, T. V. and MacIntyre, I. (1968). *Nature (London)*, **220,** 984
187. Rittel, N., Brugger, M., Kamber, B., Riniker, B. and Sieber, P. (1968). *Helv. Chim. Acta*, **51,** 924
188. Guttmann, S., Pless, J., Sandrin, E., Jaquenaud, P. A., Bossert, H. and Willems, H. (1968). *Helv. Chim. Acta*, **51,** 1155
189. Sieber, P. and Iselin, B. (1968). *Helv. Chim. Acta*, **51,** 614
190. Guttmann, S., Pless, J., Huguenin, R. L., Sandrin, E., Bossert, H. and Zehnder, K. (1969). *Helv. Chim. Acta*, **52,** 1789
191. Sieber, P., Brugger, M., Kamber, B., Riniker, B., Rittell, W., Maier, R. and Staehelin, M. (1970). *Calcitonin 1969: Proc. Second Internat. Symposium*, 28 (S. F. Taylor and C. Foster, editors) (London: Heinemann)
192. Guttmann, S., Pless, J., Huguenin, R., Sandrin, E. and Zehnder, K. (1971). *Proc. Amer. Peptide, Symp., 2nd., 1970*
193. Pearse, A. G. E. (1968). *Proc. Roy. Soc. B.*, **170,** 71
194. Lee, M. R., Deftos, L. J. and Potts, J. T. Jr. (1969). *Endocrinology*, **84,** 36
195. Cooper, C. W., Deftos, L. J. and Potts, J. T., Jr. (1971). *Endocrinology*, **88,** 747
196. Deftos, L. J., Habener, J. F., Mayer, C. P., Bury, A. B. and Potts, J. T., Jr. (1972). *J. Lab. Clin. Med.*, **79,** 480
197. Care, A. D., Bates, R. F. L. and Gitelman, A. J. (1969). *J. Endocrinol.*, **43,** 25
198. Cooper, C. W. and Deftos, L. J. (1970). *Fed. Proc. (Fed. Amer. Soc. Exp. Biol.)*, **29,** 253
199. Gray, T. K. and Munson, P. L. (1969). *Science*, **166,** 512
200. Deftos, L. J., Bury, A. B., Habener, J. F., Singer, F. R. and Potts, J. T., Jr. (1971). *Metabolism*, **20,** 1129
201. Sturtridge, W. C. and Kumar, M. A. (1968). *J. Endocrinol.*, **42,** 501
202. Raisz, L. G., Au, W. Y. W., Friedman, J. and Nieman, I. (1967). *Amer. J. Med.*, **43,** 684
203. Deftos, L. J., Lee, M. R. and Potts, J. T., Jr. (1968). *Proc. Nat. Acad. Sci. USA*, **60,** 293
204. Singer, F. R., Green, E., Godin, P. and Habener, J. F. (1972). *Abst. Fourth Internat. Congress Endocrinal.*, **450,** 180 (Amsterdam: Excerpta Medica)
205. deLuise, M., Martin, T. J. and Melick, R. A. (1970). *J. Endocrinol.*, **48,** 181
206. Singer, F. R., Habener, J. F., Godin, P., Greene, E. and Potts, J. T., Jr. (1972). *Nature New Biol.*, **237,** 269

207. Marx, S. J. and Aurbach, G. D. (In preparation)
208. O'Riordan, J. L. H. and Aurbach, G. D. (1968). *Endocrinology*, **82**, 377
209. Parsons, J. A. and Robinson, C. J. (1968). *Calcitonin: Symposium on Thyrocalcitonin and the C Cells*, 256 (S. F. Taylor, editor) (London: Heinemann)
210. Aliapoulios, M. A., Goldhaber, P. and Munson, P. C. (1966). *Science*, **151**, 330
211. Friedman, J. and Raisz, L. G. (1965). *Science*, **150**, 1465
212. Tashjian, A. H. (1965). *Endocrinology*, **77**, 375
213. Martin, T. J., Robinson, C. J. and MacIntyre, I. (1966). *Lancet*, **1**, 900
214. Klein, D. C. and Talmage, R. V. (1968). *Proc. Roy. Soc. Exp. Biol. Med.*, **127**, 95
215. Wasserman, R. H., Carradino, R. A. and Taylor, A. N. (1968). *J. Biol. Chem.*, **243**, 3978
216. Deftos, L. G., Watts, E. G. and Copp, D. H. (1972). *Abst. Fourth Internat. Congres. Endocrinal.*, **452**, 180 (Amsterdam: Excerpta Medica)
217. Murad, F. M., Brewer, H. B., Jr. and Vaughan, M. (1970). *Proc. Nat. Acad. Sci. USA* **65**, 446
218. Gaillard, P. J. (1967). *Koninkl. Ned. Akod. Wettenschop.* **70**, 309
219. Haddad, J., Birge, S. and Avioli, L. V. (1970). *N. Engl. J. Med.*, **238**, 549
220. Woodhouse, N., Bodier, P. and Fisher, M. (1971). *Lancet*, **1**, 1139
221. Shai, F., Baker, R. K. and Wallach, S. J. (1971). *J. Clin. Invest.*, **50**, 1927
222. Bijvoet, O. C. M., Van der Sluys, J. and Jansen, A. P. (1968). *Lancet*, **1**, 876
223. Foster, G. V., Joplin, G. F., MacIntyre, I., Melvin, K. E. W. and Slack, E. (1966) *Lancet*, **1**, 107
224. Lund, J., DeLuca, H. F. (1966). *J. Lipid Res.*, **7**, 739
225. DeLuca, H. F. (1969). *Fed. Proc. (Fed. Amer. Soc. Exp. Biol.)*, **28**, 1678
226. Mandel, H. G. and Weiss, W. P. (1965). *The Pharmacological Basis of Therapeutic.* (L. S. Goodman and A. Gillman, editors) (New York: Macmillan)
227. Belsey, R., DeLuca, H. F. and Potts, J. T., Jr. (1971). *J. Clin. Endocrinol.*, **33**, 554
228. Haddad, J. G. and Chyu, K. J. (1971). *J. Clin. Endocrinol.*, **33**, 992
229. DeLuca, H. F. (1970). *Recent Progress in Hormone Research*, **27**, 479 (G. B. Eastwood editor) (New York: Academic Press)
230. Holick, J. F., Schnoes, H. K., DeLuca, H. F., Suda, T. and Cousins, J. (1971). *Bio chemistry*, **10**, 2799
231. Omdahl, J., Holick, M., Suda, T., Tanaka, Y. and De Luca, H. F. (1971). *Biochemistry* **10**, 2939
232. Tanaka, Y., DeLuca, H. F., Omdahl J. and Holick, H. F. (1971). *Proc. Nat. Acad Sci. USA*, **68**, 1286
233. Gray, R. W. and DeLuca, H. F. (1971). *Arch. Biochem. Biophys.*, **145**, 276
234. Tanaka, Y. and DeLuca, H. F. (1971). *Arch. Biochem. Biophys.* **146**, 574
235. Raisz, L. G., Trummel, C. L., Holick, M. L. and DeLuca, H. F. (1972). *Science*, **175** 768
236. Holick, M. F., Garabedian, M. and DeLuca, H. F. (1972). *Science*, **176**, 1247
237. Omdahl, J. L. and DeLuca, H. F. (1971). *Science*, **174**, 949
238. Boyle, I. T., Gray, R. W. and DeLuca, H. F. (1971). *Proc. Nat. Acad. Sci. USA*, **68** 2131
239. Fairbanks, B. W. and Mitchell, H. H. (1936). *J. Nutrition*, **11**, 551
240. Rottensten, K. V. (1938). *Biochem. J.*, **32**, 1285
241. Drescher, D. and DeLuca, H. F. (1971). *Biochemistry*, **10**, 2302
242. Drescher, D. and DeLuca, H. F. (1971). *Biochemistry*, **10**, 2308
243. DeLuca, H. F. (1972). *Proc. Fourth Parathyroid Conf., Calcium, Parathyroid Hormone and the Calcitonins*, 221 (R. V. Talmage and P. L. Munson, editors) (Amsterdam Excerpta Medica)
244. Garabedian, M., Holick, M. F., DeLuca, H. F. and Boyle, I. T. (1972). *Proc. Nat Acad. Sci. USA*, **69**, 1673
245. Rasmussen, H., Wong, M., Bikle, D. and Goodman, D. B. P. (1972). *J. Clin. Invest.* **51**, 2502
247. Robinson, C. J., Rafferty, B., Parsons, J. A. (1972). *Clin. Sci.*, **42**, 235
248. Robertson, W. G., Peacock, M., Atkins, D. and Webster, L. A. (1972). *Clin. Sci.*, **43** 715
249. Nicolaysen, R., Eeg-Larsen, N. and Malm, O. J. (1953). *Physiol. Rev.*, **33**, 424
250. MacCallum, W. G. and Voegtlin, C. (1909). *J. Exp. Med.*, **11**, 118

251. Berstein, D., Kleeman, C. R. and Maxwell, M. H. (1963). *Proc. Soc. Exp. Biol. Med.*, **112,** 353
252. Singer, F. R., Keutmann, H. T., Neer, R. M., Potts, J. T., Jr. and Krane, S. M. (1972). *Calcium, Parathyroid Hormone and the Calcitonins* (R. V. Talmage and P. C. Munson, editors), 89 (Amsterdam: Excerpta Medica)
253. Chase, L. R. and Aurbach, G. D. (1970). *J. Biol. Chem.*, **245,** 1520

5
Biochemical Studies on the Receptor Mechanisms involved in Androgen Actions

S. LIAO
The University of Chicago

5.1	INTRODUCTION	154
5.2	STEROID STRUCTURES AND ANDROGENIC ACTIVITIES	155
5.3	ANDROPHILIC PROTEINS IN BLOOD	157
5.4	UPTAKE AND RETENTION OF ANDROGENS BY TARGET TISSUES	159
5.5	SELECTIVE RETENTION OF DIHYRDOTESTOSTERONE BY PROSTATE CELL NUCLEI	160
5.6	ANDROGEN-BINDING PROTEINS IN TARGET CELLS	162
5.7	STRUCTURAL REQUIREMENTS AND NATURE OF THE BINDING OF ANDROGENS TO A RECEPTOR PROTEIN	165
5.8	RETENTION OF A DIHYDROTESTOSTERONE–PROTEIN COMPLEX BY ISOLATED PROSTATE NUCLEI	167
5.9	NUCLEAR ACCEPTOR ACTIVITY FOR THE RETENTION OF A DIHYDROTESTOSTERONE–PROTEIN COMPLEX	168
5.10	RIBONUCLEOPROTEIN BINDING OF A DIHYDROTESTOSTERONE–PROTEIN COMPLEX	170
5.11	SOME SPECULATIVE VIEWS ON THE RECEPTOR MECHANISMS INVOLVED IN GENE EXPRESSION IN TARGET CELLS	171
5.12	ANTI-ANDROGENS AND DIHYDROTESTOSTERONE-BINDING PROTEINS	175

5.13 DIHYDROTESTOSTERONE AND ITS BINDING PROTEINS IN ANDROGEN
DYSFUNCTION   176

5.14 ANDROGEN ACTIONS NOT ATTRIBUTED TO DIHYDROTESTOSTERONE 177

ACKNOWLEDGEMENTS   179

5.1 INTRODUCTION

This review is prepared mainly for those who are interested in the recent efforts of biochemists to search for the chemical bases for androgen actions in male accessory genital organs. Much of the information reviewed came from work on the rat ventral prostate. As will be shown later, it may not be applicable to all androgen-sensitive tissues.

The androgenic activities of steroid hormones[1-8] are often characterised by their stimulatory action on the growth and functions of male accessory genital organs, such as the prostate and seminal vesicles of vertebrates and chick combs. Most of these androgenic steroids also stimulate muscle growth or nitrogen retention. Such action is sometimes called myotropic or anabolic activity. The effects of androgens on these tissues are rather striking; most other vertebrate tissues, however, are also affected to some extent. Liver, kidney, exorbital lacrimal gland and bone marrow, as well as other more typical androgen-sensitive tissues such as levator ani muscle, preputial gland, hair follicle, testis and epididymis, have been the subjects of biochemical studies of androgen actions. Many of the effects of androgens are stimulatory with respect to their action on the growth or functioning of tissues. Androgens also exert certain inhibitory effects on lymphoid tissues and on the hypothalamic mechanism responsible for gonadotropin production. Aside from the stimulatory or inhibitory actions of androgens on well-differentiated tissues (which have been categorised as excitational or activational effects), some androgenic steroids can influence the differentiation of the male genital tract and hypothalamic structures during limited periods of early development. This class of action has been called inductive, organisational, or morphogenetic[2,8].

The primary molecular processes by which steroid hormones elicit their effects are still not clear. One of the most widely used approaches to study this problem has been to search for the earliest effect of a hormone *in vivo* and to hope to show such an effect in an *in vitro* system. However, the *in vivo* time sequences for visible hormone effects may depend more on the relative time required to magnify each of the sequential events, detectable by available methods, than on the true order in the initial chain or processes. During the last decade, considerable attention has been given to an alternative approach in which the investigators have followed the fate of steroid hormones and hoped to find the receptor molecules that play key roles in the trigger mechanism. This approach is complicated by the fact that some steroid hormones are metabolised to many other compounds in animals; also, some of them may interact with various biological molecules that do not participate directly in the primary processes of hormone action.

The study of the mechanism of androgen action has followed these general approaches, and similar difficulties have been encountered. A major portion of this review deals with recent work on the search for androgen receptors. Since there are many similarities among steroid hormones in regard to their ways of interacting with cellular macromolecules, the reader should also consult other recent reviews on steroid receptors in general[10-17]. Various metabolic parameters of androgen action[17-28] have also been well reviewed; only a few of them are mentioned in this article. The use of the term 'steroid receptors' for specific steroid-binding proteins in target cells is controversial[29, 30] since no clear function is known for any intracellular steroid-binding protein. However, the term will be retained in some sections of this article where the usage agrees with the terminology employed by other investigators[12, 13] in the field. Proteins that have a selective affinity toward androgens are also called *androphilic proteins*[18].

5.2 STEROID STRUCTURES AND ANDROGENIC ACTIVITIES[5,6,18]

Practically all of the potent androgens are derivatives of androstanes. The structures of androstenedione and testosterone, two major testicular androgens produced by the testis and circulated in the blood, as well as of 5α-dihydrotestosterone (DHT), are shown in Figure 5.1. Non-steroidal compounds, such as diethylstilboestrol, having high oestrogenic activity are

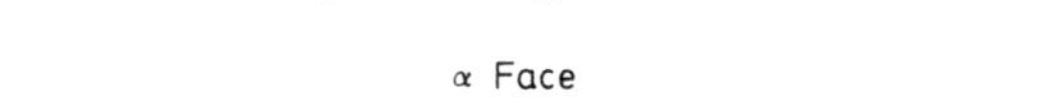

5α -Androstan-17β -OL

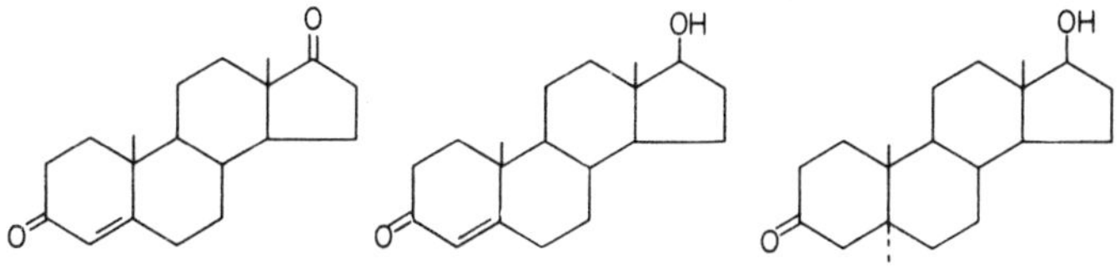

Androstenedione   Testosterone   5α -Dihydrotestosterone (DHT)

Figure 5.1 Chemical structures of representative androgens. Symbols for substitutions: 'a' and 'e', axial and equatorial; '- - - -' and '——', at α- and β-faces of a steroid respectively. (From Liao and Fang[18], by courtesy of Academic Press.)

known, but this is not the case for androgens. All carbon rings (A, B, C and D) in the steroid nuclear structures of potent androgens are *trans*-fused. Steroids with *cis*-fused rings are either inactive or show very weak potency[31-35]. Addition of one extra carbon (homosteroid) to rings B, C and D is permissible in some cases but A-homosteroids are often inactive[36-39].

The presence of oxygen at C-3 of ring A is common to the natural steroid hormones (Table 5.1). Whether a steroid with a 3-keto group is more effective than the one with a 3-hydroxy group may be dependent on the hydroxysteroid dehydrogenases in the target cells[40], or on the need for the direct action of these steroids at different sites of cellular processes. Active compounds that have no oxygen at C-3 are known[41-44]. For example, 5α-androstan-17β-ol (Figure 5.1) has been shown to be a potent androgen in chick comb[44] and in rats[42]. The myotropic activity of a steroid with a methylene group fused to C-2 and C-3 is known to be comparable to that of testosterone propionate[45]. Since the methanosteroid may not be converted to a C-3 oxygenated steroid, the finding also supports the suggestion that a C-3 oxygen is not required for myotropic action. However, the possibility of an *in vivo* conversion of these steroids to 3-hydroxysteroids has not been completely excluded. The requirement of an oxygen function at C-17 is almost absolute. Androgens having a 17β-hydroxy group are generally much more active than 17α-hydroxylated isomers. The androgenic activity of 17-keto compounds varies in different assay systems. This may be due to the availability of 17β-hydroxysteroid dehydrogenase which generates the 17β-hydroxysteroids for target cells.

DHT is at least as active as testosterone[1, 46, 47] for the growth of rat seminal

Table 5.1 Relative androgenic and anabolic activities of some representative androstanes and androstenes*. (From Liao and Fang[18], by courtesy of Academic Press.)

Androgens	Chick comb	VP	Rat SV	LA	Rat EL
Testosterone	100	100	100	100	100
5α-Dihydrotestosterone	228	268	158	152	74
17α-Methyltestosterone	300	103	100	108	162
17α-Methyl-5α-dihydrotestosterone	480	254	78	107	—
Androst-4-ene-3,17-dione	121	39	17	22	14
5α-Androstane-3,17-dione	115	33	13	11	—
5α-Androstane-3α,17β-diol	75	34	24	30	238
Androst-4-ene-3β,17β-diol	—	124	133	95	—
5α-Androstane-3β,17β-diol	2	—	10	—	5
Androst-4-en-3-on-17α-ol	—	8	2	3	—
5α-Androstan-3α-ol-17-one	115	53	8	10	46
19-Nortestosterone	—	—	10	180	52
19-Nordihydrotestosterone	118	—	—	—	—
17α-Methyl-19-nortestosterone	—	25	25	60	81
Testosterone propionate	—	161	146	187	195
5α-Androstan-17β-ol	128	—	—	—	5

* Testosterone as 100. Rat test are by injection; comb test by inunction. VP: prostrate; SV: seminal vesicle; LA: levator ani muscle; EL: exorbital lacrimal gland tests.

vesicles, prostate, or chick comb. Therefore, the androgenic action of testosterone in these tissues may not be dependent on unsaturation at C-4. The presence of unsaturation at C-1 decreases the androgenic activity of DHT in these male accessory organs. Nevertheless, the dehydro compound was found to be 2–3 times more active than the saturated one with respect to myotropic activity[48]. Similar experiments with 5α-androstane-17β-ol derivatives suggested that a high electron density at C-2 and C-3 is a factor strongly promoting high myotropic activity[45, 49, 50]. It was pointed out that the introduction of more than one sp^2-hybridised carbon atom into ring A results in a pronounced flattening of the ring from a cyclohexane chair form to a more planar conformation in which the active steroid may be better able to rest on a receptor surface, with a concomitant increase in the degree of orbital overlap. However, the need for the formation of a π complex with the receptor site if hormone action is to occur has been questioned, since active compounds having little possibility of forming sp^2 hybridisation are known[51].

By studying the androgenic activity of various alkyl-substituted androstanes, one may visualise the manner in which androgens interact with their receptors. For example, from the fact that 17α-ethyltestosterone and 1α-methyl-19-nor-DHT are much less active than the unsubstituted steroids one may suggest[52] that these axial-substituted alkyl groups interfere with receptor binding of the androgen from the α face (Figure 5.1). The difficulty of drawing such a conclusion is indicated by arguments concerning the angular methyl group on C-10. Based on the finding that the removal of this methyl group results in a moderate loss of androgenic activity, it was suggested that the angular methyl group may participate in the receptor binding from the β face of the steroid[45]. It was argued, however, that if the receptor is to bind from the β face the methyl group should interfere with the binding, and that the removal of the methyl group should, therefore, enhance rather than diminish the androgenic activity[37, 52].

In many instances, the structure–activity relationship is not straightforward[18]. For example, while 6α-(equatorial)-methyl-substituted testosterone or DHT is less androgenic than the 6β-(axial)-methyl isomers[54], 6α-chlorotestosterone is 4 and 15 times more androgenic and myotropic respectively than 6β-chlorotestosterone[55]. In general, no simple structure–activity relationship can be arrived at by such a semi-empirical approach comparing chemical structure and end-point activity. The results are probably complicated by the structural requirements for the selection processes (absorption, transport, transformations), which undoubtedly involve interactions of steroids with various specific macromolecules and do not simply reflect the relationships between steroids and the ultimate *true* receptors at the functional sites.

5.3 ANDROPHILIC PROTEINS IN BLOOD

The rate of production (by the testis) and plasma concentration of androgens are subject to individual, diurnal and seasonal variations and to many other factors[11, 18, 22, 25]. The major blood androgens are testosterone and androstenedione. Testosterone concentration increases after puberty. In young adult male mammals, the blood concentration of testosterone is of the order of 0.1–0.5 µg/100 ml[18] or *ca.* 10^{-8} mol l^{-1}. Recent studies have revealed that the concentration of DHT in adult human peripheral plasma is about 10–30% of that for blood testosterone[56,57]. These androgens are metabolised by enzymes in various tissues, especially in liver; some are conjugated as glucuronide.

The binding of steroid hormones by blood protein has been studied extensively and has been reviewed in detail[18,58,59]. Since albumin is the most abundant protein constituent of plasma, it may bind a large portion of blood steroids. Although steroids may bind preferentially to albumin through specific portions of the molecules such as the Δ^4-3-keto group[58,60,61], generally there is no great specificity involved in such an interaction and the association constants are usually low (of the order of only 10^4 l mol^{-1}). On the other hand, studies with human plasma have revealed that certain plasma globulins may bind testosterone and certain steroid hormones more firmly[62-72]. Although these proteins are present in small quantities (*ca.* 10^{-7} mol l^{-1})[64], they have high affinity constants (about 10^9 l mol^{-1})[63], and therefore can compete well with plasma albumin for the binding of androgens.

The amounts of testosterone-binding globulin(s) increase in pregnancy[64-66], in patients with prostate cancer who are receiving oestrogen therapy[64] and are higher in the female than in the male[67]. In general, these globulin(s) show high, but not absolute, specificity toward 17β-hydroxy-androstanes or -androstenes with a ring-A oxygen group, such as DHT, testosterone and 5α-androstan-3β,17β-diol. They also strongly bind 17β-oestradiol and, to a lesser extent, 4-androstenedione and oestrone[68-72]. The binding of steroids (including androgens) by erythrocytes is not significant and does not appear to have a major role in the transport or metabolism of androgens[73,74]. In contrast with the non-covalent binding of steroids by blood proteins described above, some protein binding of oestrogens (the complex was called 'oestroprotein'[75]) and androgens[76] has been characterised as very stable, and possibly indicates covalent binding. Since no information is available on the structure of the steroid moieties in these complexes, the studies cannot be assessed further. It should be noted that enzymatic oxidation of hydroxylated steroids can result in steroid radicals which react readily, but indiscriminately, with other macromolecules.

The biological role of the blood steroid-binding proteins is not clear[58]. Suggestions that cortisol-binding[77] and progesterone-binding proteins[78] in blood may act as protectors to keep these steroids from enzymatic inactivation may be applicable to androgen-binding plasma proteins. Since the binding by both the plasma albumin[73,79,80] and globulin[72] of the conjugated metabolites (17β-glucuronides and sulphates) with high aqueous solubility is poor, the proteins may be responsible for the slower clearance of the unconjugated steroids which they bind. This may enable androgens to be taken up by target tissues for a prolonged period of time after their production in the testis. The possibility that these blood steroid-binding proteins may participate in the mechanisms involved in steroid uptake at the cell membrane of the target cells cannot be excluded at this time. Such an idea has not been favoured because of the lack of evidence to disprove the assumption that steroids penetrate cell membranes by a simple diffusion process.

5.4 UPTAKE AND RETENTION OF ANDROGENS BY TARGET TISSUES[18,56]

The classic finding by Jensen, Jacobson and Gupta[81-83] that oestrogens can be retained against concentration gradients from blood by their target tissues, such as uterus and ovary, led to the discovery of specific oestradiol–receptor proteins which are largely responsible for the phenomena[14, 16, 84, 85]. Studies of this type for the androgens were difficult, however, because of the rapid and multiple transformations[22-27, 86] of androgens by the liver and also in the target tissues themselves. Earlier investigations were further complicated by the use of large doses of androgens with low specific radioactivity[18]. Nevertheless, the ability of rat ventral prostate and seminal vesicles to accumulate radioactive androgens has been indicated in these investigations[87, 88]. Studies by Pearlman and Pearlman[89] and Harding and Samuels[90] in rats indicated clearly that, while large amounts of conjugated metabolites exist in blood and liver, the prostate accumulates unconjugated androgen metabolites by a selective process. This is supported by a more recent study by Tveter and Aakvaag[91], who showed that the ratios of acetone-insoluble radioactivity (conjugated fraction) to total radioactive material were about 0.64 for liver, 0.22 for epididymis, 0.13 for dorsal prostate, 0.09 for coagulating gland and below 0.02 for seminal vesicle, ventral prostate and lateral prostate. These studies are of great importance since the differences in the concentration of blood androgens (threshold) needed to elicit androgen responses in different tissues are likely to be related to the differences in the selective processes for the uptake of various androgen metabolites and metabolic activities (including conjugation) in target tissues.

A clear demonstration of the selective accumulation of radioactive androgens by male accessory sex organs became possible with the availability of testosterone of high specific radioactivity[91-98]. In rats, clear-cut retention of radioactive androgens by ventral and dorsal prostate, seminal vesicles and coagulating gland can be observed at 1–3 h from the time of ^{3}H-testosterone injection[18, 92, 96-99]. The radioactivity in blood, spleen, lung, thymus and diaphragm is rapidly cleared within the first hour. In the prostate and seminal vesicles, radioactive steroids can be detected even 16 h after injection. However, more than 50% of the well-retained androgens appear to leave the tissues within 6 h.

The uptake and localisation of radioactive androgens in the rat accessory sex glands were also studied by autoradiographic techniques[100-102]. Within a few minutes after injection of ^{3}H-testosterone in rats, radioactivity could be detected in these target tissues. Selective nuclear labelling was seen from 0.5 to 7 h following injection (Figure 5.2). The epithelial cells of different prostate follicles of rats injected with radioactive testosterone were labelled unevenly[102]. The flattened epithelium was labelled weakly or not at all. In both the prostate and the seminal vesicles of rats, the connective tissue appeared unlabelled. In the ventral prostate, the stromal tissue contained less radioactive material than the epithelium and very few silver grains were in the glandular lumina.

At 1–7 h after the injection of ^{3}H-testosterone, the nuclear concentration of radioactivity could be seen in the epithelial cells of rat ventral prostate, seminal vesicles, coagulation gland and epididymis, but not in the liver or

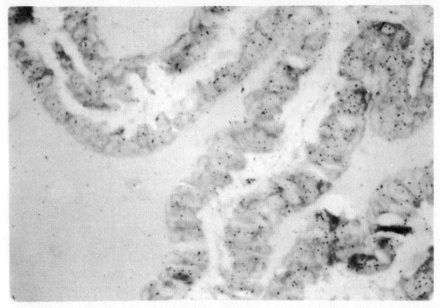

Figure 5.2 Autoradiograms of ventral prostate 1 h after a subcutaneous injection of 4 μg of 7α-³H-testosterone (10 Ci mmol⁻¹) into a 3-month old male rats castrated 72 h previously. (See Sar et al.[100], by courtesy of Lippincott.)

diaphragm[100-102]. A similar nuclear concentration of radioactive androgens appeared to occur in the muscle cells of seminal vesicles, coagulation gland and epididymis, but not in the muscle cells of the prostate. The autoradiograms showed that the nuclear radioactive steroids are not concentrated at nucleolar regions (Figure 5.2) but dispersed to all areas containing nuclear chromatins. Smear autoradiograms[100] of nuclei isolated from minced ventral prostate previously incubated with ³H-testosterone at 37 °C revealed the retention of radioactivity in the nuclei which was in agreement with *in vivo* experiments.

5.5 SELECTIVE RETENTION OF DIHYDROTESTOSTERONE BY PROSTATE CELL NUCLEI

Prior to the autoradiographic studies described above, a number of investigators had focused their attention on the association of radioactive steroids *in vivo* with nuclear components of target cells. In many of these earlier studies[18, 103-105], the radioactive steroids were not identified. In 1968, Bruchovsky and Wilson[97] reported that although DHT, androstane-3,17-diol and testosterone could be shown in prostatic cytoplasm within 1 min after the administration of ³H-testosterone to rats, only DHT and testosterone were found in the isolated prostatic nuclei for as long as 2 h. Independent experiments by Anderson and Liao[95] also revealed that nuclei of both ventral prostate and seminal vesicles can selectively retain ³H-DHT. Such retention could be demonstrated at the time when prostatic cytosol contained many other radioactive steroids. The retention of ^{3}H-DHT by prostate nuclei was tissue specific: in liver, where many metabolites (including DHT) of testosterone were demonstrable, there was no selective retention of DHT by liver nuclei. The selective binding of DHT by prostate cell nuclei *in vivo* could be reproduced *in vitro* by incubating minced prostate glands with ^{3}H-testosterone or ^{3}H-androstenedione[95]. After incubation, about 80–90% of the radioactivity associated with isolated nuclei was found to be DHT. The nuclear retention of DHT *in vitro* is also tissue-specific. When other rat tissues, such as liver, brain thymus and diaphragm, were minced and incubated with ^{3}H-testosterone, there was no clearcut nuclear retention of DHT (Table 5.2). The nuclear retention of DHT *in vivo* and *in vitro* is also steroid-specific. No distinct retention of radioactive 17β-oestradiol or cortisol could be observed.

Table 5.2 Retention of radioisotopes by isolated nuclei of minced rat tissues incubated with ^{3}H-testosterone for 30 m at 37°C. (From Anderson and Liao[95], by courtesy of the Editor of *Nature*.)

Tissues	Radioactivity of isolated nuclei (dpm/100 μg DNA)	Distribution of radioactivity*		
		Testosterone (%)	DHT (%)	Androstenedione (%)
Brain	105	96.2	2.8	0.0
Thymus	78	93.5	0.0	2.1
Diaphragm	53	78.2	2.1	17.6
Liver	60	73.4	0.0	4.5
Ventral prostate	1250	17.3	78.2	1.0

* Steroid extracted from nuclei were subjected to thin-layer chromatography, and the percentage distribution of radioactivity associated with the three steroid spots was calculated

The possibility that such nuclear retention is due to cytoplasmic contamination or non-specific adsorption of ^{3}H-DHT has been eliminated. In these experiments, the nuclei were prepared by centrifugation through a hypertonic (2.2 mol l^{-1}) sucrose solution to remove cytoplasmic particles. Electron micrographs of such preparations showed attachment of small amounts of endoplasmic reticulum to the outer nuclear membranes, but washing of nuclei with 0.5% Triton X-100, a non-ionic detergent, removed these membranes[107] without any loss of radioactivity[18, 95, 96]. Moreover, the addition of 100-fold excess of unlabelled testosterone or DHT during the homogenisation or isolation of nuclei did not influence the amount of radioactivity retained by the nuclei. This strongly suggests that ^{3}H-DHT was retained firmly inside the cell nuclei *in vivo* and not during the preparation of nuclear fractions.

It has been found that DHT is a more potent androgen than testosterone in a number of androgenic bioassay systems (Table 5.1). This fact, together with the new finding that DHT can be retained by prostate nuclei whose functions are under the control of androgenic steroids, strongly supports the idea that DHT may be an 'active form' of testosterone in some, but not necessarily in all, androgen-responsive cells. The conversion of testosterone to DHT is carried out by a NADPH: Δ^4-3-ketosteroid 5α-oxidoreductase (5α-reductase), which has been located in microsomal[26, 108] and nuclear[95, 97] fractions. The enzyme appears to be located in nuclear membranes[110] and is very sensitive to low concentrations (0.04%) of non-ionic detergent such as Triton X-100[109].

5.6 ANDROGEN-BINDING PROTEINS IN TARGET CELLS

When prostate cell nuclei that retain ^{3}H-DHT were extracted with a buffered salt solution (0.4–1.0 mol l^{-1} KCl or NaCl) and analysed by sucrose-gradient centrifugation or Sephadex gel filtration, the radioactive testosterone was found to be associated with protein fractions[18, 96, 111-117]. According to Fang and Liao[18, 96, 114-119], the sedimentation constant of the nuclear DHT-protein complex, as measured in a gradient medium containing 0.4 mol l^{-1} KCl, is 3S$_{20,w}$. The DHT–protein complex aggregates rapidly if the salt concentration is lower than 0.1 mol l^{-1} KCl; such aggregation occurs without loss of the bound androgen. The aggregation appears to involve other nuclear components (see below). It is generally agreed that the DHT-binding protein is not a histone-type basic protein but rather an acidic protein.

The cytoplasmic-soluble (cytosol) fraction of ventral prostate of rats injected with ^{3}H-testosterone contains ^{3}H-DHT–protein complexes which can also be prepared by mixing the prostate cytosol fraction of castrated rats with ^{3}H-DHT. Liao and co-workers[18, 96, 114-119] showed that these complexes have a sedimentation constant of about 3.5S (Figure 5.3) which gradually aggregates to larger forms. Analogous to the 17β-oestradiol–receptor complex in rat uterus[14, 16], ^{3}H-DHT–protein complexes having 8–9S have been detected by Unhjem et al.[120-122], Mainwaring[123-125] and Baulieu and Jung[127] (see also Figure 5.3). The 8S complex, as well as larger aggregates, dissociate in buffered 0.4 mol l^{-1} KCl solution and, as a result, ^{3}H-DHT–protein complexes can be detected in the area of 3–4S.

While gradient centrifugation and gel chromatography are convenient techniques in the detection of DHT binding proteins in crude extracts, their use in quantitative measurements and in the study of the physical properties of DHT-binding proteins needs extreme care. For example, based on agarose column chromatography and on the sedimentation constant of 8S, a molecular weight of about 200 000 could be suggested for the DHT–protein complex[124] whereas that of the 3S steroid protein complex should be of the order of 30 000[115]. In fact, the sedimentation constant of an unstable protein (which has a high tendency to aggregate) in a preparation less than 0.001 % purity is expected to be very dependent on the conditions of the preparation and analysis. Recent studies in the author's laboratory have shown that altering the pH, salt concentrations and temperatures in crude extracts results in sedimentation constants ranging from 3–12S for ^{3}H-DHT–protein complex(es) of prostate cytosol. It is not certain at this time which forms of the complexes actually represent those of the *in vivo* situation.

Only limited studies have been reported concerning the androgen-binding proteins in tissues other than rat ventral prostate. Most of these studies are based on the detection of 3–4S and 8S androgen-binding proteins by gradient

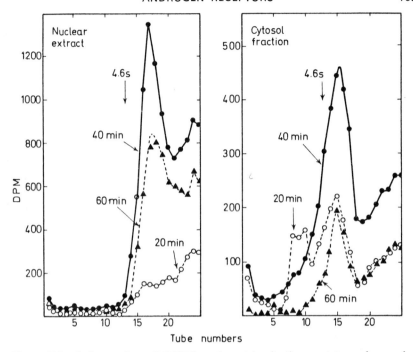

Figure 5.3 Sedimentation of DHT-bound proteins in the prostate nuclear and cytosol fractions of rats injected with ^{3}H-testosterone (50 µCi/0.34 µg per 450 rat). Adult male rats were castrated 3 days previously. Rats were killed 20, 40 and 60 min after the injections. Nuclear extracts and cytosol fractions were prepared and analysed by gradient centrifugations. KCl(0.4 mol l^{-1}) was included in the sucrose gradient solution used for the nuclear extracts but not in the one for the cytosol fractions. Bovine albumin (4.6S) was used as a reference standard. Tubes are numbered from the bottom of the centrifuge tube. (From Fang et al.[96], by courtesy of the American Society of Biological Chemists.)

centrifugations of tissue cytosols. Ritzen et al.[128] found a 4S complex in rat epididymis which binds DHT with about twice the affinity it has for testosterone and 17β-oestradiol. Androstenedione and 5α-androstane-3α,17β-diol bind to a much lesser extent. Earlier, Baulieu et al.[127] reported that the 8–10S DHT-binding proteins of rat ventral prostate also bind testosterone to a significant extent, but not 17β-oestradiol or 5α-androstane-3α(or 3β),17β-diol. Rat seminal vesicles also contain DHT-binding protein(s) which migrates as 7S in lower concentrations KCl solutions, but as 3S in 0.4 mol l^{-1} KCl solution[118]. McCann et al.[130] found a 9S DHT–protein complex in calf uterus extracts which can be distinguished from 17β-oestradiol-binding proteins. Similar androgen-binding proteins have been described in other androgen-sensitive organs such as the preen gland and cloacal gland of birds[103, 131, 132], testis[133-135], kidney[136, 137], androgen-sensitive tumours[138-140], hair follicles[141] and sebaceous and preputial glands[152, 250].

The purification of a cellular androgen-binding protein has to date not been very successful. Some of the obstacles are: (a) the amount of the target tissues as well as the quantity of the receptor proteins in the tissues are very small; (b) these receptor proteins are unstable, especially in the absence of the steroid hormones; (c) the tendency of the steroid binding proteins to form aggregates; (d) the assay methods are based on the binding of the radioactive steroids to the proteins and not on the function of the protein (or the steroid–protein complex); thus it is difficult to determine whether the 'purified' binding proteins are, in fact, the unaltered receptor molecules.

At least two proteins in the cytosol of rat ventral prostate bind DHT. These proteins are designated as α- and β-proteins[115,118]. When they form a complex with DHT, they are called Complex I and Complex II, respectively[115-119]. Complex II or β-protein can be precipitated by bringing the whole cytosol fraction of rat prostate to 40% saturation with respect to ammonium sulphate. For the precipitation of Complex I and α-protein, higher concentrations (55–70% saturation) of ammonium sulphate are required. Complex I, with a sedimentation constant of 3–3.5S (with or without 0.4 mol l^{-1} KCl), is relatively stable up to 50 °C for 10 min and has no tendency to aggregate to a larger complex. On the other hand, Complex II, with a sedimentation constant in the vicinity of 3S in the absence of KCl gradually aggregates to larger forms. The aggregates dissociate in 0.4 mol l^{-1} KCl and Complex II can be recovered as the 3S complex. Although the nature of the aggregation is not clear, recentrifugation of the 3S complex obtained from sucrose-gradient centrifugation tends to minimise the extent of aggregation, suggesting that other components are involved. Purification of Complex II by hydroxyapatite gel chromatography also appears to remove such components. The purification of the cytoplasmic DHT-binding proteins has also been attempted by the use of electrofocusing techniques[125,135], and by chromatography on cellulose phosphate[142] or DNA–cellulose columns[125]; however, no great success has been recorded.

The α-protein fraction of prostate cytosol has protein(s) that bind DHT and, to a lesser extent, progesterone, 17β-oestradiol and testosterone, but not cortisol[115]. On the contrary, the β-protein fraction exhibits an extremely high specificity toward DHT. This specificity and the nature of the steroid–protein interaction will be discussed in the next section. Complex II appears to be much less stable at 40 °C than Complex I. A criterion for distinguishing Complex II from Complex I is that the former, but not the latter, is retainable by prostate cell nuclei[115-119]. Our recent studies also show that certain DHT–protein complexes in the Complex II fraction cannot be retained by prostate nuclei. Whether this is due to the altered forms of Complex II or to another class of androgen receptors remains to be investigated[145].

A ^{3}H-DHT–protein complex can be obtained from microsomal fractions of rat prostates injected with radioactive testosterone, or the minced prostate incubated with ^{3}H-DHT[18,59,116]. Such a complex is also formed when ^{3}H-DHT is added directly to the microsomal fraction. The microsomal ^{3}H-DHT-binding protein binds tightly to particles that are insoluble in a buffered solution of low ionic strength (<0.1 mol l^{-1} KCl), but can be released readily if the KCl concentration is raised to 0.4 mol l^{-1}. The amount of the microsomal DHT-binding protein is comparable to the amount of total cytosol β-protein in a cell. The microsomal protein may be partially responsible for the retention of DHT by the microsomal fraction of the prostate[143]. Microsomal androgen-binding protein is very similar to cytosol β-protein in its steroid specificity, heat stability, sedimentation constant and ability to aggregate. These properties are also shared by the nuclear DHT-binding protein, suggesting that they may be closely related, if not identical.

The DHT-binding protein in the prostate nuclear and microsomal fractions is apparently not 5α-reductase. This conclusion is based on the observations that (a) Triton X-100 can remove or inactivate all the reductase activity of prostate nuclei without loss of DHT-binding protein from the nuclei[109]; (b) the loss or gain in binding protein by the prostate nuclei, in response to androgen manipulation, is not accompanied by a comparable change in the reductase activity[18, 95, 96]; (c) the cytoplasmic DHT-binding proteins and reductase activity do not behave in a parallel manner during the fractionations;[144] and (d) DHT at high concentrations (5×10^{-6} mol l^{-1}) does not inhibit the reductase activity, showing no firm binding of the product to the active sites of the enzyme[118].

5.7 STRUCTURAL REQUIREMENTS AND NATURE OF THE BINDING OF ANDROGENS TO A RECEPTOR PROTEIN

The binding of androgens by the β-protein is highly specific. At concentrations below 1×10^{-9} mol l^{-1}, 5β-DHT, testosterone, androst-4-ene-3,17-dione, 5α-androstane-3,17-dione, 3α(or 3β), 17β-dihydroxy-5α-androstane, progesterone, cortisol, or 17β-oestradiol do not compete well with the binding of ^{3}H-DHT to β-protein[115]. Thus, β-protein can act as a specific selector in deciding which natural steroids can or cannot be retained by the target cells. Available information indicates that there is a good correlation between those structural elements of the steroid required for androgenic activity in rat prostate and for binding to β-protein (or metabolism to a binding derivative)[59, 115, 117]. This correlation, as well as the selective retention of the β-protein (in complex with DHT) by the target cell nuclei and the fact that these processes are antagonised by anti-androgens (Section 5.12), strongly support the contention that β-protein is a receptor protein for DHT in the target cells.

From studies of the ability of various steroids to compete with ^{3}H-DHT for protein binding[153], one can deduce that C-3 should not have a hydroxy group and C-17 should not have a carbonyl group on the steroid molecule. Whether an oxygen function at both C-3 and C-17 is required for the binding is yet to be tested with deoxysteroids. Since 5β-DHT cannot bind to β-protein, a flat steroid carbon skeleton appears to fit better to the binding site. At C-7, methyl substitution at the equatorial (7β) position greatly hinders the binding, whereas a methyl group at the axial (7α) position enhances the binding ability in some cases. Some restrictions on the peripheral sides of the steroid molecules are therefore apparent[117-119, 153].

Since the affinity of DHT for β-protein is much higher than that of testosterone, one might suspect that the local electronic structures associated with the unsaturated bond in ring A are not desirable. This is apparently not so, since potent androgens such as 7α-17α-dimethyl-19-nortestosterone which have unsaturated bonds at C-4 and are not reduced under the conditions of our study can bind to β-protein with 3–5 times the affinity of DHT[59, 117-119]. It is possible that these androgens, having no 19-methyl group substitution, are thin enough to fit into the binding site on the β-protein. The androgen-binding site on the β-protein may be inside a narrow hole, which can accommodate flat molecules like DHT better than testosterone which is about 1 Å thicker. In fact, it is very likely that a large portion of the steroid molecule is physically enveloped in the receptor protein molecule[59, 117-119, 153]. This suggestion is in accord with the observation that ^{3}H-DHT, once bound to the β-protein (with an apparent affinity constant much higher than 10^{10} l mol^{-1}), does not exchange readily with a several-hundred-fold excess of non-radioactive DHT while maintained for many hours at 0 °C. Such exchange can be performed however by freezing and thawing the complex in the presence of the competitor[115, 153].

Androgen - receptor complex

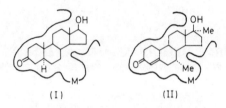

(I) (II)

Oestrogen - receptor complex

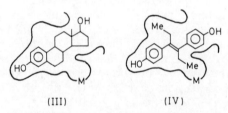

(III) (IV)

Figure 5.4 A schematic representation of the interaction of sex steroids with their receptor protein. See text for explanation. Steroid structures shown are for DHT (I), 7α,17α-dimethyl,19-nortestosterone (II), 17β-oestradiol (III) and diethylstilboestrol (IV). (From Liao et al.[153].)

The increase in receptor-binding affinity due to the substitution of a 7α-methyl group on 19-nortestosterone is in accord with the increase in the androgenicity due to the same substitution[43, 147]. It is possible that a specific binding site (M site) for the binding of the 7α-methyl group[117, 153] is present in the β-protein (Figure 5.4), and that both the binding affinity and the androgenicity are thus enhanced many fold. The substitution of the 7α-methyl group to DHT or 17α-methyl-19-Nor-DHT does not significantly enhance the binding affinity or the biological activity. In these compounds the 7α-methyl group orientates slightly further from C-3 of a ring A and, as a result, may not fit perfectly on the M site. It is interesting to imagine that the terminal methyl groups on diethylstilboestrol, a potent synthetic oestrogen, may also behave like the 7α-methyl group on the 19-nortestosterone in enhancing its ability to bind to the oestrogen receptor and also its oestrogenicity[117]. This implies a similarity between the androgen receptor and oestrogen receptor in their hormone-binding sites and their importance in biological activities. If this is so, then the M site must play only an accessory role and is not absolutely necessary for the functioning of sex steroids, since the natural steroids do not have the methyl group[153].

5.8 RETENTION OF A DIHYDROTESTOSTERONE–PROTEIN COMPLEX BY ISOLATED PROSTATE NUCLEI

The prostate nuclear DHT-binding protein appears to originate in the cytoplasm[18, 96]. This is evident from the fact that the prostate nuclei of castrated rats contain very few proteins that bind DHT[115]. Incubation of the prostate nuclei with radioactive DHT also did not result in the formation of the DHT–protein complex. However, if ^{3}H-Complex II was prepared first and then mixed with prostate nuclei, one could detect the ^{3}H-DHT–receptor complex

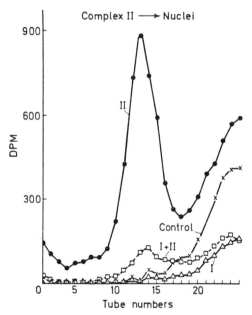

Figure 5.5 Selective retention of the prostate ^{3}H-DHT–protein complex by isolated nuclei. ^{3}H-Complex I and ^{3}H-Complex II, either alone or in mixture (I + II) were incubated with prostate nuclear preparations. Nuclei were re-isolated, extracted and the ^{3}H-DHT–protein complex in the extract was analysed by sucrose-gradient centrifugation. Tubes are numbered from the bottom of the centrifuge tube. The control tube contained only the nuclear preparation. (From Fang and Liao[115], by courtesy of the American Society of Biological Chemists.)

in the nuclear extract of the re-isolated nuclei. The Complex I (or α-protein) was not retained by the nuclei and also strongly reduced the nuclear retention of Complex II (Figure 5.5). Apparently the cytosol Complex II itself is retained as the nuclear complex[115].

It has been proposed that steroid-binding proteins may function as carriers of steroids and thus regulate the availability of the hormones at various cellular sites. However, one may also visualise[96, 115-119] the binding of a steroid to a specific protein as a necessary step in removing a protein from a specific sub-cellular site and translocating it to another sub-cellular site, for example, to the cytoplasmic membranes or to cell nuclei[115, 119]. An experiment designed to test this possibility has been reported. In this study, the amount of β-protein (not bound to steroid) retainable by prostate nuclei was found to be very limited. However, addition of ^{3}H-DHT to the β-protein fraction enhances β-protein retention more than threefold. Since DHT retention by prostate cell nuclei is highly steroid- and tissue-specific, the retention of DHT–receptor complex is expected to have similar specificities. In fact, when various ^{3}H-steroids were mixed with a whole cytosol preparation or a β-protein fraction and then incubated with prostate nuclei, only ^{3}H-DHT formed the complex with a protein that could be retained by the nuclei whereas radioactive testosterone, 17β-oestradiol progesterone and cortisol were ineffective[115, 116, 164]. The nuclei from liver, thymus, brain and diaphragm of rats were not able to retain as much of the prostate DHT–receptor complex as prostate nuclei. Similar tissue specificities can be observed if crude nuclear chromatins are made from tissues and assayed for their ability to associate with prostate ^{3}H-DHT–receptor complexes[106, 148]. The cytosol fractions of adult liver, thymus and diaphragm provide very little DHT–protein complex that can be retained by prostate nuclei[115].

The microsomal DHT–protein complex could be extracted first and mixed with purified prostate nuclei to show its retention by the nuclei[59].

5.9 NUCLEAR ACCEPTOR ACTIVITY FOR THE RETENTION OF A DIHYDROTESTOSTERONE–PROTEIN COMPLEX

Since the nuclei of the ventral prostate but not of other rat tissues can effectively retain ^{3}H-Complex II in a specific manner, it was suggested that prostate nuclei may contain specific nuclear 'acceptor' substance(s)[116-119]. Under favourable conditions, with an excess of ^{3}H-Complex II, each nucleus of rat ventral prostate in *in vivo* or in *in vitro* tissue immersion experiments can retain about 2000–3000 molecules of the DHT–receptor complex[116, 118, 124]. Higher values of 6000[99]–10 000[117] molecules have been given, but it appears that there are a limiting number of available 'acceptor' sites in the nucleus. Since the nuclear 'acceptor' activity, i.e., the ability to retain ^{3}H-Complex II, could be inactivated by heating for 10 min at temperatures above 50 °C, or by treatment of the nuclei with trypsin or pronase, it was suggested that the 'acceptor' substance(s) is at least in part protein(s)[115-119].

For a further experimental study of the receptor–'acceptor' interaction, one may consider the possibility that the 'acceptor' substance(s) is constituted of more than one macromolecule[149, 150]. Some of these molecules ('acceptor' factors) may be responsible for specifying the site where a steroid–receptor complex is to be retained by a larger macromolecular complex ('acceptor') whose biological function is to be regulated. In an effort to search for the 'acceptor' factor (f) and the 'acceptor' (A) for the DHT–receptor complex in rat ventral prostate, Tymoczko and Liao[149] designed a simple assay system. For this purpose, nuclei that might contain the hypothetical 'acceptor' factors were extracted with various KCl concentrations. The nuclear extracts were then tested to determine whether they contained 'acceptor' factors which were necessary for the retention of the steroid–receptor complex by the reconstructed nucleoprotein aggregates. According to the work described in the previous section, the 'acceptor' activity should be dependent on (a) the DHT–receptor complex, but not DHT alone, and also on (b) heat-labile protein-factors. Such retention should also be (c) steroid- and receptor-specific; and one should be able (d) to release the retained steroid–receptor complex from the 'acceptor' molecules in 0.4 mol l^{-1} KCl and (e) to identify the complex by gradient centrifugation. These properties therefore have been used to design the assay methods.

In typical experiments, prostate nuclei which had not been exposed to DHT were extracted with 0.4 mol l^{-1} KCl. The protein extracts were then incubated in 0.1 mol l^{-1} KCl with ^{3}H-Complex II preparations in the presence of calf-thymus DNA or synthetic polyribonucleotides. The mixture was passed through a Millipore filter to retain the aggregates formed. After washing with 0.1 mol l^{-1} KCl, the receptor-bound ^{3}H-steroid was extracted with 0.4 mol l^{-1} KCl and analysed by gradient centrifugation. When assayed in this manner, the liver nuclear extracts are usually much less active than the the prostate nuclear extracts. Heat treatment of the nuclear extract rapidly destroys the 'acceptor' activity. The heat-labile factors needed for the retention of DHT–receptor complex are not dialysable. They can be fractionated by ammonium sulphate and ethanol or precipitated from the solution at pH 4.0–4.5. The acceptor factors are destroyed by treatment with trypsin.

The 'acceptor' activity of the nuclear extract can be stimulated several-fold by the addition of an adequate amount of rat prostate or calf-thymus DNA. One striking observation is that such an effect is abolished if DNA is denatured by heating. Excess DNA is inhibitory, possibly due to the non-specific association of the 'acceptor' factors of ^{3}H-Complex II with nucleic acids. In the presence of DNA alone (without an 'acceptor' fraction), a large amount of ^{3}H-Complex II is retained by a Millipore filter, showing its binding to DNA. The retained radioactivity cannot be extracted with 0.4 mol l^{-1} KCl, however. It appears that in the assay one is measuring the portion of the androgen–receptor complex which binds to the nucleoprotein sites associated with heat-labile protein(s), and that the ternary complex formed dissociates in 0.4 mol l^{-1} KCl. The stimulatory effect of DNA can be imitated by liver ribosomal RNA, poly (G), or poly (A), while poly (U) and poly (C) are less effective[149].

The possibility that steroid–receptor complexes may bind directly to DNA has been investigated by King[151] for oestrogen and by O'Malley for progesterone[17] receptors. By gradient centrifugation, these female hormones in complex with receptor proteins were found to bind to both single- and double-stranded DNA of their target tissues as well as DNA of other non-target tissues of the animals used. Mainwaring and Mangan[125] chromatographed ³H-DHT–receptor complexes of rat prostate on DNA bound to cellulose columns. More androgen–receptor complexes were retained by the columns when DNA was prepared from prostate instead of from liver and kidney of rats, calf thymus, or *Escherichia coli*. To what extent this is due to differential alterations of DNA during the preparations of DNA is not known. It was also considered that certain specific prostate proteins which remain attached to prostate DNA may be responsible for the specificity[99, 125]. If this is the case, the proteins appear to be heat-insensitive and stable with respect to the drastic treatments needed for the DNA preparation. Therefore they may not be identical with the heat-labile 'acceptor' factors described above.

5.10 RIBONUCLEOPROTEIN BINDING OF A DIHYDROTESTOSTERONE–PROTEIN COMPLEX

For the identification of the nuclear components which are present in the KCl extracts of prostate nuclei and responsible for the 'acceptor' activity, a gradient centrifugation technique was employed[155]. Cell nuclei were first extracted with KCl solutions; the extracts were then fractionated and mixed with ³H-steroid–receptor preparations. The mixtures were subjected to gradient centrifugation to see whether or not new radioactive peaks could be detected. By this method, we found that some of the prostate nuclear ribonucleoprotein (RNP) particles could bind ³H-Complex II to form 60–80S particles[155]. Some of these particles were present in the 0.4 mol l⁻¹ KCl nuclear extract, but many of them appeared to bind tightly to the chromatin aggregate even after extensive extraction with 1 mol l⁻¹ KCl. The chromatin-bound RNP particles could be freed by incubating the chromatin with pancreatic DNase followed by deoxycholate treatment. To demonstrate the association

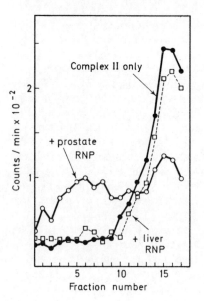

Figure 5.6 Interaction of a ³H-DHT–protein complex of rat ventral prostate with nuclear ribonucleoprotein particles of the prostate and rat liver. ³H-Complex II (1800 c/min/0.17 μg protein) alone or in the presence of 0.1 absorbancy unit (at 260 nm) of nuclear RNP was layered on the top of a centrifuge tube containing sucrose-gradient (10–30%) solution. (From Liao *et al.*[155], by courtesy of the Editor of *Nature*.)

of the nuclear particles and ^{3}H-Complex II, they were mixed and incubated at 0 °C for 30 min. Following the incubation, the mixture was analysed by gradient centrifugation (Figure 5.6). The steroid–receptor binding activity of the RNP particles was heat-sensitive (10 min at temperature above 60 °C) and was abolished by treatment with pancreatic RNase and T_1-RNase, but not with DNase-I. Apparently the interaction of the DHT–receptor complex and the prostate RNP particle did not require DNA. The RNP particle or the ternary complex of the particle and the steroid–receptor complex may bind to DNA in a specific manner, however. The steroid–receptor complex could be dissociated from the ternary complex by 0.4 mol l^{-1} KCl or RNase treatment. Apparently the 80S peak was formed by a reversible association of nuclear RNP particles with the androgen–protein complex and possibly with other proteins in the ^{3}H-Complex II fraction. RNP preparations isolated in the same manner from rat liver were usually much less active in binding the ^{3}H-Complex II (Figure 5.6). The ^{3}H-progesterone and ^{3}H-17β-oestradiol-binding proteins of calf uterus did not bind to RNP isolated from the chromatin of rat ventral prostate. In the presence of excess ^{3}H-Complex II, less than 10% of the RNP particles could bind to the ^{3}H-DHT–receptor complex. It is possible that only those RNP particles having the heat-labile 'acceptor' factor(s) can associate with the androgen–receptor complex in a specific manner[154, 155]. Some possible implications of these observations will be discussed in the next section. Similar studies on the binding of ^{3}H-17β-oestradiol-receptor complex by nuclear RNP of calf uteri have also been carried out[155].

5.11 SOME SPECULATIVE VIEWS ON THE RECEPTOR MECHANISMS INVOLVED IN GENE EXPRESSION IN TARGET CELLS

Many steroid hormones can transform target cells into a form with biochemically and morphologically distinctive characteristics. It is not known how these hormones trigger such effects, but it is obvious that their actions are closely tied to the regulation of gene expression, including the synthesis of proteins and RNA[14-18, 20, 23, 156-159].

One of the early effects of androgens in male accessory reproductive glands is their ability to enhance RNA synthesis[18, 23]. This can be shown by ^{32}P-pulse labelling of tissue RNA *in vivo*[160, 161], or by measuring the RNA-synthesising (polymerase) activity of nuclei isolated from castrated and testosterone-injected rats[18, 163, 165]. Effects can be observed within 1 h after the administration of androgens[163, 165]. Although androgens appear to stimulate the production of all types of RNA, including messenger-RNA[166, 167], biochemical[168-170] and autoradiographic studies[171] have shown that the effect is most dramatic on the synthesis of RNA in the nucleolar chromatin regions[169]. The RNA produced in these chromatin regions is rich in guanine and cytosine (and certain dinucleotide sequences), a characteristic of ribosomal RNA. At low concentrations, actinomycin D injected into rats or added to the isolated nuclear systems *in vitro*, can selectively inhibit the androgen-induced enhancement of nucleolar RNA synthesis, without significantly affecting the production of RNA in the non-nucleolar chromatin regions.

The enhancement of ribosomal RNA synthesis is known to be a general feature of a number of growth-promoting hormones[157]. Since DHT does not appear to be retained in significant amount at nucleolar sites[100], the action of DHT–receptor complex may be considered to be extranucleolar. Alternatively the DHT–receptor complex may interact with nucleolar components without being retained firmly by them. A more sophisticated assessment cannot be made because our understanding of the relationship between extranucleolar activity and nucleolar RNA synthesis is very limited[173-176].

After rats have been castrated, the RNA polymerase activity of rat prostate cell nuclei regresses rapidly (Figure 5.7)[18]. The rate of regression (50% in 10–15 h) is somewhat slower than, but comparable to, the rate of disappearance of androgens from the tissues (50% in 5–7 h). During this period, the

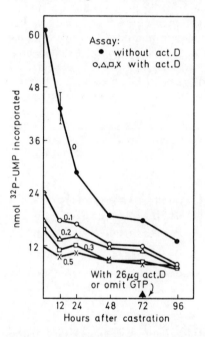

Figure 5.7 Selective regression of RNA polymerase activity at chromatin highly sensitive to actinomycin D. Prostate nuclei were isolated from normal rats and rats castrated for different time intervals and were assayed for the polymerase activity in the presence and absence of different concentrations of actinomycin D (the numbers are the amount of the antibiotic, in µg per 0.5 ml of reaction mixture used). (From Liao and Fang[18], by courtesy of Academic Press.)

prostate nuclei appear to contain excess polymerase protein that can form RNA on purified calf-thymus DNA[18]. It is unlikely, therefore, that androgens mediate the synthesis of the catalytic units of RNA polymerase. In 1967, Liao and Lin[169] proposed that RNA polymerase may be sequestered at certain regions of cells and that androgen action in the cell can result in the translocation of polymerase proteins. It was suggested that the restricted distribution of certain nucleotide sequences may serve an important function in directing polymerase to a specific area of the DNA template or in regulating the rate of RNA synthesis, and that the action of steroid hormones may involve a conformational fit of molecules required for the initiation of RNA synthesis. (Regulatory factors for RNA polymerase have been described recently in bacterial systems[177]). Following this line of thinking, it has been suggested that the steroid–protein complex may recognise certain nucleotide sequences in DNA and or RNA, and that such a process plays a key role in the selective synthesis as well as the utilisation or protection of RNA[18]. The experimental indication that the association of prostate DHT–receptor complex with reconstructed nuclear chromatin and isolated ribonucleoprotein particles may involve a selective recognition of certain species of polynucleotides (see Sections 5.9 and 5.10) may be pertinent to such a view.

The disappearance of androgens from rat ventral prostate after castration is accompanied by a loss of the receptor protein from the prostate nuclei (Section 5.8). Since the amounts of the cytoplasmic receptor proteins for androgens are not changed until a much later time[118], the nuclear loss of the receptor protein is probably due to the inability of the receptor protein (in the absence of DHT) to bind to the nuclear sites[118, 119]. Even in the presence of androgens, the nuclear retention of androgen–receptor complexes may be a dynamic process. Also, the release of the steroid–receptor complex from the target cell nuclei may be associated with a process related to androgen action. Since certain nuclear RNP particles can bind a DHT–receptor complex in a specific manner, it is fascinating to consider that some of these RNP particles may indeed be involved in such a functional process[150, 154, 155]. The function of the RNP particles that bind steroid–receptor complexes in prostate (and uterus) has yet to be identified. It is possible that the nuclear RNP particles may participate in the regulation of gene expression by being (a) a regulator of gene transcription[178-179]; (b) an RNP particle of the type identified in electron micrographs of mammalian cell nuclei[180], perhaps related to the biochemically-defined informosomes[181] or informofers[182]; or (c) a ribosome precursor particle[183, 184].

From the fragmentary information available, it is also possible to consider a model[150, 154] for the intracellular cycling of a steroid–receptor complex and its function (Figure 5.8). In such a model, a steroid hormone forms a complex with a receptor protein in the cytoplasm. After a conformational change, the complex enters the cell nucleus and becomes involved in the regulation of RNA synthesis. During such a process, the steroid–receptor complex and other proteins or nucleoproteins (including 'acceptor' factors) may bind to a specific RNA synthesised. The steroid-receptor-bound RNP may be processed

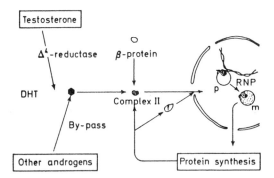

Figure 5.8 A model for androgen–receptor cycling. See text for explanation. (From Liao et al.[154], by courtesy of Plenum Press.)

to a mature form and enter the cytoplasm and the RNP particle may then participate in protein synthesis. The role of the steroid–receptor complex is seen to provide the structural specificity needed for the formation, processing and also function of the RNP.

In the above model, the receptor protein may lose its ability to bind to the RNP at different stages of processing and utilisation, especially if the steroid hormone of the cell is depleted. Both the receptor proteins and the 'acceptor' factors may re-associate with these RNP particles when the steroid hormone is replenished. Thus, the recycling process and its functions may be re-initiated by the steroid hormone at various points (in the nucleus or in the cytoplasm) of the receptor cycle. This would suggest that the importance of gene transcription (RNA synthesis) in relation to gene translation (protein synthesis) for the overall function of a steroid hormone in the target cells may be dependent on the amounts of RNP particles at different stages of processing and on their RNA and protein constituents in the target cells at the time when the hormone is supplied. If the target cells contain sufficient amounts of these RNA and protein constituents of the RNP, the earlier actions of the hormone may be simply dependent on the processing and the utilisation of the RNP and not dependent on the RNA synthesis. This may explain some androgen actions that have been reported to be actinomycin-D-insensitive[185-187].

Examples in other animal cells showing that nuclear activities are controlled by certain cytoplasmic proteins are not rare[188-191]. There are also indications that, in the uterus, oestrogen-sensitive nuclear-RNA synthesis requires a continuous supply of proteins, possibly synthesised in the cytoplasm[192]. This finding, taken together with the findings of other investigators[14, 16, 158, 193, 194], can be interpreted to indicate that oestrogen, in its early stage of action, enhances the synthesis of certain messenger-RNAs. The proteins synthesised from these messenger-RNAs may be utilised for other oestrogen effects, including a more general stimulation of RNA synthesis. Several investigators have claimed recently that, in cell-free systems containing uterine nuclei, 17β-oestradiol together with uterine cytosol protein fractions can enhance the RNA-synthesising activity of the uterine nuclei[195-198]. The detailed biochemical characterisation of the systems may open a new approach to our understanding of the role of steroid hormones in controlling the nuclear-RNA synthesis in the target tissues. Attempts to show this in an androgen–prostate system have not been successful, but it has been claimed that DHT (but not testosterone) in the absence of added cytoplasmic proteins, can stimulate the incorporation of radioactive nucleosides into RNA fractions[199, 200].

Since certain steroid hormones can selectively increase the amounts of certain proteins in target cells[156-159, 201, 202], it has been suggested that steroids may act as specific de-repressors by binding to gene repressors[159]. If the cytoplasmic steroid-binding proteins are indeed gene repressors, it is hard to understand the need for the steroid–repressor complexes to enter nuclei and be retained there. It is possible of course that steroid–receptor complexes rather than the free steroids are de-repressor molecules. Tomkins[156, 203, 204] proposed the existence of a labile post-transcriptional repressor that both inhibits messenger-RNA translation and enhances messenger degradation. The steroid–receptor complex is assumed to antagonise these actions of the repressor either by interacting directly with the repressor to antagonise its action, or by inhibiting its synthesis or speeding its degradation. Similarly, Ohno[205] suggested that the receptor protein for an androgen may act in the absence of the androgen, as a 'translational block' by binding to certain messenger-RNA. Androgen is considered to bind to the suppressive protein; as a result, the messenger-RNA is released and translated. Ohno believes that the same androgen–receptor complex enters the prostate nuclei and activates nucleolar RNA polymerase.

Although the concept that certain steroid hormones act by direct activation of limited numbers of gene is attractive, it is entirely possible that steroids may act by a more general mechanism that is not specific for particular genes. In the latter case, the sequential or specific gene expressions in target cells may follow their own intrinsic programme. This type of hormone action appears to occur in a number of cells influenced by some non-steroid hormones which use adenosine 3',5'-mononucleotide (cAMP) as a second messenger[206-208]. The administration of cAMP can produce testosterone-like enhancement of certain enzyme activities in orchdectomised and immature rats[209, 210]; however, there is no convincing evidence to support the suggestion that steroid actions are mediated by cAMP. Although certain polypeptide hormones may regulate the local nuclear concentrations of cAMP produced by prostate nuclei[211], testosterone or DHT do not stimulate the adenylcyclase activity[211, 212] nor the histone phosphokinase activities[213] of rat prostate preparations. Polyamines have also been mentioned as possible mediators of androgen-induced change in RNA and protein synthesis[214, 215]. However, a careful comparison of the changes in RNA polymerase activity[18, 163] and in the level of polyamines[216, 217] in rat prostate, in response to androgen manipulation, shows that this may not be so. Polyamines can probably support androgen actions on protein or RNA synthesis by stabilising nucleic acids, ribosomes, cellular organelles, or even nuclear structures which are critical for RNA polymerase activities[218, 219].

As discussed in other sections, the role of a steroid may be that of transforming a structurally-incompatible protein to a form structurally 'fit' to an 'acceptor' site. The steroid molecule may not necessarily participate directly in the function of the protein. It is reasonable to speculate that, in tissues not sensitive to steroid hormones, proteins similar to the steroid–receptor proteins may function in the same manner, but since they would be compatible with the functional sites they would not need prior interactions with steroid hormones.

5.12 ANTI-ANDROGENS AND DIHYDROTESTOSTERONE–BINDING PROTEINS

Many steroids or steroid-like compounds can suppress the effectiveness of androgens in animals. For example, oestrogens are well known for their anti-androgenic effect on the growth of prostate[220, 221] and chick comb[222]. Although some of these effects may be due to the alteration of gonadotropin function which results in a lower output of androgens by the testis, a more direct mechanism is not unlikely since oestrogen inhibition has been also observed in hypophysectomised animals[223]. Oestrogens appear not to be selectively retained by prostate nuclei and antagonise the binding of DHT to β-protein only poorly[18,96,115]. In the prostate as well as the uterus, the specific 17β-oestradiol-binding protein appears to be distinct from the DHT-binding protein[130]. Large doses of oestrogenic compounds ($>10^{-5}$ mol l^{-1}) are reported to inhibit NADPH-dependent enzymic reduction of testosterone to DHT by cell-free preparations of prostate[108].

Many progestogens and weak androgens are antagonists of the more potent androgens. In high concentrations, some of these compounds may be expected to bind to androgen receptors and thus to inhibit the receptor binding of the more potent androgens such as DHT. This may be one of the reasons for the reduced uptake of ^{3}H-testosterone by prostates and seminal vesicles of rats injected with non-radioactive progesterone[91]. Progesterone does not bind firmly to β-protein of rat ventral prostate[18,96,115], but prostate has another protein that binds progesterone firmly[96,224]. Recently, Voigt et al.[225] found that progesterone is a very potent inhibitor of the 5α reduction of testosterone by microsomal preparations of human skin. The inhibition is due to substrate competition during which progesterone is reduced to 5α-pregnane-3,20-dione. Progesterone thus can reduce the concentration and the retention of DHT in the target tissues.

Among the most active anti-androgens is cyproterone (6-chloro-17α-hydroxy-1,2α-methylene-4,6-pregnadiene-3,20-dione) and its 17α-acetate[226,227]. These anti-androgens can prevent the accumulation of radioactive androgens by the rat seminal vesicle and ventral prostate *in vivo*[18,114,228-230] or in tissue incubation experiments[18,114]. The reduced uptake of androgens by these target tissues is also accompanied by a decrease in the retention of DHT by their cell nuclei[114]. This is apparently due to the ability of cyproterone or cyproterone acetate to inhibit the binding of DHT to β-protein[114,115,126]. The possibility that these anti-androgens may also act on cell membranes to prevent the uptake of androgens cannot be excluded at this time. These compounds, however, are known to have no inhibitory action on the enzyme system responsible for the formation of DHT from testosterone[59,114,231].

5.13 DIHYDROTESTOSTERONE AND ITS BINDING PROTEINS IN ANDROGEN DYSFUNCTION

The levels of DHT in its target cells are unquestionably important for the androgen responses. To test the importance of the DHT formation, Wilson and Gloyna[232,233] studied the rate of conversion of testosterone to DHT in tissue slices from sex-accessory tissues of a number of species, and in human skin from a variety of anatomical sites. The results showed that DHT formation is not an essential feature of all androgen actions. Siiteri and Wilson[234] measured the androgen content in 15 normal and 10 hypertrophic human prostates by using a double-isotope derivative technique. They found that the DHT content was significantly greater in the hypertrophic tissues than in normal glands. These differences were apparently not due to the difference in the rate of DHT formation. The DHT content was also 2–3 times greater in the periurethral area, where prostatic hypertrophy usually commences, than in the outer regions of the gland. A similar study was made in dogs[235]. The concentration of DHT at hypertrophic areas may be due to DHT-binding proteins which may serve as 'storage receptors'[236]. In some cases of prostatic carcinoma[237, 238], conversion of testosterone to DHT was reported to be very small. Hansson et al.[239] have detected androphilic proteins which bind radioactive androgens originating from ^{3}H-testosterone in human prostate with benign hyperplasia. The formation of the androgen–protein complexes was reduced by anti-androgens such as cyproterone and 17α-methyl-β-nortestosterone (SK and F 7690).

A lowered rate of conversion of testosterone to DHT has been suggested[240, 241] as a possible reason for the androgen-insensitivity of testicular feminisation syndrome[242-244]. Maurais-Jarris et al.[240] have shown that patients with the androgen-insensitivity syndrome are about 10 times more sensitive to exogenous oestrogens than normal males in regard to the increase in the level of the plasma testosterone-binding proteins and the fall in urinary DHT and 5α-androstanediol. The higher plasma-protein binding of testosterone may reduce the supply of testosterone to peripheral tissues. The androgen insensitivity in these patients cannot be due simply to the reduced supply of DHT to tissues that are normally androgen-responsive, since DHT formation appears to be normal in some cases[245, 247, 264]. Moreover, the administration of DHT, like that of testosterone, does not fully restore the nitrogen retention or other typical effects of androgens[246, 247]. It has been suggested that the androgen insensitivity may be due to an alteration in the normal functions of the cellular androgen-binding proteins and/or of the nuclear 'acceptor' molecules[247].

A strain of male pseudo-hermaphroditic rats developed by Stanley and Gumbreck was characterised by a female phenotype and a lack of androgen-dependent differentiation. Testosterone or DHT injected into these rats produced very little androgen effect in preputial glands and other tissues[248, 249]. The preputial tissue can convert testosterone to DHT but does not retain DHT, suggesting that the insensitivity of this tissue may be due to the lack of the ability of preputial nuclei to concentrate DHT[250]. Diminished DHT uptake by liver nuclei and lower retention of radioactive androgens (80% testosterone) by kidney nuclei have been observed in these androgen-insensitive rats[137]. Since the in vivo uptake of radioactive androgens by these tissues is similar to that in normal rats, the defect in nuclear retention may be due to an abnormality in the androgen-binding proteins or nuclear-acceptor substances. In mice with a testicular-feminisation (Tfm) mutation[251], these androgenic steroids also failed to induce hypertrophy of kidney or alcohol dehydrogenase and β-glucuronidase characteristic of the normal animals[252]. Gehring et al.[136] showed that the amount of the ^{3}H-DHT-binding protein in the cytoplasm and nuclei of kidney is distinctly less in Tfm than in the normal mouse.

5.14 ANDROGEN ACTIONS NOT ATTRIBUTED TO DIHYDROTESTOSTERONE

The suggestion that DHT is an active form of androgen in target cells is supported by the finding that similar DHT-binding proteins are present in a variety of androgen-sensitive tissues (see Section 5.6). However, the specific

DHT-binding proteins have not been clearly identified in skeletal and levator ani muscles which are also major androgen target tissues. In these tissues[122, 147] testosterone itself or metabolites other than DHT may function by binding to another class of receptor proteins.

The possibility that different androgen metabolites may function differently was elegantly shown by Huggins et al. in 1954[40]. It was found that 3α-hydroxy-androstanes and androstenes as well as the corresponding 3-keto compounds cause preferential stimulation of the superficial cells of vaginal epithelium to produce mucus, whereas the deeper layers respond to compounds containing 3β-hydroxy groups. These differential physiological effects may be due either to a differential distribution of two (3α and 3β) hydroxysteroid dehydrogenases in different types of cells, or to direct action of these steroid isomers at different sites of cellular processes.

Baulieu, Lasnitzki and Robel[53] reported that, in organ cultures of rat prostate, testosterone was more effective than DHT in maintaining epithelial height and secretory activity, whereas DHT had a greater effect on the induction of epithelial hyperplasia. These investigators suggested that other metabolites of testosterone, such as androstenediols, may play certain roles differing from that of DHT in prostates. Mann et al.[253] found recently that, although DHT is capable of stimulating the growth of the bovine seminal vesicle to approximately the same extent as testosterone in vivo, the secretory output of fructose and citric acid of the DHT-stimulated seminal vesicle is distinctly less than that of the testosterone-stimulated organ. Furthermore, the possibility that both testosterone and DHT may serve as intranuclear effectors of androgen action in mouse kidney was suggested by Mowszowicz et al.[254].

Testicular androgens can also regulate gonadotropin secretion[2], influence oestrous behaviour in females[2] and induce hypothalamic virilisation[255]. Since 5α-reductase activity is present in mammalian hypothalamus[256, 257], these effects may depend on the reduction of testosterone to DHT in situ. However, testosterone administered to animals appears to be far more active than DHT in some of these effects, especially in maintaining androgen-dependent sexual behaviour in male rats[258, 259]. In these systems, there is a suggestion that certain products of the aromatisation of testosterone are partly responsible for the phenomena[260, 261].

One of the most distinct differences in the action of testosterone and DHT in intact animals is the inability of DHT to mimic the action of testosterone in inducing anovulatory (androgen) sterility during perinatal days[146, 262]. Testosterone is believed to inhibit the differentiation of neural cells which are later involved in gonadotropin control mechanisms. As a consequence, female rats treated perinatally with testosterone do not exhibit gonadotropin control of feminine ovulatory or behavioural patterns in adulthood. DHT does not show such an inhibitory effect. Since DHT appears to pass the blood-brain barrier[259], the findings suggest that androgens other than DHT act in these neural cells. Recently, Schultz and Wilson[263] also presented indications that male differentiation of the Wolffian ducts in the female embryo can be induced by testosterone when 5α-reductase activity is not present. This also suggests the possibility that testosterone itself or another foetal androgen besides DHT is the active hormone.

Acknowledgements

The author thanks Mrs. Diane K. Howell for her help during the preparation of the manuscript. Work from this laboratory has been supported by Grants AM 09461 and HD 07110 of U.S. National Institutes of Health.

References

1. Dorfman, R. I. and Shipley, R. A. (1956). *Androgens.* (New York: Wiley)
2. Young, W. C. (editor) (1961). *Sex and Internal Secretions*, Vol. 1 and 2 (Baltimore: Wilkins)
3. Vollmer, E. (editor) (1963). *Biology of the Prostrate and Related Tissues, U.S. National Cancer Institute Monograph, No. 12*
4. Mann, T. (1964). *The Biochemistry of Semen and the Male Reproductive Tract*, (London: Methuen)
5. Kruskemper, H. L. (1968). *Anabolic Steroids*, (New York: Academic Press)
6. Vida, J. A. (1969). *Androgens and Anabolic Agents*, (New York: Academic Press)
7. Griffiths, K. and Pierrepoint, C. G. (editors) (1970). *Some Aspects of the Aetiology and Biochemistry of Prostatic Cancer*, (Cardiff: Alpha Omega Alpha Publishing)
8. Eik-Nes, K. B. (editor) (1970). *The Androgens of the Testis*, (New York: Marcel Dekker)
9. Harris, G. W. (1964). *Endocrinology*, **75,** 627
10. Hechter, O. and Halkerston, I. D. K. (1964). *The Hormones*, Vol. 5, 697 (G. Pincus, K. V. Thimann and E. B. Astwood, editors) (New York: Academic Press)
11. Pincus, G., Nakao, T. and Tait, J. F. (editors) (1966). *Steroids Dynamics*, (New York: Academic Press)
12. Raspe, G. (editor) (1971). *Adv. Biosciences*, **7**
13. Pasqualini, J. R. and Scholler, R. (editors) (1972). *First International Symposium, J. Steroid Biochemistry*, **3,** No. 3, (Oxford: Pergamon Press)
14. Gorski, J., Shyamala, G. and Toft, D. (1969). *Curr. Topics Develop. Biol.*, **4,** 149
15. Williams-Ashman, H. G. and Reddi, A. H. (1971). *Ann. Rev. Physiol.*, **33,** 31
16. Jensen, E. V. and DeSombre, E. R. (1972). *Ann. Rev. Biochem.*, **41,** 203
17. O'Malley, B. W., Spelsberg, T. C., Schrader, W. T., Chytil, F. and Steggles, A. W. (1972). *Nature (London)*, **235,** 141
18. Liao, S. and Fang, S. (1969). *Vitam. Horm.*, **27,** 17
19. Huggins, D. (1947). *Harvey Lectures*, **42,** 148
20. Kochakian, C. D. (1965). *Mechanism of Hormone Action*, 192 (P. Karlson, editor) (Stuttgart: Thieme)
21. Talalay, P. (1965). *Ann. Rev. Biochem.*, **34,** 347
22. Dorfman, R. I. and Ungar, F. (1965). *Metabolism of Steroid Hormones*, (New York: Academic Press)
23. Williams-Ashman, H. G. (1965). *Cancer Res.*, **25,** 1096
24. Baulieu, E.-E., Corpechot, C., Dray, M. F., Emiliozzi, R., Lebeau, M. C., Mauvais-Jarvis, P. and Robel, P. (1965). *Rec. Prog. Horm. Res.*, **21,** 411
25. Lipsett, M. B., Wilson, H., Kirschner, M. A., Korenman, S. G., Fishman, L. M., Sarfaty, G. A. and Bardin, C. W. (1966). *Rec. Prog. Horm. Res.*, **22,** 245
26. Ofner, P. (1968). *Vitam. Horm.*, **26,** 271
27. Samuels, L. T. and Eik-Nes, K. B. (1968). *Metabolic Pathways*, Vol. II, 169 (D. M. Greenberg, editor) (New York: Academic Press)
28. Williams-Ashman, H. G. and Reddi, A. H. (1972). *Biochemical Actions of Hormones*, Vol. II, 257 (G. Litwack, editor) (New York: Academic Press)
29. Wurtman, R. J. (1968). *Science*, **159,** 1261
30. Jensen, E. V. (1968). *Science*, **159,** 1261
31. Butenandt, A. and Poschmann, L. (1944). *Chem. Ber.*, **77,** 394
32. Segaloff, A. and Gabbard, R. B. (1960). *Endocrinology*, **67,** 887
33. Shoppee, C. W. (1964). *Chemistry of the Steroids*, (London: Butterworth)
34. Fieser, L. F. and Fieser, M. (1959). *Steroids*, (New York: Reinhold)

35. Djerassi, C., Manson, A. J. and Bendas, H. (1957). *Tetrahedron*, **1**, 22
36. Turner, R. B., Helling, R., Meier, J. and Heusser, H. (1955). *Helv. Chim. Acta*, **38**, 411
37. Ringold, H. J. (1960). *J. Amer. Chem. Soc.*, **82**, 961
38. Herrmann, M. and Goslar, H. G. (1963). *Experientia*, **19**, 76
39. Goldberg, M. W. and Monnier, R. (1940). *Helv. Chim. Acta*, **23**, 840
40. Huggins, C., Jensen, E. V. and Cleveland, A. S. (1954). *J. Exptl. Med.*, **100**, 225
41. Kochakian, C. D. (1952). *Proc. Soc. Exptl. Biol. Med.*, **80**, 386
42. Huggins, C. and Jensen, E. V. (1954). *J. Exptl. Med.*, **100**, 241
43. Segaloff, A. (1963). *Steroids*, **1**, 299
44. Dorfman, R. I., Rooks, W. H., II., Jones, J. B. and Leman, J. D. (1966). *J. Med. Chem.*, **9**, 930
45. Wolff, M. E., Ho, W. and Kwok, R. (1964). *J. Med. Chem.*, **7**, 577
46. Huggins, C. Mainzer, K. (1957). *J. Exptl. Med.*, **105**, 485
47. Saunders, F. J. (1963). *U.S. Natl. Cancer Inst. Monograph.*, **12**, 139
48. Nutting, E. F., Klimstra, P. D. and Counsell, R. E. (1966). *Acta Endocrinol.*, **53**, 627 and 635
49. Bowers, A., Cross, A. D., Edwards, J. A., Carpio, H., Calzada, M. C. and Denot, E. (1963). *J. Med. Chem.*, **6**, 156
50. Irmscher, K., Kraft, H. G. and Bruckner, K. (1964). *J. Med. Chem.*, **7**, 345
51. Klimstra, P. D., Zigman, R. and Counsell, R. E. (1966). *J. Med. Chem.*, **9**, 924
52. Ringold, H. J. (1961). *Mechanism of Action of Steroid Hormones*, 200 (C. A. Villee and L. L. Engel, editors) (Oxford: Pergamon Press)
53. Baulieu, E. E., Lasnitzki, I. and Robel, P. (1968). *Nature (London)*, **219**, 1155
54. Ringold, H. J., Batres, E. and Rosenkranz, G. (1957). *J. Org. Chem.*, **22**, 99
55. Cross, A. D., Carpio, H. and Ringold, H. J. (1963). *J. Med. Chem.*, **6**, 198
56. Ito, T. and Horton, R. (1970). *J. Clin. Endocrinol. Metab.*, **31**, 362
57. Tremblay, R. R., Beitins, I. Z., Kowarski, A. and Migeon, C. J. (1970). *Steroids*, **16**, 29
58. Westphal, U. (1970). *Biochemical Actions of Hormones*, Vol. 1, 209 (Litwack, editor) (New York: Academic Press)
59. Liao, S., Fang, S., Tymoczko, J. L. and Liang, T. *Structure and Function of Male Sex Accessory Organs*, (D. Brandes, editor) (New York: Academic Press), in press
60. Westphal, U. (1961). *Mechanisms of Action of Steroid Hormones*, 33 (C. A. Villee and L. L. Engel, editors) (Oxford: Pergamon Press)
61. Westphal, U. and Ashley, B. D. (1962). *J. Biol. Chem.*, **237**, 2763
62. Chen, P. S., Mills, I. H. and Bartter, F. C. (1961). *J. Endocrinol.*, **23**, 129
63. Mercier-Bodard, C., Alfsen, A. and Baulieu, E. E. (1970). *Acta Endocrinol. Suppl.*, **147**, 204
64. Pearlman, W. H. and Crepy, O. (1967). *J. Biol. Chem.*, **242**, 182
65. Pearlman, W. H. (1970). *Acta Endocrinol. Suppl.*, **147**, 225
66. Rivarola, M. A., Snipes, C. A. and Migeon, C. J. (1968). *Endocrinology*, **82**, 115
67. Forest, M. G., Rivarola, M. A. and Migeon, C. J. (1968). *Steroids*, **12**, 323
68. De Moor, P., Heyns, W., Van Baelan, H. and Steeno, O. (1968). *Proc. 3rd Intern. Congr.*, *Endocrinol* 159 (Amsterdam: Excerpta Med. Found.)
69. Deakins, S. and Rosner, W. (1968). *Proc. 3rd Intern. Congr. Endocrinol.*, 159 (Amsterdam: Excerpta Med. Found.)
70. Bardin, C. W., Hembree, W. C. and Chvambach, A. (1968). *Proc. 3rd Intern. Congr. Endocrinol.*, 160 (Amsterdam: Excerpta Med. Found.)
71. Kato, T. and Horton, R. (1968). *J. Clin. Endocrinol. Metab.*, **28**, 1160
72. Lebeau, M. C. and Baulieu, E. E. (1970). *J. Clin. Endocrinol. Metab.*, **30**, 166
73. West, C. D., Tyler, F., Brown, H. and Samuels, L. T. (1951). *J. Clin. Endocrinol. Metab.*, **11**, 897
74. Sandberg, A. A. and Slaunwhite, W. R., Jr. (1958). *J. Clin. Endocrinol. Metab.*, **18**, 253
75. Roberts, S. and Szego, C. M. (1955). *Ann. Rev. Biochem.*, **24**, 543
76. Sandberg, A. A., Slaunwhite, W. R., Jr. and Antoniades, H. N. (1957). *Rec. Prog. Horm. Res.*, **13**, 209
77. Matsui, N. and Plager, J. E. (1966). *Endocrinology*, **78**, 1159
78. Hoffmann, N., Forbes, T. R. and Westphal, U. (1969). *Endocrinology*, **85**, 778
79. Bischoff, F. and Pilhorn, H. R. (1948). *J. Biol. Chem.*, **174**, 663
80. Eik-Nes, K., Schellman, J. A., Lumry, R. and Samuels, L. T. (1954). *J. Biol. Chem.*, **206**, 411

81. Jensen, E. V. and Jacobson, H. I. (1960). *Biological Activities of Steroids in Relation to Cancer*, 161 (G. Pincus and E. P. Vollmer, editors) (New York: Academic Press)
82. Jensen, E. V. and Jacobson, H. I. (1962). *Rec. Prog. Horm. Res.*, **18**, 387
83. Gupta, G. N. (1960). *Ph.D. Dissertation*, University of Chicago
84. Talwar, G. P., Seqal, S. J., Evans, A. and Davidson, O. W. (1964). *Proc. Nat. Acad. Sci. USA*, **52**, 1059
85. Toft, D. and Gorski, J. (1966). *Proc. Nat. Acad. Sci. USA*, **55**, 1574
86. Bruchovsky, N. (1971). *Endocrinology*, **89**, 1212
87. Greer, D. S. (1959). *Endocrinology*, **64**, 898
88. Resko, J. A., Goy, R. W. and Phoenix, C. H. (1967). *Endocrinology*, **80**, 490
89. Pearlman, W. H. and Pearlman, M. R. J. (1961). *J. Biol. Chem.*, **236**, 1321
90. Harding, B. W. and Samuels, L. T. (1962). *Endocrinology*, **70**, 109
91. Tveter, K. J. and Aakvaag, A. (1969). *Endocrinology*, **85**, 683
92. Tveter, K. J. and Attramadal, A. (1968). *Acta Endocrinol.*, **59**, 218
93. Tveter, K. J. and Unhjem, O. (1969). *Endocrinology*, **84**, 963
94. Tveter, K. J. (1970). *Acta Endocrinol.*, **63**, 489
95. Anderson, K. M. and Liao, S. (1968). *Nature (London)*, **219**, 227
96. Fang, S., Anderson, K. M. and Liao, S. (1969). *J. Biol. Chem.*, **244**, 6584
97. Bruchovsky, N. and Wilson, J. D. (1968). *J. Biol. Chem.*, **243**, 2012
98. Belham, J. E., Neal, G. E. and Williams, D. C. (1969). *Biochim. Biophys. Acta*, **187**, 159
99. Mainwaring, W. I. P. and Peterken, B. M. (1971). *Biochem. J.*, **125**, 285
100. Sar, M., Liao, S. and Stumpf, W. E. (1970). *Endocrinology*, **86**, 1008
101. Tveter, K. J. and Attramadal, A. (1969). *Endocrinology*, **85**, 350
102. Stumpf, W. E., Baerwaldt, C. and Sar, M. (1971). *Basic Actions of Sex Steroids in Target Organs*, (P. O. Hubinont and F. Leroy, editors) (Basel: S. Karger), in press
103. Wilson, J. D. and Loeb, P. M. (1965). *J. Clin. Invest.*, **44**, 1113
104. Sluyser, M. (1966). *J. Mol. Biol.*, **22**, 411
105. Mangan, F. R., Neal, G. E. and Williams, D. C. (1968). *Arch. Biochem. Biophys.*, **124**, 27
106. Steggles, A. W., Spelsberg, T. C., Glasser, S. R. and O'Malley, B. W., (1971). *Proc. Nat. Acad. Sci., USA*, **68**, 1479
107. Anderson, K. M., Lee, F. H. and Miyai, K. (1970). *Exptl. Cell Res.*, **61**, 371
108. Shimazaki, J., Kurihara, H., Ito, T. and Shida, K. (1965). *Gunma J. Med. Sci.* **14**, 326
109. Anderson, K. M. (1969). *Ph.D. Dissertation*, University of Chicago
110. Moore, R. J. and Wilson, J. D. (1972). *J. Biol. Chem.*, **247**, 958
111. Bruchovsky, N. and Wilson, J. D. (1968). *J. Biol. Chem.*, **243**, 5953
112. Mainwaring, W. I. P. (1969). *J. Endocrinol*, **44**, 323
113. Unhjem, O. (1970). *Acta Endocrinol.*, **63**, 69
114. Fang, S. and Liao, S. (1969). *Mol. Pharmacol.*, **5**, 428
115. Fang, S. and Liao, S. (1971). *J Biol. Chem.*, **246**, 16
116. Liao, S. and Fang, S. (1970). *Some Aspects of Aetiology and Biochemistry of Prostate Cancer*, 105. (K. Griffiths and C. G. Pierrepoint, editors) (Cardiff: Alpha Omega Alpha Publishing)
117. Liao, S., Liang, T. and Tymoczko, J. L. (1972). *J. Steroid Biochem.*, **3**, 401
118. Liao, S., Tymoczko, J. L., Liang, T., Anderson, K. M. and Fang, S. (1971). *Adv. Bioscience*, **7**, 155
119. Liao, S., Tymoczko, J. L. and Fang, S. (1971). *Hormonal Steroids*, 434 (V. H. T. James and L. Martini, editors) (Amsterdam: Excerpta Medica)
120. Unhjem, O., Tveter, K. J. and Aakvaag, A. (1969). *Acta Endocrinol.*, **62**, 153
121. Tveter, K. J., Unhjem, O., Attramadal, A., Aakvaag, A. and Hansson, V. (1971). *Adv. Bioscience*, **7**, 193
122. Aakvaag, A., Tveter, K. J., Unhjem, O. and Attramadal, A. (1972). *J. Steroid Biochem.* **3**, 375
123. Mainwaring, W. I. P. (1969). *J. Endocrinol.*, **45**, 531
124. Mainwaring, W. I. P. (1971). *Hormonal Steroids*, 368. (V. H. T. James and L. Martini, editors) (Amsterdam: Excerpta Medica)
125. Mainwaring, W. I. P. and Mangan, F. R. (1971). *Adv. Bioscience*, **7**, 165
126. Baulieu, E. E. and Jung, I. (1970). *Biochem. Biophys. Res. Commun.*, **38**, 599
127. Baulieu, E. E., Jung, I., Blondeau, J. P. and Robel, P. (1971). *Adv. Bioscience*, **7**, 179
128. Ritzen, E. M., Nayfeh, S. N., French, F. S. and Dobbins, M. C. (1971). *Endocrinology*, **89**, 143

129. Danzo, B. J., Orgebin-Crist, M. C. and Strott, C. A. (1972). *Abstracts for Fourth International Congress for Endocrinology* 80, (Amsterdam: Excerpta Medica)
130. McCann, S., Gorlich, L., Janssen, V. and Jungblut, P. W. (1970). *Excerpta Med. Inter. Congr. Ser.*, **210**, 150
131. Wilson, J. D. (1966). *Proc. 2nd Intern. Congr. Hormonal Steroids*, 45 (Amsterdam: Excerpta Medica)
132. Wilson, J. D. and Loeb, P. M. (1965). *Developmental and Metabolic Control Mechanisms and Neoplasia*, 375 (Baltimore: Williams and Wilkins)
133. Ritzen, E. M., Dobbins, M. C., French, F. S. and Nayfeh, S. N. (1972). *Abstracts for Fourth International Congress of Endocrinology*, 79 (Amsterdam: Excerpta Medica)
134. Vernon, R. G., Dorrington, J. H. and Fritz, I. B. (1972). *Abstracts for Fourth International Congress of Endocrinology*, 79 (Amsterdam: Excerpta Medica)
135. Mainwaring, W. I. P. (1972). *Abstracts for Fourth International Congress of Endocrinology*, 80 (Amsterdam: Excerpta Medica)
136. Gehring, U., Tomkins, G. M. and Ohno, S. (1971). *Nature (New Biol.)*, **232**, 106
137. Ritzen, E. M., Nayfeh, S. N., French, F. S. and Aronin, P. A. (1972). *Endocrinology*, **91**, 116
138. Mainwaring, W. I. P. Personal communication
139. Bruchovsky, N., (1972). *Biochem. J.*, **127**, 561
140. Rennie, P. and Bruchovsky, N. (1972). *Abstracts for Fourth International Congress of Endocrinology*, 79 (Amsterdam: Excerpta Medica)
141. Fazekas, A. G. and Sandor, T. (1972). *Abstracts Fourth International Congress of Endocrinology*, 80 (Amsterdam: Excerpta Medica)
142. Rennie, P. and Bruchovsky, N. (1972). *J. Biol. Chem.*, **247**, 1546
143. Kowarski, A., Shalf, J. and Migeon, C. J. (1969). *J. Biol. Chem.*, **244**, 5269
144. Mainwaring, W. I. P. (1970). *Biochem. Biophys. Res. Commun.*, **40**, 192
145. Unpublished observations
146. Whalen, R. E. and Luttge, W. G. (1971). *Endocrinology*, **89**, 1320
147. Segaloff, A. and Gabbard, R. B. (1962). *Endocrinology*, **71**, 949
148. Baulieu, E. E. and Jung, I. (1972). *Nature (New Biol.)*, **237**, 24
149. Tymoczko, J. L. and Liao, S. (1971). *Biochim. Biophys. Acta*, **252**, 607
150. Liao, S., Tymoczko, J. L., Howell, D. K., Lin, A. H., Shao, T. C. and Liang, T. (1972). *Fourth International Congress of Endocrinology*, (Amsterdam: Excerpta Medica), in press
151. King, R. J. B., Beard, V., Gordon, J., Pooley, A. S. Smith, J. A., Steggles, A. W. and Vertes, M. (1971). *Adv. Bioscience*, **7**, 21
152. Adachi, K. and Kano, M. (1972). *Steroids*, **19**, 567
153. Liao, S., Liang, T., Fang, S., Castaneda, E. and Shao, T-C. (1973). *J. Biol. Chem.*, in press
154. Liao, S., Liang, T., Shao, T. C. and Tymoczko, J. L. (1973). *Receptors for Reproductive Hormones*, 232 (B. W. O'Malley and A. R. Means, editors) (New York: Plenum Press)
155. Liao, S., Liang, T. and Tymoczko, J. L. (1972). *Nature (London)*, in press
156. Tomkins, G. M. and Gelehrter, T. D. (1972). *Biochemical Actions of Hormones*, Vol. 2, 1 (G. Litwack, editor) (New York: Academic Press)
157. Tata, J. R. (1970). *Biochemical Actions of Hormones*, Vol. 2. 89 (G. Litwack, editor) (New York: Academic Press)
158. Hamilton, T. H. (1968). *Science*, **161**, 649
159. Karlson, P. (1963). *Angew, Chem. Int. Ed. Engl.*, **2**, 175
160. Wicks, W. D., Greenman, D. L. and Kenney, F. T. (1965). *J. Biol. Chem.*, **240**, 4414
161. Greenman, D. L., Wicks, W. D. and Kenney, F. T. (1965). *J. Biol. Chem.*, **240**, 4420
162. Hancock, R. L., Zelis, R. F., Shaw, M. and Williams-Ashman, H. G. (1962). *Biochim. Biophys. Acta*, **55**, 257
163. Liao, S., Leininger, K. R., Sagher, D. and Barton, R. W. (1965). *Endocrinology*, **77**, 763
164. Jung, I. and Baulieu, E. E. (1971). *Biochemie*, **53**, 807
165. Mainwaring, W. I. P., Mangan, F. R. and Peterken, B. M. (1971). *Biochem. J.*, **123**, 619
166. Liao, S. and Williams-Ashman, H. G. (1962). *Proc. Natl. Acad. Sci. USA*, **48**, 1956
167. Liao, S. (1965). *J. Biol. Chem.*, **240**, 1236
168. Liao, S., Barton, R. W., and Lin, A. H. (1966). *Proc. Natl. Acad. Sci. USA*, **55**, 1593
169. Liao, S. and Lin, A. H. (1967). *Proc. Natl. Acad. Sci. USA*, **57**, 379
170. Liao, S., Lin, A. H. and Barton, R. W. (1966). *J. Biol. Chem*, **241**, 3869

171. Liao, S. and Stumpf, W. E. (1968). *Endocrinology*, **83**, 629
172. Liao, S. (1968). *Amer. Zoologist*, **8**, 233
173. Muramatsu, M., Shimada, N. and Higashinakagawa, T. (1970). *J. Mol. Biol.*, **53**, 91
174. Petrov, P. and Sekeris, C. E. (1971). *Expt. Cell Res.*, **69**, 393
175. Ringborg, U., Daneholt, B., Edstrom, J. E., Egyhazi, E. and Rydlander, L. (1970). *J. Mol. Biol.*, **51**, 679
176. Tamaoki, T. and Mueller, G. C. (1963). *Biochem. Biophys. Res. Commun.* **11**, 404
177. Burgess, R. R. (1971). *Ann. Rev. Biochem.* **40**, 711
178. Paul, J. (1970). *Curr. Topics Develop. Biol.*, **5**, 317
179. Britten, R. J. and Davidson, E. H. (1969). *Science*, **165**, 349
180. Stevens, B. J. and Swift, H. (1966). *J. Cell. Biol.*, **31**, 55
181. Spirin, A. S. (1969). *Eur. J. Biochem.*, **10**, 20
182. Samarina, O. P., Lukanidin, E. M., Molnar, J. and Georgiev, G. P. (1968). *J Mol. Biol.*, 33
183. Burdon, R. H. (1971). *Prog. Nucleic Acid Res. Mol. Biol.*, **11**, 33
184. Kumar, A. and Warner, J. R. (1972). *J. Mol. Biol.*, **63**, 233
185. Talwar, G. P., Modi, S. and Rao, K. N. (1965). *Science*, **150**, 1315
186. Frieden, E. H., Harper, A. A., Chin, F. and Fishman, W. H. (1964). *Steroids*, **4**, 777
187. Frieden, E. H. and Fishel, S. S. (1968). *Biochem. Biophys. Res. Commun.*, **31**, 515
188. Gurdon, J. B. and Woodland, H. R. (1970). *Curr. Topics Develop. Biol.*, **5**, 39
189. Thompson, L. R. and McCarthy, B. J. (1968). *Biochem. Biophys. Res. Commun.*, **30**, 166
190. Laskey, R. A. and Gurdon, J. B. (1970). *Nature (London)*, **228**, 1332
191. Harris, H. (1967). *J. Cell. Sci.*, **2**, 23
192. Nicolette, J. A., Lemahieu, M. A. and Mueller, G. C. (1968). *Biochim. Biophys. Acta*, **166**, 403
193. Katzenellenbogen, B. S. and Gorski, J. (1972). *J. Biol. Chem.*, **247**, 1299
194. Baulieu, E. E. (1972). *Fourth International Congress of Endocrinology* (Amsterdam: Excerpta Medica), in press
195. Raynaud-Jammet, C. and Baulieu, E. E. (1969). *Compt. Rend.*, **268D**, 3211
196. Arnaud, M., Beziat, Y., Guilleux, J. C., Hough, A., Hough, D. and Mousseron-Canet M. (1971). *Biochim. Biophys. Acta*, **232**, 117
197. Arnaud, M., Beziat, Y., Guilleux, J. C. and Mousseron-Canet, M. (1971). *Compt. Rend.*, **272D**, 635
198. Mohla, S., DeSombre, E. R. and Jensen, E. V. (1972). *Biochem. Biophys. Res. Commun.*, **46**, 661
199. Bashirelahi, N., Chader, G. J. and Villee, C. A. (1969). *Biochem. Biophys. Res. Commun.*, **37**, 976
200. Bashirelahi, N. and Villee, C.A. (1970). *Biochim. Biophys. Acta*, **202**, 192
201. Shaw, C. R. and Koen, A. L. (1963). *Science*, **140**, 70
202. Means, A. R., Comstock, J. P., Rosenfeld, G. C. and O'Malley, B. W. (1972). *Proc. Nat. Acad. Sci. USA*, **69**, 1146
203. Tomkins, G. M., Martin, D. W. Jr., Stellwagen, R. H., Baxter, J. D. Mamont, P. and Levinson, B. B. (1970). *Cold Spring Harbor Symposia on Quantitative Biology*, **35**, 635
204. Tomkins, G. M. and Martin, D. W., Jr. (1970). *Ann. Rev. Genet.*, **4**, 91
205. Ohno, S. (1971). *Nature (London)*, **234**, 134
206. Robinson, G. A., Butcher, R. and Sutherland, E. W. (971). *Cyclic AMP* (New York: Academic Press)
207. Robinson, G. A., Butcher, R. W. and Sutherland, E. W. (1968). *Ann. Rev. Biochem.*, **37**, 149
208. Butcher, R. W., Robinson, G. A. and Sutherland, E. W. (1972). *Biochemical Actions of Hormones*, Vol. 2, 21 (G. Litwack, editor) (New York: Academic Press)
209. Singhal, R. L., Vijayvarigiya, R. and Ling, G. M. (1970). *Science*, **168**, 261
210. Singhal, R. L., Parulekar, M. R. and Vijayvargiya, R. (1971). *Biochem. J.*, **125**, 329
211. Liao, S., Lin, A. H. and Tymoczko, J. L. (1971). *Biochim. Biophys. Acta*, **230**, 535
212. Rosenfeld, M. G. and O'Malley, B. W. (1970). *Science*, **168**, 253
213. Reddi, A. H., Ewing, L. L. and Williams-Ashman, H. G. (1971). *Biochem. J.*, **122**, 333
214. Caldarera, C. M., Moruzzi, M. S., Barbiroli, B. and Moruzzi, G. (1968). *Biochem. Biophys. Res. Commun.*, **33**, 266

215. Moultan, B. C. and Leonard, S. L. (1969). *Endocrinology*, **84**, 1461
216. Pegg, A. E., Lockwood, D. H. and Williams-Ashman, H. G. (1970). *Biochem. J.*, **117**, 17
217. Williams-Ashman, H. G., Pegg, A. E. and Lockwood, D. H. (1969). *Adv. Enzyme Regul.*, **7**, 291
218. Tabor, H. and Tabor, C. W. (1964). *Pharmacol. Rev.*, **16**, 245
219. Russell, D. H., Levy, C. C. and Taylor, R. L. (1972). *Biochem. Biophys. Res. Commun.*, **47**, 212
220. Huggins, C. and Clark, P. J. (1940). *J. Exp. Med.*, **72**, 747
221. Huggins, C. and Russell, P. S. (1946). *Endocrinology*, **39**, 1
222. Hoskins, W. H. and Koch, F. C. (1939). *Endocrinology*, **25**, 266
223. Goodwin, D. A., Rasmussen-Taxdal, D. S., Ferreira, A. A. and Scott, W. W. (1961). *J. Urology*, **86**, 134
224. Karsznia, R., Wyss, R. H., Heinrichs, W. M. L. and Herrmann, W. L. (1969). *Endocrinology*, **84**, 1238
225. Voigt, W., Fernandez, E. P. and Hsia, S. L. (1970). *J. Biol. Chem.*, **245**, 5594
226. Bridge, R. W. and Scott, W. W. (1964). *Invest., Urology*, **2**, 99
227. Neumann, F. and Von Berswordt-Wallrabe, R. (1966). *J. Endocrinol.*, **35**, 363
228. Geller, J., Damme, O. V., Garabieta, G., Loh, A., Rettura, J. and Seifter, E. (1969). *Endocrinology*, **84**, 1330
229. Stern, J. M. and Eisenfeld, A. J. (1969). *Science*, **166**, 233
230. Whalen, R. E., Luttge, W. G. and Green, R. (1969). *Endocrinology*, **84**, 217
231. Belham, J. E. and Neal, G. E. (1971). *Biochem. J.*, **125**, 81
232. Wilson, J. D. and Gloyna, R. E. (1970). *Rec. Prog. Horm. Res.*, **26**, 309
233. Gloyna, R. E. and Wilson, J. D. (1969). *J. Clin. Endocrinol., Metab.*,
234. Siiteri, P. K. and Wilson, J. D. (1970). *J. Clin. Invest.*, **49**, 1737
235. Gloyna, R. E., Siiteri, P. K. and Wilson, J. D. (1970). *J. Clin. Invest.*, **49**, 1746
236. Haltmeyer, G. C. and Eik-Nes, K. B. (1972). *Acta. Endocrinol.*, **69**, 394
237. Giorgi, E. P., Stewart, J. C., Grant, J. K. and Scott, R. (1971). *Biochem. J.*, **123**, 41
238. Giorgi, E. P., Stewart, J. C., Grant, J. K. and Shirley, I. M. (1972). *Biochem. J.*, **126**, 107
239. Hansson, V., Tveter, K. J., Attramadal, A. and Torgersen, O. (1971). *Acta Endocrinol* **68**, 79
240. Mauvais-Jarvis, P., Bercovici, J. P., Crepy, O. and Gauthier, F. (1970). *J. Clin. Invest.*, **49**, 31
241. Northcutt, R. C., Island, D. P. and Liddle, G. W. (1969). *J. Clin. Endocrinol. Metab.*, **29**, 422
242. Money, J., Ehrhardt, A. A. and Masica, D. N. (1968). *Johns Hopkins Med. J.*, **123**, 105
243. Simmer, H. H., Pion, R. J. and Dignam, W. J. (1965). *Testicular Feminization: Endocrine Function of Feminising Testes: Comparison with Normal Testes*, (Springfield, Illinois: Thomas)
244. Federman, D. D. (1967). *Abnormal Sexual Development*, (Philadelphia: W. B. Saunders)
245. Wilson, J. D. and Walker, J. D. (1969). *J. Clin. Invest.*, **48**, 371
246. Strickland, A. L. and French, F. S. (1969). *J. Clin. Endocrinol. Metab.*, **29**, 1284
247. Rosenfield, R. L., Lawrence, A. M., Liao, S. and Landau, R. L. (1969). *J. Lab. Clin. Med.* **74**, 1003
248. Bardin, C. W., Allison, J. E., Stanley, A. J. and Gumbreck, J. G. (1969). *Endocrinology*, **84**, 435
249. Bardin, C. W., Bullock, L., Schneider, G., Allison, J. E. and Stanley, A. J. (1970). *Science*, **167**, 1136
250. Bullock, L. and Bardin, C. W. (1970). *J. Clin. Endocrinol. Metab.*, **31**, 113
251. Lyon, M. F. and Hawkes, S. G. (1970). *Nature (London)*, **227**, 1217
252. Dofuku, R., Tettenborn, U. and Ohno, S. (1971). *Nature (New Biol.)*, **232**, 5
253. Mann, T., Rowson, L. E. A., Baronos, S. and Karagiannidis, A. (1971). *J. Endocrinol.*, **51**, 707
254. Mowszowicz, I., Bullock, L. and Bardin, C. W. (1972). *Abstracts for Fourth International Congress of Endocrinology*, 19 (Amsterdam: Excerpta Medica)
255. Barraclough, C. A. (1967). *Neuroendocrinology*, Vol. 2, 61 (L. Martini and Ganong, W. F., editors) (New York: Academic Press)
256. Jaffe, R. B. (1969). *Steroids*, **14**, 483

257. Perez-Palacios, G., Castaneda, E., Gomez-Perez, F., Perez, A. E. and Gual, C. (1970). *Biol. Reprod.*, **3**, 205
258. McDonald, P., Beyer, C., Newton, F., Brien, B., Baker, R., Tan, H. S., Sampson, C., Kitching, P., Greenhill, R. and Pritchard, D. (1970). *Nature (London)*, **227**, 964
259. Whalen, R. E. and Luttge, W. G. (1971). *Hormones and Behaviour*, **2**, 117
260. Beyer, C., Vidal, N. and Mijares, A. (1970). *Endocrinology*, **87**, 1386
261. Beyer, C., Morali, G. and Cruz, M. L. (1971). *Endocrinology*, **89**, 1158
262. Luttge, W. G. and Whalen, R. E. (1970). *Hormones and Behaviour*, **1**, 265
263. Schultz, F. M. and Wilson, J. D. (1972). *Abstracts for Fourth International Congress of Endocrinology*, 77 (Amsterdam: Excerpta Medica)
264. Perez-Placios, G., Morato, T., Perez, A. E., Castaneda, E. and Gual, C. (1971). *Steroids*, **17**, 471

6
The Mode of Action of the Female Sex Steroids

B. W. O'MALLEY and A. R. MEANS
Baylor College of Medicine, Houston

6.1	PROGESTERONE	187
	6.1.1 *Physiologic responses to progesterone*	187
	6.1.2 *Tissue uptake and metabolism of progesterone*	189
	6.1.3 *Binding of progesterone to target cell receptors*	191
	6.1.4 *Biochemical sequence of events in the action of progesterone*	192
6.2	OESTROGEN	196
	6.2.1 *Introduction*	196
	6.2.2 *Regulation of uterine transcription*	197
	6.2.3 *Regulation of uterine translation*	199
	6.2.4 In vitro *effects of oestrogen on the uterus*	201
	6.2.5 *Oestrogenic control of transcription in oviduct*	202
	6.2.6 *Oestrogenic control of translation in oviduct*	204
	6.2.7 *Oestrogenic control of specific oviduct mRNA*	205

6.1 PROGESTERONE

6.1.1 Physiologic responses to progesterone

Progesterone has been established as the most important steroid mediator of pregnancy[1-3]. The major physiological effect of progesterone appears to be the induced transformation of uterine endometrial cells so that implantation of the developing blastocyst is permitted. Other likely effects of this steroid hormone are listed as follows: (a) myometric activity may be suppressed, aiding in retention of the embryo during implantation and growth prior to normal parturition; (b) this steroid is thought to support mammary development; (c) progesterone acts as a direct antagonist to oestrogen stimulation of numerous metabolic activities; (d) finally, a number of metabolic parameters are altered which appear to have no major physiological effect on maintenance or termination of pregnancy or reproductive processes in general. Progesterone has been thought to be needed for alveolar development in the breast tissue of some species; this chapter, however, will be limited to a discussion of progesterone action in uterus and oviduct.

The effects of progesterone on the myometrial cell have been studied by a number of investigators[4]. Uterine muscle functions normally only if these cells possess, as a result of oestrogen stimulation, sufficient amounts of contractile proteins. In addition, the excitable membrane of the uterine muscle cell must be capable of undergoing periodic changes between rest and activity. The spontaneous rhythmic activity of the individual cell membrane must be transferred to the contractile system of the myoplasm through an effective excitation. In the rabbit, progesterone exerts a 'blocking effect' so that an excitation wave cannot spread from one region to another. This steroid has been reported to increase the membrane potential to an extent where spontaneous activity is suppressed[4], and a gradual reduction in spike discharge occurs. This 'quietening' effect on uterine myometrial contractions could conceivably play a physiological role to aid implantation and retention of the embryo. Nevertheless, this phenomenon is not a universal one for, in the guinea-pig, progesterone has been shown not to be a myometrial-blocking agent[5]. A similar role in the primates has not yet been adequately investigated. Therefore, a general concept for the role of progesterone in the regulation of gestational length and determination of the onset of labour is at present unconfirmed.

Unquestionably, the most important function which appears in animals in response to progesterone is the ability of the uterine endometrial cells to accept the developing blastocyst. Under the influence of oestrogen, the endometrium proliferates and becomes dense. Progesterone inhibits further endometrial proliferation and the epithelium becomes secretory. The endometrium now has the capacity to accept, retain and nourish the blastocyst. This 'sensitivity' for implantation must certainly reflect a series of, as yet unproved, earlier intracellular biochemical events.

The deciduoma reaction has been used as a uterine model for implantation of the blastocyst into the endometrium[6]. Physical or chemical irritation of the endometrial cells of a progesterone-treated 'receptive' uterus leads to a proliferative reaction which resembles the uterine reaction to the normal blastocyst as evidenced by both light and electron microscopy[6]. The uterus of a pregnant (or pseudopregnant) animal is 'receptive' to implantation (or decidualisation) on the fourth day but not the third or fifth day of pregnancy. The decidual reaction has been found to be determined by progesterone alone, since the castrate immature rat will respond with resultant decidual growth when treated only with progesterone. Progesterone also exerts a profound stimulatory influence on decidual RNA and protein synthesis in untreated or oestrogen-treated castrates. Furthermore, the response is specific for progestational agents (natural or synthetic) but not androgens, glucocorticoids, or mineralocorticoids. Actinomycin D blocks the decidual reaction, suggesting that the response is dependent on new RNA synthesis*.

The paucity of suitable tissue models to investigate a 'positive' progesterone effect has thwarted our progress in understanding the mechanism of action of

* Glasser, S. R., personal communication.

this important steroid hormone. Unfortunately, we cannot as yet specifically define the newly synthesised proteins responsible for the 'uterine sensitivity' which leads to the deciduoma reaction.

The chick oviduct has proved to be a useful model system in which to study the biochemistry of progesterone action in target tissues. Hormonal changes that occur in this tissue have been characterised in terms of both the long-term developmental and the early biochemical effects caused by oestrogens and progestins[7,8]. Oestrogenic substances are known to stimulate oviduct growth, and after oestrogen treatment three new cell-types emerge from the primitive mucosa[9]. Two of these, the goblet cells and the tubular gland cells, will subsequently synthesise tissue-specific proteins. After a single exposure to progesterone, the goblet cells of the oestrogen-stimulated oviduct synthesise the protein, avidin[8,10].

Avidin then serves as a specific intracellular biochemical marker for the action of progesterone in reproductive target tissue of the chicken. The protein can easily be quantified and serves as a definite end-point for progesterone action. Events prior to avidin induction can then be related in a time-sequence manner. The studies performed in the chick oviduct will be summarised later in this chapter.

6.1.2 Tissue uptake and metabolism of progesterone

Apart from the uterus and breast, it is not clear exactly which tissues are physiologic targets for progesterone. For instance, it has been demonstrated that progesterone can alter hepatic function, e.g. increase conversion of amino acids into urea resulting in reduced concentration of plasma amino acids[11]. Progesterone can influence the renal permeability to sodium leading to a natriuresis[11]. The physiological importance of these observations to the total body economy is not understood, and may simply represent pharmacological effects of progesterone.

A commonly used approach to determine 'target' tissues for hormones involves the *in vivo* injection of tracer amounts of radiolabelled hormone followed by monitoring of the distribution of radioactivity. Tissues demonstrating a capacity to concentrate the hormone relative to blood can then be considered as potential target tissues. Such studies have demonstrated the 'target tissue' nature of the uterus in a number of species[12-15]. It has also been shown that priming animals with oestrogen results in greater amounts of progesterone uptake by the uterus and longer retention times[15]. This latter finding is interesting in light of evidence which reveals a specific induction of progesterone–receptor protein in target tissue by oestrogen[14,16-18]. Similarly, progesterone uptake and retention by vaginal tissue was increased by pretreatment with oestradiol[19].

The potential importance of progesterone uptake by the hypothalamus and pituitary[12] lies in the fact that progesterone can inhibit gonadotropin secretion[20], and could presumably be part of a feed-back mechanism in the control of LH release during the reproductive cycle[21]. Examination of the possibility has generated conflicting experimental results. It was thus concluded that in the rat specific progesterone-binding receptors do not exist in the hypothalamus and pituitary gland[22], but in subsequent studies, in which rats were also used, the results suggested the existence of limited capacity binding sites for progesterone in the mesencephalone, diencephalone and in the pituitary[23].

During tissue retention, steroid hormones are subject to various enzyme systems which transform them into a variety of metabolites. Once progesterone enters a 'target' cell it may effect a particular response by direct action or through the formation of a specific metabolite(s). It is not as yet clear whether progesterone metabolism is related solely to a local disposal mechanism or if the formation of a metabolite might also act as a mediator of a specific biochemical or biological effect. It is even conceivable that different metabolites perform independent functions such as has been demonstrated for testosterone metabolites in androgen-dependent tissue[24]. There are at least 26 different metabolites of progesterone that can be formed without introducing additional oxygen functions. The problem is thus a complex one, and it is not clear as to why a cell should form so many different metabolites. If, however, the importance in the formation of any particular metabolite is to be understood, it will be necessary to examine each metabolite as it is identified for some biological or biochemical activity. Such an approach has been initiated using several animal model systems. Some preliminary results are now available for the chick oviduct[10] and the rat deciduoma reaction[6].

Because it has been previously shown that in the chick oviduct there is a specific cytoplasmic receptor for progesterone which appears to function in transporting progesterone into the nucleus where it acts[16], the metabolism of progesterone associated with the cytoplasmic and nuclear receptors was examined.

Several metabolites, in addition to unreacted progesterone, have been positively identified by recrystallisation to constant specific activity. These compounds include: 5α-pregnane-3,20-dione; 5β-pregnane-3,20-dione; 3α-hydroxy-5β-pregnan-20-one; 3β-hydroxy-5α-pregnan-20-one; 3β-hydroxy-5β-pregnan-20-one; 20β-hydroxy-4-pregnen-3-one; 20β-hydroxy-5α-pregnan-3-one. No radioactivity was recovered when 20α-hydroxy-4-pregnen-3-one was recrystallised, suggesting that chick oviduct tissue may be lacking in 20α-hydroxysteroid dehydrogenase activity*.

In terms of radioactive incorporation, 5α-pregnane-3,20-dione and progesterone were the major metabolites recovered from nuclear chromatin, a finding similar to that of a previous report[25]. After 30 min incubation, 5α-pregnane-3,20-dione and progesterone were present in essentially equal amounts for as long as 4 h. This finding prompted the quantification of avidin synthesis after giving 5α-pregnane-3,20-dione *in vivo* as carried out with progesterone[10]. It was found that this metabolite could induce avidin synthesis with a potency equal to that of progesterone. This observation is analogous to the conversion of testosterone to 5α-dihydrotestosterone in the rat prostate[26] and the finding that 5α-dihydrotestosterone has important biological activity in that system[27].

The question thus arises: In the chick oviduct, does progesterone induce avidin synthesis directly or through the formation of 5α-pregnane-3,20-dione, or are both compounds active? In addition, it is possible that further metabolism of 5α-pregnane-3,20-dione is important. Of course, avidin synthesis

* Strott, C. A., unpublished observations.

is a measurable biological response, and a metabolite of progesterone may effect some biochemical change that may or may not lead to measurable biological activity. It is also not known how the cell ultimately disposes of the progesterone molecule which it originally extracted from the extracellular fluid.

The pattern of metabolites associated with the cytosol receptor protein was similar to that found in the nucleus. At the end of a 4 h incubation period, however, the greatest amount of unbound radioactivity was found as an unknown polar steroid metabolite*. It is conceivable that this polar material is the ultimate disposal product of progesterone. It is of interest that a similar observation in an *in vivo* study has been reported for another animal species[2].

Several metabolites have been identified following incubation of rat deciduoma tissue with ^{3}H-progesterone, but in contrast to the chick oviduct very little 5α-pregnane-3,20-dione was formed. Instead, the major metabolite in this animal model system appears to be 3α-hydroxy-5α-pregnan-20-one. This was also the case when highly purified nuclear preparations were extracted. At all stages examined, progesterone was at least 3–4 times more abundant than any metabolite. The control horn and the horn containing the deciduoma appeared to metabolise progesterone similarly.

Thus, an interesting species difference exists with respect to 5α-pregnane-3,20-dione. It was the major *in vitro* metabolite in the chick oviduct and had biological activity in that system, at least equivalent to progesterone. In contrast, it was a relatively minor *in vitro* metabolite in the rat uterus, and, unlike progesterone, was unable to support the deciduoma reaction or maintain pregnancy in the ovariectomised rat†.

Similar experimental evidence is required which can be used to correlate tissue metabolism of progesterone with the biological activity of the individual metabolites in many different species. In this way, the complex metabolic pattern of this steroid may become more meaningful in relation to progesterone physiology.

6.1.3 Binding of progesterone to target cell receptors

A number of investigators have reported binding of ^{3}H-progesterone to target cells. Both mouse vagina[28] and guinea-pig uterus[15] have been reported to bind ^{3}H-progesterone following *in vivo* injections of the hormone. The guinea-pig progesterone receptor has been subsequently isolated and shown to be a heat-labile protein with a K_a value of 2×10^{-9} mol l^{-1} for the steroid and a sedimentation value of 6.75 on sucrose gradient analysis[29]. The presence of specific progesterone-binding proteins has also been demonstrated in both cytosol and nuclear fractions of rabbit uteri[30]. In this tissue, the progesterone-binding protein is distinct from transcortin (CBG) and the concentration of progesterone-binding sites in castrate rabbit uterus can be increased by prior oestrogen treatment[30]. Finally, specific progesterone-binding sites have been shown to be present in human endometrium[30]‡.

---

* Strott, C. A., unpublished observations.
† Glasser, S. R., personal communication.
‡ Eaton, W. L., unpublished data.

Preferential binding of progestins has also been reported for rat uterine tissue[31]. Earlier, published studies failed to show the progesterone-binding capacity to be distinct from rat transcortin[33]. However, a more recent report has confirmed the existence of specific progesterone-binding proteins in both rat and mouse uterus[32].

In the different species studied to date, i.e. rat[31], guinea-pig[29], rabbit[30], chicken[16, 17] and human[30], these uterine and oviduct progesterone receptors have consistently demonstrated specificity for progesterone when compared to other classes of biologically-active steroids. Thus, cortisol, oestradiol-17β and testosterone have been shown not to compete effectively for the receptor. In addition, various kinds of steroids (precursors, metabolites and synthetic forms) have been tested. So far only those synthetic compounds known to have progestational biological activity have been shown to bind to the receptor. Few of the naturally occurring metabolites of progesterone have been tested, but one in particular has proved to be of some interest, i.e. 5α-pregnane-3,20-dione. This is the only metabolite to date that has demonstrated significant binding to the receptor, and in this respect there is remarkable species variability. In the chicken, it competes equally well with progesterone for the receptor, in the guinea-pig about half as well. In the rat, rabbit and human, however, it competes very little. In the chicken, 5α-pregnane-3,20-dione is known to have substantial biological activity whereas in the rat it does not (the rabbit and human have not been tested). Also, in the chicken, the 5β-pregnane-3,20-dione isomer has been shown to bind very little to the receptor, and likewise has been found to have little biological activity. Only a limited amount of work has been done to date on the relation of progesterone metabolites in target tissue to the progesterone receptor and biological activity. Studies of species variability in this respect will be important in order to understand the nature of progesterone metabolism in target tissue and the relationship of this process to its mechanism of action.

6.1.4 Biochemical sequence of events in the action of progesterone

Since we have not yet elucidated the complete sequence of steps in steroid hormone action, we have often tended to place emphasis on the earliest events occurring in target cells following exposure to steroids. One of the earliest detectable intracellular events in steroid hormone action is thought to be the binding of the steroid hormone to a specific macromolecular protein component of the target tissue cytoplasm. Since only target cells have this binding protein, the distribution of these 'receptor' molecules* might determine tissue specificity. Present evidence suggests that it is the hormone–receptor complex rather than the free hormone which exerts a regulatory influence on RNA and protein metabolism in target tissues.

The chick oviduct has been the subject of the bulk of the studies on molecular mechanisms of progesterone action. Following an *in vivo* injection

---

* At present, the steroid-binding protein cannot unequivocally be defined as a receptor in the strictest pharmacological sense of the word since the subsequent functional significance of this steroid–protein complex has not been proved.

of ³H-progesterone to an oestrogen-pretreated chick, the tritiated progestin is found distributed in both the cytoplasm and nucleus of the oviduct. The cytoplasmic radioactivity appears to exist bound to a macromolecular complex that does not dissociate on passage of the cytosol (105 000 g supernatant extract) through a Sephadex G-200 column[8]. Both cytosol and nuclear extracts can be centrifuged through 5–20% sucrose gradients and shown to contain steroid-binding proteins. A small amount of ³H-progesterone is bound in the cytoplasm as early as 1 min after injection, and maximum binding occurs after 25 min; the percentage of total steroid partitioned into the nucleus progressively increases during this same period. The cytosol and nuclear progesterone-binding macromolecules have sedimentation coefficients of 4S during sucrose gradient centrifugation in the presence of 0.3 mol l⁻¹ KCl but the cytosol-binding molecule aggregates to 5S and 8S forms under low salt (no KCl) conditions[16, 34]. The oviduct-binding component shows a high affinity for progesterone ($Kd \times 10^{-10}$ at 4 °C).

Because of the tight affinity displayed by transcortin for progesterone, tissue contamination by the plasma protein must be ruled out, especially since it also sediments at about 4S in high-salt sucrose gradients. However, the cytosol progesterone-binding macromolecule has been unequivocally distinguished from plasma transcortin by agarose gel chromatography, discontinuous polyacrylamide electrophoresis, isoelectric gradient chromatography and protamine sulphate precipitation[16, 34].

The progesterone-binding molecule can be destroyed by proteolytic enzymes but not RNase or DNase. Thus, the active molecular binding site for the steroid may be proteinaceous in nature. Furthermore, the receptor is a heat-sensitive, non-dialysable and ammonium sulphate-precipitable protein with an isoelectric point at pH 4.0. Indirect physicochemical calculations suggest that the molecule can exist in the shape of a prolate ellipsoid with a monometric molecular weight of approximately 90 000 daltons*.

The cytosol-binding protein of the oviduct shows very little affinity for oestrogens (oestradiol, oestrone), mineralocorticoids (aldosterone), glucocorticoids (cortisol) or progesterone precursors and inactive metabolites. However, 5a-pregnane-3,20-dione does bind to the progesterone receptor. The tissue concentration of progesterone-binding protein can be markedly increased by prior oestrogen treatment[18]. This oestrogen-mediated stimulation of progesterone-binding capacity correlates quite closely with the oestrogen-induced enhancement of the oviduct progesterone response, i.e. avidin synthesis. In addition, the progesterone-binding protein is found only in progesterone-responsive target tissue. For these reasons, the evidence can be considered as compatible with the concept that the progesterone-binding protein may indeed be a 'physiological receptor' for progesterone. The receptor could thus interact initially with the steroid upon its entrance into the target cell and play a major role in the subsequent events involved in the action of the hormone.

Since progesterone is thought to act in the nucleus to influence gene transcription, it was of interest to establish whether the hormone is also bound to a receptor in the nucleus of the target cell. Purification of oviduct nuclei and extraction with salt revealed the presence of such a nuclear receptor

* Unit of atomic mass: $\frac{1}{16}$ of the mass of the oxygen atom.

which appeared almost identical to that which is found in the cytoplasm of the oviduct[16]. No progesterone receptor could be detected in the nucleus prior to exposure to the hormone. However, upon incubation of tissue slices with ^{3}H-progesterone at 37°C for short periods of time, it was possible to detect and monitor the actual transfer of the receptor–progesterone complex from the cytosol to the nucleus of the oviduct target cell[16]. This same transfer process can be accomplished under more defined cell-free conditions by incubating cytosol–receptor preparations labelled with ^{3}H-progesterone together with purified nuclei[16]. This *in vitro* binding of receptor to oviduct nuclei showed specificity requirements not only for oviduct cytosol but also for oviduct nuclei. Both cytosol and nuclear preparations from non-target tissues were ineffective in these cell-free recombination studies. The intra-nuclear progesterone–receptor complex was subsequently shown to be contained on the chromatin. This series of events involving the progesterone receptor of chick oviduct is quite similar to that noted for oestrogens[14, 35, 36-40], androgens[41-43], aldosterone[44-46] and glucocorticoids[47, 48] in their respective target tissues.

Since there is a specific requirement for oviduct nuclei in the binding of progesterone–receptor complex to chromatin, we began a search for specific 'acceptor' sites on the nuclear chromatin-DNA for the receptor–hormone complex. Indeed, incubation of cytosol–receptor with purified oviduct chromatin revealed sites capable of specifically binding the steroid–receptor complex[16, 17]. Much less binding was noted when non-target chromatins were used. Determination of the nature of these chromatin sites was considerably more difficult. Experiments were performed in which a progesterone–receptor interaction with chromatin was monitored during selective removal of histones, non-histone proteins and chromosomal-RNA. Further experiments were carried out using reconstituted 'hybrid chromatins', containing the histones from one organ or species and the non-histone acidic proteins and DNA from another organ or species[50]. These rather complex experiments can be summarised by stating that the receptor–hormone complex appeared to interact with the genome at specific DNA sites which are determined by the chromatin acidic proteins[50]. Histones appeared to be not involved in the interaction of the hormone–receptor complex with target cell chromatin. In subsequent experiments, these chromatin non-histone 'acceptor' proteins have been partially purified and characterised[51].

More recent studies have shown that the progesterone receptor may contain two non-identical sub-units. One of these sub-units has a specific affinity for the non-histone chromosomal 'acceptor' proteins while the second sub-unit appears to react directly with target-cell-DNA[52, 53]. The speculation is that these DNA sites are activator sites in the genes regulating functional responses to progesterone.

The experimental results described above are thus compatible with the following sequence of events suggested for the mechanism of the action of progesterone in the chick oviduct: (a) steroid permeation of the cell membrane; (b) binding of progesterone to a specific cytoplasmic receptor; (c) transfer of the steroid–receptor complex to the nucleus; and (d) binding of the steroid–receptor complex to specific sites on the chromatin-DNA.

If the specific interaction of progesterone with the target-tissue-DNA is related to gene activation, it would be expected that new gene transcription products (RNA) should appear in response to progesterone. Some effort, therefore, has gone towards attempts to demonstrate that progesterone induces the synthesis of specific types of RNA prior to the synthesis of avidin.

Experimental evidence from studies of the progesterone induction of avidin in the presence of known inhibitors of DNA, RNA and protein synthesis suggested that new DNA-dependent RNA synthesis and continued protein synthesis are required for induction to occur. New DNA synthesis itself is not necessary since hydroxy-urea, a specific inhibitor of DNA synthesis, did not affect the induction process at low concentrations[8]. Further confirmation of this point came from the fact that ^{3}H-thymidine incorporation into DNA was not enhanced during induction. However, actinomycin D, which inhibits DNA-dependent RNA synthesis, caused a 90% reduction of avidin synthesis without blocking general protein synthesis, whereas cycloheximide, which blocks protein synthesis by inhibiting the transfer of activated amino acids from transfer-RNA to the growing polypeptide chain, caused a complete inhibition of avidin induction[8].

In order to confirm the importance of RNA in the induction process, RNA-polymerase activity was assessed at various intervals. After the administration of progesterone[54], the activity of this enzyme increased and this increase preceded the initial rise in avidin synthesis. Similarly, the pattern of rapidly-labelled nuclear-RNA synthesis was altered. These results point to an effect of progesterone at the transcriptional level. In addition, analysis of RNA synthesised *in vitro* from oviduct chromatin shows that there was a qualitative change in the dinucleotide composition of nuclear-RNA after progesterone administration[8].

New species of hybridisable rapidly-labelled nuclear-RNA (repeating sequences) were also found following administration of progesterone, but prior to induction of avidin[55-57]. That this new RNA contains the messenger-RNA for avidin is probable, but proof for such a hypothesis requires the development of a cell-free assay system that can detect a specific RNA template controlling the synthesis of this protein.

In a recent report, we have demonstrated that the oviduct contains RNA capable of coding for the *in vitro* synthesis of the specific protein, avidin[58]. This RNA can be extracted from either whole-cell homogenates or polysomal preparations and has the molecular weight and sedimentation value expected for directing synthesis of a 15 000 molecular weight peptide sub-unit. The RNA is present only in progesterone target tissue (oviduct) and the intracellular concentration is directly dependent on the state of hormonal stimulation. Progesterone alone induces accumulation of avidin mRNA in the oviduct target cells. Taken together, the studies summarised in this chapter seem to indicate that progesterone acts at the transcriptional level of protein synthesis.

In attempting to explain how progesterone transforms the mammalian endometrium into a receptive state for the new blastocyst, we might postulate two major mechanisms. Progesterone may suppress the synthesis of an agent which inhibits implantation. More likely, the steroid induces the synthesis of a protein molecule(s) which facilitates implantation. We feel the latter mechanism to be more probable for the following reasons: (a) the cell structure is primarily composed of protein and all aspects of cell metabolism (i.e. carbohydrate, lipid, protein) are ultimately controlled by enzyme proteins; (b) progesterone has been shown to stimulate the synthesis of RNA and protein in the best defined model systems (deciduoma and chick oviduct); (c) finally, other steroid hormones are generally thought to exert effects in their target

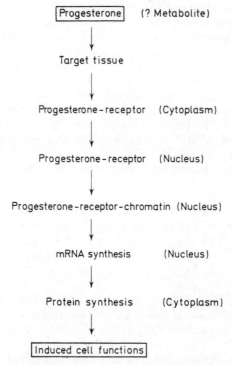

Figure 6.1 Summary of the biochemical sequence of events occurring during progesterone action in the cells of target tissues

tissues by first stimulating production of RNA and protein. Thus, the temporary acquisition of the capacity for the uterine endometrium to accept the new blastocyst during the reproductive process would then seem to be regulated most efficiently by the induction of the synthesis of a limited number of enzymes. A hypothetical scheme, based on current knowledge, for the series of biochemical events occurring in progesterone action in target tissue is listed in Figure 6.1.

6.2 OESTROGEN

6.2.1 Introduction

The rat uterus has proved to be a very popular model system in which to investigate the mechanism of action of oestrogenic hormones. It was from studies utilising this tissue that the concept of steroid hormone receptors was initially elucidated. Thus, it was demonstrated that uterus, vagina and anterior pituitary selectively bound oestradiol-17β to intracellular macromolecules[61]. Upon fractionation of the uterine cell, these oestradiol-binding proteins could be isolated from the cytoplasm and characterised by their distinctive sedimentation profile upon sucrose gradient centrifugation analysis[36]. Further experiments revealed that when rat uterine slices were incubated in the presence of ^{3}H-oestradiol-17β, the steroid formed an initial complex with its receptor in the cytoplasm of the target cell. However, upon continued incubation, the steroid–receptor complex relocated to the nuclear compartment[39]. This concept was supported by other studies which demonstrated that, following *in vivo* administration of ^{3}H-oestradiol to rats, the hormone could be found complexed to the nuclear chromatin of uterine target tissue[62]. Finally, recent studies have revealed the presence of specific 'acceptor sites' on target-tissue chromatin which bind this steroid hormone–receptor complex upon its entrance into the nucleus[63]. These initial events apparently set the stage for the subsequent early effects of oestrogen upon transcription and translation[8, 64, 70].

6.2.2 Regulation of uterine transcription

The role of RNA synthesis as a primary event in the early action of oestrogen has been well established[65]. Two minutes following injection of oestrogen into the ovariectomised adult rat, a 40% stimulation in the nuclear synthesis of rapidly-labelled RNA is observed[66]. The magnitude of this response reaches a peak of 400–500% over control after 20 min and then begins to decline. However, the oestrogen-stimulated value remains higher than control for at least 24 h[67]. On the other hand, the increased incorporation of ^{3}H-uridine into RNA was accompanied by an increased uterine uptake of the nucleotide. Therefore, labelling techniques could not distinguish whether the oestrogen-induced variation in RNA specific activity was a cause or a consequence of the fluctuation of precursor uptake. Recent evidence, from *in vivo* and *in vitro* studies, however, suggests that the increased labelling of uterine-RNA with ^{3}H-uridine is a result of enhanced synthesis of both nucleotides and nucleic acid, but not of increased nucleoside transport[68, 69].

The transcriptional components regulated by oestrogen in its early action on the uterus have been investigated in some detail. Thus, oestrogen has been shown to stimulate the activity of RNA-polymerase assayed in isolated nuclei, the template capacity of uterine chromatin and the production of new species (repeating sequences) of hybridisable nuclear-RNA. The order in which these various events were stimulated, however, is open to question (see Refs. 70, 71 for review). Gorski[72] first demonstrated that oestradiol increased Mg^{2+}-dependent RNA-polymerase activity of a crude uterine nuclear pellet (now termed Polymerase I) within 1 h of injection of immature rats. This observation was subsequently confirmed using purified uterine nuclei[67, 73]. Moreover, a second polymerase activity which required the presence of Mn^{2+} and 0.4 mol l^{-1} $(NH_4)_2SO_4$ (Polymerase II) was shown to be stimulated by oestradiol administration (*vide supra*). However, treatment for 12 h was required in order to observe an oestrogen stimulation. Subsequently, it was revealed by radioautography that the Mg^{2+}-dependent Polymerase I was localised in the nucleolus whereas the Mn^{2+}-dependent enzyme (II) was restricted to the nucleoplasm[74]. Barry and Gorski[75] have further investigated the effect of oestrogen on polymerase activity employing methods which were capable of distinguishing between an oestrogen-mediated increase in the number of growing nucleotide chains and the rate of chain elongation. The results obtained suggested that oestrogen brings about an increased rate of chain elongation within 1 h of administration but does not affect the number of growing chains. The implication of such experiments is that oestrogen stimulates the activity but not the number of polymerase molecules. Thus, the authors suggested that the oestrogen effect at this early stage is only minimally due to transcription of additional template.

RNA-polymerase activity in response to oestrogen has recently been reevaluated by Glasser, Chytil and Spelsberg[76] utilising rigorously-controlled kinetic experiments. By performing the assays at low temperature and under proper kinetic conditions it has been possible to demonstrate a rapid but transient increase in Polymerase II activity within 15 min of oestrogen administration. This increase occurs in concert with the stimulation of the synthesis of uterine rapidly-labelled nuclear-RNA, as previously reported[66]. Moreover, this increase in Polymerase II activity occurs prior to any detectable change in the chromatin template activity. Polymerase I activity was first increased 1 h after oestrogen, and reached maximum levels within 4 h.

Baulieu et al.[98] have also found that oestrogen increased RNA-polymerase activity within 1 h and suggested that this stimulation represented Polymerase I activity due to the fact that the enhancement was not sensitive to α-amanatin. It was also found that cycloheximide would block the oestrogen-mediated increase in Polymerase I activity. This suggested that the continuous synthesis of protein is necessary to support the oestrogen-activated RNA synthesis. Similar effects of inhibitors of protein synthesis on oestrogen-induced changes in RNA synthesis which occur after 1 h have been previously reported[91, 98, 125, 128]. On the other hand, the early stimulation of rapidly-labelled RNA synthesis by oestrogen is not blocked by cycloheximide[66, 82]. Baulieu et al. suggest that synthesis of a protein factor with an apparent half-life of 15 min is required for the activation of nucleolar Polymerase I. Moreover, the synthesis of this protein factor is apparently dependent on a short-lived mRNA, thus supporting the importance of transcriptional events for the early effects of oestrogen.

Teng and Hamilton[62] have described a 25% enhancement of the template capacity of rat uterine chromatin within 30 min of a single injection of oestradiol-17β into ovariectomised animals. The peak response occurred at approximately 8 h, but stimulation was noted for as long as 72 h. Similar observations have been made by Church and McCarthy[77] using chromatin isolated from the endometrium of the ovariectomised rabbit. In these studies, a significant stimulation was revealed within 10 min of oestrogen injection and remained demonstrable for at least 2 h. Finally, Barker and Warren[78] reported an effect of oestrogen upon rat uterine chromatin at 2 h post-injection. This, however, was the earliest time period investigated. Chromatin template activity was first stimulated 1 h after the administration of oestrogen in the studies of Glasser, Chytil and Spelsberg[76]. It is interesting to note that the changes in template activity assayed with bacterial polymerase appeared to correlate in a temporal fashion with changes in uterine-RNA Polymerase I activity. In studies in which chemical composition was carefully monitored, the ratio of non-histone chromosomal protein to DNA varied in direct proportion to changes in template activity[62, 76]. No such correlation was found for the ratio of histone to DNA. Thus, it seems certain from work performed in four separate laboratories that oestrogen-mediated nuclear-RNA synthesis occurs under experimental conditions where fluctuations in uterine precursor uptake are simultaneously monitored.

Church and McCarthy[77] have employed DNA–RNA competition methods to examine the effect of oestrogen on populations of nuclear-RNA sequences. Unlabelled uterine nuclear-RNA from untreated or 1-h oestrogen-treated rabbits was used to compete with ^{3}H-RNA, isolated from uteri of ovariectomised rabbits for DNA binding sites, as measured by annealing to the DNA. The unlabelled RNA from untreated animals did not compete effectively with the uterine-RNA isolated from rabbits treated for 1 h with oestrogen. These hybridisation competition experiments provide evidence that oestrogen, within 1 h of injection, induces synthesis of different populations of RNA molecules. It should be pointed out, however, that these experiments were performed using rapid re-annealing kinetics and thus cannot distinguish between unique and repetitive sequences of DNA[79]. At any rate the studies presented demonstrate conclusively an important role for genetic transcription in the early action of oestrogen on the uterus of hormone-deficient animals.

6.2.3 Regulation of uterine translation

Oestrogen stimulates the rate of incorporation of amino acids into protein in a manner which is consistent with a prior requirement for nuclear-RNA synthesis[80-82]. Indeed the earliest demonstrable effects of oestrogen upon the overall synthesis of uterine proteins occurs after 2–4 h[82, 83], a time interval which is closely paralleled by that required for the appearance of newly formed ribosomes in the cytoplasm[84, 85]. Pulse-labelling experiments with several ^{3}H-amino acids have revealed a depression in the rate of protein synthesis in all uterine sub-cellular fractions (i.e. nuclear, mitochondrial, microsomal and soluble) after 30 min of oestrogen action. Furthermore, over the period 2–4 h there is a preferential stimulation in the rate of labelling of proteins in the nuclear fraction which anticipates a marked rise in cytoplasmic protein synthesis[82, 83]. The depression of labelling of uterine protein synthesis following oestrogen administration to the adult ovariectomised rat occurs at a time when the rate of rapidly-labelled nuclear-RNA synthesis is maximally stimulated[82]. It is interesting to note that a similar observation has been reported by Barnea and Gorski[86] and by Eilon and Gorski[87] although the significance of these observations remains unknown.

Oestrogen regulates the synthesis of newly formed uterine ribosomes. Stimulation of ribosome synthesis monitored by the incorporation of ^{3}H-uridine into ribosomal-RNA is first demonstrable after 4 h and then rises at a linear rate for at least 12 h following oestrogen administration[84]. Not only is the synthesis of ribosomes under the influence of oestrogen but several reports have demonstrated that this steroid also regulates the capacity of functional ribosomes to synthesise protein. The first cell-free system for protein synthesis prepared from rat uterus was described by Greenman and Kenney[85]. These authors demonstrated that ovariectomy lowered the activity of uterine ribosomes. Associated with this lowered activity, however, was an increased response to the exogenous synthetic messenger, polyuridylic acid. On the other hand, treatment with oestrogen for 4 h increased the natural message activity of isolated ribosomes but decreased the ability to respond to Poly U. Results were interpreted to indicate that both the number and capacity of uterine ribosomes were oestrogen-dependent. The conclusions drawn by Greenman and Kenney[85] concerning the effects of oestrogen on polysomal protein synthesis have been confirmed and extended by Teng and Hamilton[88,89]. An initial stimulation of the amino acid incorporating ability of uterine polysomes was noted after 2 h with further increases for 8–12 h after the administration of oestrogen; concomitant with this increased activity is an oestrogen-mediated accumulation of polysomes in the cytoplasm. However, if animals did not receive an additional injection of oestrogen after 24 h, the level of cytoplasmic polysomes decreased. The results prompted the authors to suggest that the polysomes formed under the influence of oestrogen were relatively unstable.

Recent observations by Suvatte and Hagerman[90] suggest that oestrogen may stimulate the initiation of polypeptide synthesis on uterine ribosomes within 3 h following injection of castrated guinea-pigs. This inference was drawn from studies measuring the release of ^{3}H-peptidylpuromycin from uterine ribosomes. Subsequent to these early effects, oestrogen was also reported to enhance translocase activity of the preparations. Thus it was concluded that oestrogen affects both the initiation and completion of nascent protein on uterine ribosomes.

A general stimulation of uterine protein synthesis is not noted until after 2–4 h of oestrogen administration. However, Notides and Gorski[91] have shown that oestradiol injection into immature rats results within 30 min in the enhancement of the rate of synthesis of a specific uterine cytoplasmic protein. This protein was initially identified by starch gel electrophoresis of the soluble protein fraction. Further studies have revealed that this effect of oestrogen can be demonstrated in both immature and adult ovariectomised rats. Synthesis of the specific induced protein (IP) can also be detected by labelling *in vivo* or *in vitro*[86]. The synthesis of IP appears to reach a maximum 2 h after the administration of oestrogen and after 4 h has begun to decline; moreover, the results are compatible with a *de novo* synthesis of this protein as opposed to changes in conformation or subcellular compartmentation. Polyacrylamide gel electrophoresis of double-labelled soluble proteins from control and oestrogen-treated uteri has been utilised by Mayol and Thayer[92] to confirm the results of the Gorski group. Mayol and Thayer have also utilised preparative gel electrophoresis to partially purify the oestrogen-induced proteins, and report that this procedure results in the demonstration of considerable heterogeneity[92]. These results are interpreted to suggest that IP may actually be a group of acidic proteins ($pK = 3.5$–4.0) rather than a single molecular species.

Pretreatment of animals with actinomycin D blocks the oestrogen-stimulated synthesis of IP[93]. These results suggest that induction of IP is a very early oestrogen-induced macromolecular event. The extremely rapid stimulation of nuclear-RNA synthesis by oestrogen[66] is demonstrable prior to the earliest detectable effects on IP[93]. It is tempting to speculate that these two biosynthetic events are inter-related in the initial actions of oestrogen.

It is now well accepted that the nuclear non-histone (acidic) proteins participate in the regulation of organ-specific restriction of the expression of the genome[94,95]. In this regard, Teng and Hamilton[96] have suggested that these acidic proteins will counteract the inhibitory effect of histones on the template activity of isolated uterine chromatin. These same authors have recently investigated the influence of oestrogen on the synthesis of the non-histone nuclear proteins[97]. Tryptophan was employed as the labelled precursor since histones do not contain this amino acid. It was reported that oestrogen did stimulate the incorporation of ^{3}H-tryptophan into the acidic protein fraction. This response was first demonstrable after 8–12 h following hormone administration. Moreover, analysis of the acidic proteins on acrylamide–urea gels resulted in a resolution of the fraction into six major bands and the demonstration that only one of these bands was affected by oestrogen. A specific oestrogen-induced protein has also been found in the nucleus by Baulieu et al.[98]. Moreover, these workers have confirmed the cytoplasmic synthesis of IP, and suggest that IP and the specific nuclear acidic protein have the same apparent molecular weight. Oestrogen stimulation is first observed by an increase of the cytosol protein and only subsequently is the effect on the nuclear protein demonstrable. Baulieu et al.[98] have used these data to postulate that IP, after its cytoplasmic synthesis, may be transferred into the nucleus and become associated with the uterine chromatin.

Barker[99] has also described the oestrogen-induced synthesis of a specific non-histone protein in the uterus. This study reports increased specific activity of a nuclear protein fraction within 30 min of oestrogen administration to adult ovariectomised rats. This protein fraction co-chromatographed with the F3 histone on Sephadex G-100 but could be separated by gel electrophoresis, thus demonstrating its acidic nature. The kinetics of the effect of oestrogen on the labelling of this protein and the method of separation were considerably different from that for the nuclear acidic protein described by Teng and Hamilton[97] or by Baulieu et al.[98]. It is unlikely therefore that these acidic protein fractions represent a single molecular species.

6.2.4 *In vitro* effects of oestrogen on the uterus

Raynaud-Jammet and Baulieu[100] were the first to demonstrate an *in vitro* effect of oestradiol-17β on uterine-RNA synthesis in isolated nuclei. A stimulation of RNA synthesis was obtained by pre-incubating oestradiol-17β with calf uterine cytosol, then adding isolated nuclei and assaying for RNA-polymerase activity. Pre-incubation of the oestrogen with cytosol was an absolute necessity in order for the stimulation of RNA-polymerase activity to occur. Presumably this allowed the interaction of the steroid with its specific receptor protein[39]. More recently these observations were confirmed by Arnaud et al.[101]. These workers suggested that only the 5S form of the oestrogen–receptor complex would stimulate uterine-RNA synthesis *in vitro* and acted specifically on nucleolar-RNA Polymerase I[101]. Furthermore, the cAMP-dependent phosphorylation of the 5S receptor was reported to enhance the ability of this complex to stimulate RNA synthesis. Arnaud et al.[102] use these data to help to explain the role of increased cAMP in the uterine response to oestradiol as reported by Szego[103]. Finally, a 40–60 % stimulation of nuclear-RNA-polymerase activity has also been reported by Mohla et al.[104]. Again, in order to demonstrate the stimulation, oestrogen was pre-incubated with uterine cytosol under conditions which would convert the 8S receptor to the presumably active 5S form.

Incubation of immature rat uteri with 10^{-9} mol l^{-1} oestradiol has been reported by Katzenellenbogen and Gorski[105] to result in the induction of IP. Synthesis of this specific protein was blocked by actinomycin D, but under optimal conditions oestradiol brought about a 75 % increase in labelling. The time-dependence of stimulation was similar to that reported *in vivo* with a 50 % response after 20–30 min. Coincidence of IPs induced *in vitro* and *in vivo* was shown by co-electrophoresis on polyacrylamide gels. Finally, this *in vitro* response of oestradiol was both tissue- and steroid-specific. Independent confirmation of the *in vitro* induction of IP has been obtained by Wira and Baulieu[106]. Kinetics of the response and specificities were identical to those reported in the Katzenellenbogen and Gorski study[105]. Again this response was blocked by pretreatment of the tissue with actinomycin D. It appears, therefore, that it is now possible to obtain some of the specific biochemical responses to oestrogen in an *in vitro* system.

In summary, it appears that oestradiol-17β in its early action in the uterus first binds to a cytoplasmic receptor protein. This complex is then modified in some manner and transferred into the nucleus. A stimulation of synthesis of nuclear-RNA ensues within minutes following hormone injection. This stimulation then leads to the enhancement of the synthesis of proteins, some of which appear in the cytoplasm (IP) and some in the nucleus[70, 71, 98, 107]. It is possible that this induction phase may lead to an amplification phase; that is, increased synthesis of nuclear-RNA which is transported to the cytoplasm for regulation of overall uterine protein synthesis.

The oviduct of the immature chick has also provided a good model system for the study of the biochemical mechanism of oestrogen action. Moreover, it is also a desirable system in which to study hormone-mediated tissue differentiation. Administration of oestrogen to the immature chick results in differentiation of three distinct cell types from the homogeneous population of primitive mucosal cells[8, 9]. The resulting tubular gland cells synthesise at least two specific proteins which are easily measured by biochemical and immunochemical assays[108, 109]. Thus ovalbumin and lysozyme synthesis by the tubular gland cells can be utilised as biochemical end-point markers both for the molecular action of oestrogens and oviduct cytodifferentiation[9, 110].

6.2.5 Oestrogenic control of transcription in oviduct

Transcriptional events have been demonstrated to precede the synthesis of cell-specific proteins which are first identifiable between 2–4 days after oestrogen treatment. Thus a single injection of oestrogen was shown to cause a rapid and prolonged stimulation of nuclear-RNA-polymerase activity which reached a maximum after 12–24 h[54]. This information suggested that oestrogenic substances act at the level of gene transcription to promote oviduct protein synthesis and cytodifferentiation. Further support for this suggestion was obtained by the demonstration that oestrogen elicited an increase in total template activity of isolated chromatin as assayed *in vitro*[8]. Moreover, qualitative changes in the composition of the RNA synthesised from chromatin template during oestrogen-mediated differentiation were shown by nearest-neighbour base frequency analysis of the RNA[8]. Again, these changes preceded the appearance of cell-specific proteins and suggest that the steroid must promote the synthesis of new species of nuclear-RNA. In order to test this hypothesis, we examined total nuclear messenger-RNA activity during oestrogen-induced growth and differentiation[111]. For these experiments, chick oviduct nuclear-RNA activity was assayed by a modification of the procedure described by Nirenberg[112]. This assay assesses the ability of synthetic mRNA or natural DNA-like RNA from animal cells to direct the synthesis of phenylalanine-^{14}C into polyphenylalanine using ribosomes isolated from a mutant of *E. coli.* (MRE-600) defective in ribonuclease. A threefold stimulation of activity was revealed within 24 h of a single injection of diethylstilbestrol (DES) into previously untreated chicks. Maximum stimulation was obtained after 3 days of hormone treatment and by 6 days the messenger activity had begun to decline. Oestrogen also increases the synthesis of oviduct tRNA and rRNA with maximum stimulation occurring between 4–7 days of treatment[8, 71, 111]. Finally, induction of specific protein synthesis is blocked by actinomycin D[8].

Since new cell-types and cell-specific proteins appear after oestrogen administration, we would predict that this hormone in its early action might promote the synthesis of new species of RNA. To examine populations of nuclear-RNA transcribed from the repeating DNA sequences at various stages of oestrogen-mediated growth, DNA–RNA competition methods have been utilised[55]. Unlabelled nuclear-RNA from immature chicks and from subsequent stages of differentiation was used to compete with ^{3}H-RNA, isolated from differentiated tissue, for DNA binding sites, as measured by annealing to the DNA. The RNA from the nuclei of the immature chick oviduct competed quite poorly with the labelled RNA of the differentiated oviduct (20 days of DES treatment). However, a progressive increase in competing species of RNA was noted after 2, 5 and 10 days of DES administration. After 20 days of DES, full competition of unlabelled RNA with labelled RNA (also from 20-day DES-treated chicks) was noted. Similar results have also been obtained by Hahn *et al.*[113]. Our experiments demonstrate that the adult species of oviduct nuclear-RNA (transcribed from 'repeating sequences' of the genome) can be generated by continuous oestrogen administration to the immature chick; this finding is consistent with a marked specific effect of this steroid on nuclear gene transcription.

It has been proposed that the unique sequences of DNA code for the mRNAs for structural and enzymic proteins, while repeating sequences may contain regulatory genes. Recent advances in nucleic acid technology have made it possible to separate DNA into unique and repeating sequences. Therefore, our laboratory has utilised hydroxylapatite chromatography to subdivide the chick oviduct genome on the basis of nucleotide homology. Experiments have been conducted to study the effects of oestrogen on DNA complexity and RNA transcription from unique sequences[114]. No detectable differences were found in the renaturation profiles of oviduct DNA at the various times of oestrogen stimulation. These data indicate that oestrogen is not acting through major gene duplication or deletion, and suggest that new proteins required for growth may arise from differential gene transcription. Indeed it was found that unique sequence DNA is transcribed in the oviduct with 20–25% of the resulting RNA being processed into the cytoplasm. Moreover, oestrogen apparently causes an increase in the extent of unique DNA transcription. Thus, although the amount of RNA transcribed from unique sequence DNA that is processed into the cytoplasm does not appear to vary as a result of oestrogen treatment, a qualitative difference seems to exist in the cytoplasmic mRNA populations at different stages of oestrogen-induced differentiation[129].

Both quantitative and qualitative alterations in oviduct gene transcription result following administration of oestrogens; and these early molecular events occur prior to morphologic or biochemical differentiation. Available data support the following sequence of events: oestrogen, upon entering the oviduct cell, complexes to a specific receptor protein which rapidly passes into the nucleus. The receptor complex–genome interaction results in the coordinated synthesis of all species of RNA. The new RNA molecules must then be transported into the cytoplasm so that new polyribosomes may be formed which are necessary to support that synthesis of new proteins required for biochemical specialisation. In order to further test this hypothesis, we have examined in detail oviduct protein synthesis in response to oestrogen.

6.2.6 Oestrogenic control of translation in oviduct

Since oestrogen is reponsible for the growth of the oviduct, an early effect on certain key enzymes and structural proteins would be expected. Indeed, a stimulation of the activity of oviduct ornithine decarboxylase, completely blocked by cycloheximide, has been found within hours of *in vivo* or *in vitro* oestrogen administration[115]. Stimulation of this enzyme and of polyamine synthesis have also been linked with the chemical induction of growth in other cell-types and it is possible that polyamines may trigger, or more probably simply support, the reactions which lead to rapid nucleic acid and protein synthesis and subsequent cell growth.

In addition to this early stimulation of ornithine decarboxylase activity and the later induction of the cell-specific proteins, lysozyme and ovalbumin, oestrogen is known to stimulate overall protein synthesis in the oviduct[55, 109]. Since proteins are synthesised on polyribosomes, we have characterised these particles as to quantity and functional capacity during oestrogen-mediated differentiation of the oviduct[111]. For the first 7 days of hormone treatment, there is an increase in the oviduct content of ribosomes; this is followed by a decline by the tenth day when the rate of cytodifferentiation is beginning to plateau. The polysome profiles from the immature chick oviduct also reveal an effect of oestrogen which could account for the stimulation of protein synthesis[111]. In the unstimulated oviduct, monomers are the predominant form, but from oviducts treated for 4 days with hormone, the polysome profile reveals a large portion of aggregates. Withdrawal of oestrogen at any point of differentiation results in polysomal disaggregation, cessation of protein synthesis and the growth process is brought to an abrupt halt[111, 116, 117]. Oestrogen-induced morphologic differentiation of the chick oviduct is accompanied by increases in the total endogenous mRNA activity of oviduct polysomes[111, 118]. Concomitant increases are also noted in ribosome translational capacity and activity of peptide chain initiation factors. Once the differentiation process nears completion ($\simeq 7$ days of oestrogen treatment), total ribosome-bound mRNA activity decreases but the translational machinery remains very active. Analysis by polyacrylamide gel electrophoresis of peptides synthesised *in vivo* or *in vitro* in the cell-free polyribosome preparation following 4 days of oestrogen administration shows that striking changes occur in the peptides synthesised before and after the administration of oestrogen[111]. Since the effects are apparent in the *in vitro* system, it would seem that the hormone must be responsible in some way for affecting the specific messenger-RNA population of the cells of the oviduct mucosa. However, the only direct proof of effects on messenger-RNA is the ability of an isolated RNA fraction to direct the *de novo* synthesis of a specific protein in a chemically-defined cell-free system. We have, therefore, initiated studies designed to develop a protein-synthesising system capable of translating exogenous messages with fidelity in order to assess changes in specific mRNAs which occur during oestrogen-mediated oviduct development[107, 119, 120]. Since ovalbumin comprises 50–60% of total oviduct protein, we chose initially to investigate the mRNA for this protein.

6.2.7 Oestrogenic control of specific oviduct mRNA

A rapidly-labelled RNA fraction can be isolated from oviduct polyribosomes of oestrogen-stimulated chicks that has the characteristics of the mRNA for the cell-specific protein ovalbumin[121, 122]. This RNA fraction, which sediments in the 8–17S region of sucrose gradients, possesses properties suggestive of the presence of a polyadenylic acid sequence. Moreover, it has been possible to translate this mRNA with fidelity in a heterologous cell-free system derived from rabbit reticulocytes[121, 122]. The identity of the newly synthesised protein as ovalbumin was confirmed by three separate methods[122]. Translation of ovalbumin mRNA was dependent both on amount of exogenous oviduct mRNA and incubation time. Both the rate and extent of ovalbumin synthesis were stimulated by the addition of exogenous peptide chaininitiation factors[121, 122]. Finally, it was found that the ovalbumin mRNA could be partially purified (50–100-fold) on nitrocellulose filters[123].

The activity of ovalbumin mRNA is clearly dependent upon the hormonal state of the animal. Oestrogen induces activity within 24 h of administration to immature chicks and the appearance of ovalbumin mRNA significantly precedes the appearance of demonstrable oviduct ovalbumin *in vitro*[118]. Cessation of oestrogen injections results in the progressive disappearance of ovalbumin mRNA. Re-administration of this steroid to animals deprived of oestrogen causes, again within a few hours, reappearance of the ovalbumin message as measured by the cell-free synthesis of ovalbumin[118, 121-123]. The appearance of mRNA for ovalbumin is consistent with existing data outlined in this chapter which suggest that a primary consequence of oestrogen action in the chick oviduct is at the level of gene transcription. Furthermore, our data support the hypothesis that steroid hormones can act in target tissues to promote the accumulation of mRNA, and offer evidence that the rate-limiting event in the hormonal induction of cell-specific proteins such as ovalbumin is the synthesis and intracellular accumulation of the specific mRNA[129].

Based on available information, one can begin to construct a sequence of events by which oestrogen induces the synthesis of cell-specific proteins such as ovalbumin (Figure 6.2). Oestrogen first brings about the differentiation

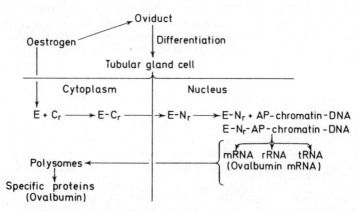

Figure 6.2 Proposed sequence of events in the mechanism of the action of oestrogen in the chick oviduct

of the tubular gland cells from the primitive oviduct mucosa. These cells contain an oestrogen-binding macromolecule which is tissue-specific, heat-labile and binds oestradiol or DES with high affinity ($K_d \simeq 10^{-10}$). Thus, upon entering the cell, oestrogen is bound to a cytoplasmic receptor which subsequently moves into the nucleus. Oestrogen then causes alterations in rapidly-labelled nuclear-RNA synthesis, chromatin template activity and RNA-polymerase activity. These alterations are paralleled by the appearance of newly synthesised tRNA and rRNA as well as a new species of mRNA. A conversion of ribosomes to polysomes then occurs presumably due to the entry of new messenger-RNA into the cytoplasm. These polyribosomes synthesise a qualitatively different population of peptides *in vitro* which reflect the earlier alterations in messenger-RNA production. Finally the end product of this action of oestrogen is the induction *de novo* of mRNAs of specific proteins such as ovalbumin and lysozyme and their subsequent translation. This sequence of events indicates, therefore, that oestrogen exerts its primary effect in the nucleus to promote selective gene activation and subsequent synthesis of new ovalbumin molecules on cytoplasmic polysomes.

In the past, steroid hormones have often been shown to influence the synthesis, distribution, degradation and function of RNA in target cells. Since it was first determined that ribonucleic acids play a central role in the control of protein synthesis in micro-organisms, a large body of experimental data has accumulated suggesting that animal hormones also regulate the amount of enzymes and structural proteins through RNA mediators[124]. In certain tissues, such as the uterus[125], male accessory sex tissue[126], liver[127] and oviduct[8,71], all major RNA fractions were found to be stimulated by the appropriate steroid hormone. These observations cast some doubt on the specificity of the role of the new RNA molecules. However, the observation that many of these changes in RNA metabolism preceded the hormonal regulation of cytoplasmic protein synthesis suggested a biologic significance for the new RNA. Our recent studies demonstrating the stimulation of the appearance of specific mRNA molecules by two steroid hormones, progesterone and oestrogen, are certainly compatible with the concept that steroid hormones do in fact act by inducing an increase in the cellular level of certain species of mRNA; these then code for the cytoplasmic synthesis of specific proteins in target tissues.

References

1. Corner, G. W. (1947). *The Hormones in Human Reproduction*, 281 (Princeton: Princeton University Press)
2. Reel, J. R., Lee, S. and Callantine, M. R. (1969). *Fifty-First Meeting of the Endocrine Society, New York, Specific Retention of an Unconjugated Polar Metabolite in the Rat Uterus Following Labeled Progesterone Administration*, 113
3. Reynolds, S. R. M. (1949). *Physiology of the Uterus*, 611 (New York: P. B. Hoeber, Inc.)
4. Csapo, S. (1961). *Brook Lodge Symposium of Progesterone*, 7 (Augusta, Michigan: Brook Lodge Press)
5. Porter, D. G. (1969). *Progesterone: Its Regulatory Effect on the Myometrium*, 79 (G. E. W. Wolstenholme and J. Knight, editors) (London: J. and A. Churchill)
6. De Feo, V. J. (1967). *Cellular Biology of the Uterus*, 191 (R. M. Wynn, editor) (New York: Appleton-Century-Crofts)
7. O'Malley, B. W. and McGuire, W. L. (1968). *Endocrinology*, **84**, 63
8. O'Malley, B. W., McGuire, W. L., Kohler, P. O. and Korenman, S. G. (1969). *Rec. Progr. Horm. Res.*, **25**, 105
9. Kohler, P. O., Grimley, P. M. and O'Malley, B. W. (1969). *J. Cell. Biol.*, **40**, 8
10. Korenman, S. G. and O'Malley, B. W. (1968). *Endocrinology*, **83**, 11
11. Landau, R. L. and Lugibihl, K. (1961). *Rec. Progr. Horm. Res.*, **17**, 249
12. Laumos, K. R. and Faroog, A. (1966). *J. Endocrinol.*, **36**, 95
13. Edwards, R., Brush, M. G. and Taylor, R. W. (1969). *J. Endocrinol.*, **45**, iii
14. Wiest, W. G. (1972). *The Sex Steroids*, 297 (K. W. McKerns, editor) (New York: Appleton-Century-Crofts)
15. Falk, R. J. and Bardin, C. W. (1970). *Endocrinology*, **86**, 1059
16. O'Malley, B. W., Sherman, M. R. and Toft, D. O. (1970). *Proc. Nat. Acad. Sci. USA*, **67**, 501
17. O'Malley, B. W., Sherman, M. R., Toft, D. O., Spelsberg, T. C., Schrader, W. T. and Steggles, A. W. (1971). *Advances in the Biosciences*, **7**, 213
18. Toft, D. O. and O'Malley, B. W. (1972). *Endocrinology*, **90**, 1041
19. Katzman, P. A., Larson, D. L. and Podratz, K. C. (1972). *The Sex Steroids*, 107 (K. W. McKerns, editor) (New York: Appleton-Century-Crofts)
20. Swerdloff, R. S. and Odell, W. O. (1966). *J. Clin. Endocrinol.*, **29**, 157
21. Ross, G. T., Cargille, C. M., Lipsett, M. B., Rayford, P. L., Marshall, J. R., Strott, C. A. and Rodbard, D. (1970). *Rec. Progr. Horm. Res.*, **26**, 1

22. Seiki, K., Miyamoto, M., Yamashita, A. and Kotani, M. (1969). *J. Endocrinol.*, **43,** 129
23. Whalen, R. E. and Luttge, W. G. (1971). *Brain Res.*, **33,** 147
24. Robel, P., Lasnitzki, I. and Baulieu, E. (1971). *Biochimie*, **53,** 81
25. Morgan, M. D. and Wilson, J. D. (1970). *J. Biol. Chem.*, **245,** 3781
26. Bruchovsky, N. and Wilson, J. D. (1968). *J. Biol. Chem.*, **243,** 2012
27. Baulieu, E. E., Lasnitzki, I. and Robel, P. (1968). *Nature (London)*, **219,** 1155
28. Podratz, K. C. and Katzman, P. A. (1968). *Fed. Proc.*, **27,** 497
29. Milgrom, E., Atger, M. and Baulieu, E. (1970). *Steroids*, **16,** 741
30. Wiest, W. G. and Rao, B. R. (1971). *Advances in the Biosciences* (Schering Workshop on Steroid Hormone 'Receptors'), **7,** 251
31. McGuire, J. L. and DeDella, C. (1971). *Endocrinology*, **88,** 1099
32. Feil, P. D., Glasser, S. R., Toft, D. O. and O'Malley, B. W., *Endocrinology*, in the press
33. Milgrom, E. and Baulieu, E. (1970). *Endocrinology*, **87,** 276
34. Sherman, M. R., Corvol,. P L. and O'Malley, B. W. (1970). *J. Biol. Chem.*, **245,** 6085
35. Jensen, E. V. (1964). *Proc. 2nd Int. Congr. Endocrinol. London*, 420 (Amsterdam: Excerpta Med. Found.)
36. Toft, D. and Gorski, J. (1966). *Proc. Nat. Acad. Sci. USA*, **55,** 1574
37. Noteboom, W. D. and Gorski, J. (1965). *Arch. Biochem. Biophys.*, **111,** 559
38. Jensen, E. V., DeSombre, E. R., Jungblut, P. W., Stumpf, W. E. and Roth, L. J. (1969). *Autoradiography of Diffusible Substances*, 81 (L. J. Roth and W. E. Stumpf, editors) (New York: Academic Press)
39. Jensen, E. V., Suzuki, T., Kawashima, T., Stumpf, W. E., Jungblut, P. W. and DeSombre, E. R. (1968). *Proc. Nat. Acad. Sci. USA*, **59,** 632
40. Toft, D. O., Shyamala, G. and Gorski, J. (1967). *Proc. Nat. Acad. Sci. USA*, **57,** 1740
41. Fang, S., Anderson K. M. and Liao, S. (1969). *J. Biol. Chem.*, **244,** 6584
42. Mainwaring, W. I. P. (1969). *J. Endocrinol.*, **45,** 531
43. Baulieu, E. E. and Jung, I. (1970). *Biochem. Biophys. Res. Commun.*, **38,** 599
44. Edelman, I. S. and Fimognari, G. (1968). *Rec. Prog. Horm. Res.*, **24,** 1
45. Herman, T. S., Fimognari, G. M. and Edelman, I. S. (1968). *J. Biol. Chem.*, **243,** 3849
46. Swaneck, G. E., Chu, L. L. H. and Edelman, I. S. (1970). *J. Biol. Chem.*, **245,** 5382
47. Beato, M., Biesewig, D., Brandle, W. and Sekeris, C. E. (1969). *Biochim. Biophys. Acta*, **192,** 494
48. Baxter, J. D. and Tomkins, G. M. (1970). *Proc. Nat. Acad. Sci. USA*, **65,** 709
49. Wira, C. and Munck, A. (1970). *J. Biol. Chem.*, **245,** 3436
50. Spelsburg, T. C., Steggles, A. W. and O'Malley, B. W. (1971). *J. Biol. Chem.*, **246,** 4188
51. Spelsberg, T. C., Steggles, A. W., Chytil, F. and O'Malley, B. W. (1972). *J. Biol. Chem.*, **247,** 1368
52. Schrader, W. T., Toft, D. O. and O'Malley, B. W. (1972). *J. Biol. Chem.*, **247,** 2401
53. Schrader, W. T. and O'Malley, B. W. (1972). *J. Biol. Chem.*, **247,** 51
54. McGuire, W. L. and O'Malley, B. W. (1968). *Biochim. Biophys. Acta*, **157,** 187
55. O'Malley, B. W. and McGuire, W. L. (1968). *Proc. Nat. Acad. Sci. USA*, **60,** 1527
56. O'Malley, B. W. and McGuire, W. L. (1968). *Biochem. Biophys. Res. Commun.*, **32,** 595
57. Hahn, W. E., Church, R. B., Gorbman, A. and Wilmot, L. (1968). *Gen. Comp. Endocrinol.*, **10,** 438
58. O'Malley, B. W., Rosenfeld, G. C., Comstock, J. P. and Means, A. R. (1972). *Nature (London)*, **240,** 45
59. Chatterton, R. T. Jr., (1972). *The Sex Steroids*, 345 (K. W. McKerns, editor) (New York: Appleton-Century-Crofts)
60. Korenman, S. G. (1969). *Steroids*, **13,** 163
61. Jensen, E. V. and Jacobson, E. P. (1966). *Rec. Prog. Horm. Res.*, **18,** 387
62. Teng, C. S. and Hamilton, T. H. (1968). *Proc. Nat. Acad. Sci. USA*, **60,** 1410
63. Steggles, A. W., Spelsberg, T. C., Glasser, S. R. and O'Malley, B. W. (1971). *Proc. Nat. Acad. Sci. USA*, **68,** 1479
64. Hamilton, T. H. (1968). *Science*, **161,** 649
65. O'Malley, B. W. (1970). *Metabolism*, **20,** 981
66. Means, A. R. and Hamilton, T. H. (1966). *Proc. Nat. Acad. Sci. USA*, **56,** 1594
67. Hamilton, T. H., Widnell, C. C. and Tata, J. R. (1968). *J. Biol. Chem.*, **243,** 408
68. Oliver, J. M. (1971). *Biochem. J.*, **121,** 83
69. Knowler, J. T. and Smellie, R. M. S. (1971). *Biochem. J.*, **125,** 605

70. Hamilton, T. H., Teng, C. S., Luck, D. H. and Means, A. R. (1972). *The Sex Steroids*, 197 (K. W. McKerns, editor) (New York: Appleton-Century-Crofts)
71. Means, A. R. and O'Malley, B. W. (1972). *Metabolism*, **21**, 357
72. Gorski, J. (1964). *J. Biol. Chem.*, **239**, 889
73. Hamilton, T. H., Widnell, C. C. and Tata, J. R. (1965). *Biochim. Biophys. Acta*, **108**, 168
74. Maul, G. G. and Hamilton, T. H. (1967). *Proc. Nat. Acad. Sci. USA*, **57**, 1371
75. Barry, J. and Gorski, J. (1971)1 *Biochemistry*, **10**, 2384
76. Glasser, S. R., Chytil, F. and Spelsberg, T. C. (1972). *Biochem. J.*, **130**, 947
77. Church, R. H. and McCarthy, B. J. (1970). *Biochim. Biophys. Acta*, **199**, 103
78. Barker, K. L. and Warren, J. C. (1966). *Proc. Nat. Acad. Sci. USA*, **56**, 1298
79. Britten, R. J. and Kohn, D. E. (1968). *Science*, **161**, 529
80. Ui, H. and Mueller, G. C. (1963). *Proc. Nat. Acad. Sci. USA*, **50**, 256
81. Talwar, G. P. and Segal, S. (1963). *Proc. Nat. Acad. Sci. USA*, **50**, 226
82. Means, A. R. and Hamilton, T. H. (1966). *Proc. Nat. Acad. Sci. USA*, **56**, 686
83. Means, A. R. and Hamilton, T. H. (1966). *Biochim. Biophys. Acta*, **129**, 432
84. Moore, R. J. and Hamilton, T. H. (1964). *Proc. Nat. Acad. Sci. USA*, **52**, 439
85. Greenman, D. L. and Kenney, F. T. (1964). *Arch. Biochem. Biophys.*, **107**, 1
86. Barnea, A. and Gorski, J. (1970). *Biochemistry*, **9**, 1829
87. Eilon, G. and Gorski, J. (1972). *Fed. Proc.*, **31**, 245 Abstr
88. Teng, C. S. and Hamilton, T. H. (1967). *Biochem. J.*, **105**, 1091
89. Teng, C. S. and Hamilton, T. H. (1967). *Biochem. J.*, **105**, 1101
90. Suvatte, A. B. and Hagerman, D. D. (1970). *Endocrinology*, **87**, 641
91. Notides, A. and Gorski, J. (1966). *Proc. Nat. Acad. Sci. USA*, **56**, 230
92. Mayol, R. F. and Thayer, S. A. (1970). *Biochemistry*, **9**, 2484
93. DeAngelo, A. B. and Gorski, J. (1970). *Proc. Nat. Acad. Sci. USA*, **66**, 693
94. Paul, J. and Gilmour, R. (1968). *J. Mol. Biol.*, **34**, 305
95. Spelsberg, T. C., Hnilica, L. S. and Ansevin, A. T. (1971). *Biochim. Biophys. Acta*, **228**, 550
96. Teng, C. S. and Hamilton, T. H. (1969). *Proc. Nat. Acad. Sci. USA*, **63**, 465
97. Teng, C. S., and Hamilton, T. H. (1970). *Biochem. Biophys. Res. Commun.*, **40**, 1231
98. Baulieu, E. E., Alberga, A., Raynaud-Jammet, C. and Wira, C. R. (1972). *Nature (London)*, **236**, 236
99. Barker, K. L. (1971). *Biochemistry*, **10**, 284
100. Raynaud-Jammet, C. and Baulieu, E. E. (1969). *C.R. Acad. Sci. Paris*, **268**, 3211
101. Arnaud, M., Beziat, Y., Guilleux, J. C., Hough, A., Hough, D. and Mousseron-Canet, M. (1971). *Biochim. Biophys. Acta*, **232**, 117
102. Arnaud, M., Beziat, Y., Borgna, J. L. Guilleux, J. C. and Mousseron-Canet, M. (1971). *Biochim. Biophys. Acta*, **254**, 241
103. Szego, C. M. (1965). *Fed. Proc.*, **24**, 1343
104. Mohla, A., DeSombre, E. R. and Jensen, E. V. (1972). *Biochem. Biophys. Res. Commun.*, **46**, 661
105. Katzenellenbogen, B. S. and Gorski, J. (1972). *J. Biol. Chem.*, **247**, 1299
106. Wira, C. and Baulieu, E. E. (1972). *C.R. Acad. Sci. Paris*, **274**, 73
107. Means, A. R. and O'Malley, B. W. (1971). *Acta Endocrinol.*, **Suppl. 153**, 318
108. Kohler, P. O., Grimley, P. M. and O'Malley, B. W. (1968). *Science*, **160**, 86
109. O'Malley, B. W. and McGuire, W. L. (1968). *J. Clin. Invest.*, **47**, 654
110. O'Malley, B. W., McGuire, W. L. and Korenman, S. G. (1967). *Biochim. Biophys. Acta*, **145**, 204
111. Means, A. R., Abrass, I. B. and O'Malley, B. W. (1971). *Biochemistry*, **10**, 1561
112. Nirenberg, M. W. (1963). *Methods in Enzymology*, **VI**, 17
113. Hahn, W. E., Church, R. B., Gorbman, A. and Wilmot, L. (1968). *Gen. Comp. Endocrinol.*, **10**, 438
114. Liarakos, C. D., Rosen, J. M. and O'Malley, B. W. (1973). *Biochemistry*, **12**, 2804
115. Cohen, S., O'Malley, B. W. and Stastny, M. (1970). *Science*, **170**, 336
116. Oka, T. and Schimke, R. T. (1969). *J. Cell. Biol.*, **43**, 123
117. Palmiter, R. D., Christensen, A. K. and Schmike, R. T. (1970). *J. Biol. Chem.*, **245**, 833
118. Comstock, J. P., Rosenfeld, G. C., O'Malley, B. W. and Means, A. R. (1972). *Proc. Nat. Acad. Sci. USA*, **69**, 2377

119. Means, A. R., Comstock, J. P. and O'Malley, B. W. (1971). *Biochem. Biophys. Res. Commun.*, **46,** 759
120. Comstock, J. P., O'Malley, B. W. and Means, A. R. (1972). *Biochemistry*, **11,** 646
121. Rosenfeld, G. C., Comstock, J. P., Means, A. R. and O'Malley, B. W. (1972). *Biochem. Biophys. Res. Commun.*, **46,** 1695
122. Means, A. R., Comstock, J. P., Rosenfeld, G. C. and O'Malley, B. W. (1972). *Proc. Nat. Acad. Sci. USA*, **69,** 1146
123. Rosenfeld, G. C., Comstock, J. P., Means, A. R. and O'Malley, B. W. (1972). *Biochem. Biophys. Res. Commun.*, **47,** 387
124. Tata, J. R. (1966). *Progr. Nucleic Acid Res. Mol. Biol.*, **5,** 191
125. Gorski, J., Noteboom, W. D. and Nicolette, J. A. (1965). *J. Cell. Comp. Physiol.*, **66,** Suppl. 1, 91
126. Williams-Ashman, H. G., Liao, S., Hancock, R. L., Jurkowitz, L. and Silverman, D. A. (1964). *Rec. Progr. Horm. Res.*, **20,** 247
127. Kenney, F. T., Wicks, W. D. and Greenman, D. L. (1965). *J. Cell. Comp. Physiol.*, **66,** Suppl. 1, 125
128. Nicolette, J. A. and Babler, M. (1972). *Arch. Biochem. Biophys.*, **149,** 183
129. Chan, L., Means, A. R. and O'Malley, B. W. (1973). *Proc. Nat. Acad. Sci.*, **70,** 1870

7
The Mode of Action of Glucocorticoids

W. D. WICKS
University of Colorado Medical Center, Denver

7.1	INTRODUCTION	212
7.2	INTRACELLULAR GLUCOCORTICOID RECEPTORS	212
	7.2.1 Liver and hepatoma cells	213
	7.2.2 Thymocytes	215
7.3	EFFECTS OF GLUCOCORTICOIDS ON THE SYNTHESIS OF SPECIFIC PROTEINS	216
	7.3.1 Liver	216
	7.3.1.1 Tyrosine transaminase	217
	7.3.1.2 Glutamylalanine transaminase	222
	7.3.1.3 Phosphoenol pyruvate carboxykinase	223
	7.3.1.4 Other liver enzymes	225
	(a) Tryptophan pyrrolase	225
	(b) Serine dehydratase	226
	7.3.2 Other tissues	227
	7.3.2.1 Glutamine synthetase—neural retina	227
	7.3.2.2 Glycerol phosphate dehydrogenase—glial cells	228
	7.3.2.3 Alkaline phosphatase—HeLa cells	229
7.4	GLUCOCORTICOIDS, CYCLIC-AMP AND THE PERMISSIVE EFFECT	230
	7.4.1 General considerations	230
	7.4.2 Evidence for or against glucocorticoid requirements for the action of cyclic-AMP and mediation of the effects of glucocorticoids by cyclic-AMP	232
7.5	REGULATION OF THE GROWTH AND FUNCTION OF LYMPHOID TISSUES BY GLUCOCORTICOIDS	233
7.6	CONCLUDING REMARKS	236
	NOTE ADDED IN PROOF	237

7.1 INTRODUCTION

The steroid hormones secreted by the inner zones of the adrenal cortex, the glucocorticoids, influence a remarkable number and variety of metabolic processes and cannot be as readily confined to a particular target organ or biological response as other steroid hormones. Because of this diversity of metabolic effects, and the frequency with which the subject has been reviewed[1-10], the present reviewer has taken the liberty of dealing with only selected aspects of the problem which, it is hoped, will highlight possible mechanisms by which glucocorticoids may exert their effects.

Glucocorticoids have traditionally been classified on the basis of their effects on carbohydrate and protein metabolism in contrast to the mineralocorticoids which act principally on sodium retention in the kidney[11]. The most common natural glucocorticoid is cortisol and, although it has weak mineralocorticoid activity, it is the most potent natural glucocorticoid[11]. Among the synthetic steroids, prednisolone (Δ^1-cortisol) is generally more potent as a glucocorticoid but less so as a mineralocorticoid, whereas dexamethasone (Δ^1,9α-fluoro,16α-methylcortisol) is one of the most potent glucocorticoids known and possesses no detectable mineralocorticoid activity[11]. The distinction between the two types of corticoid effects is somewhat arbitrary since, for example, aldosterone, the major natural mineralocorticoid does possess glucocorticoid activity[11]. (See Figure 7.1.)

It is virtually impossible to discuss the metabolic effects of glucocorticoids without also considering the intimate inter-relationship which exists between glucocorticoids, insulin and hormones that act via cyclic-AMP. In many ways overall metabolic homeostasis in higher organisms depends upon a complex interplay between these agents. As a consequence of this consideration, the involvement of insulin and cyclic-AMP as co-participants in, or antagonists of, the action of glucocorticoids will be discussed when deemed appropriate.

7.2 INTRACELLULAR GLUCOCORTICOID RECEPTORS

This subject has been reviewed extensively quite recently[3,12] and will be discussed here only briefly with special reference to possible mechanisms of action of the glucocorticoids. As will be seen, it is clear now that the primary site of recognition of external steroid input signals involves interaction of the steroid with a specific receptor in the cytosol, which is most probably a protein[3,12]. Thus, in this regard steroid hormones differ fundamentally from amino acid and polypeptide hormones which apparently interact only at the cell surface[13].

In order to fulfil specific physiological requirements, receptors should possess the properties of high affinity but low capacity of binding (i.e. can be saturated at physiological concentrations of steroid). They should also exhibit the requisite discrimination between active and inactive steroids in addition to being present only in tissues which respond to the hormone in question[12]. That such species exist for glucocorticoids in several tissues is attested to by the following paragraphs.

7.2.1 Liver and hepatoma cells

The liver appears to be a major target organ for glucocorticoids in view of the profound but selective effects they exert on its metabolism, and the fact that plasma corticosteroids are concentrated by the liver to a significant extent[14]. Radioactive cortisol accumulates rapidly in the liver reaching a maximum concentration 30–45 min after intraperitoneal injection, with the peak in the cytosol occurring prior to that in the nucleus[15]. (This time sequence correlates well with the kinetics of cortisol action on specific hepatic functions (see Section 7.3). Labelled cortisol and metabolites which form rapidly become distributed throughout the various sub-cellular compartments and are bound to a variety of macromolecules; this makes the task of determining which if any of these represent the active form of the hormone rather difficult. A number of techniques have been employed to overcome this problem, including amongst others sucrose density gradient centrifugation, gel filtration, equilibrium dialysis and charcoal adsorption[3,12,15-17]. (The fact that specifically-bound corticoids do not dissociate immediately from the receptors, especially at low temperatures, has allowed both the detection of specific binding and the use of a wide variety of physicochemical techniques for the study of the characteristics of the receptors[12,16,18]. The fact that non-specifically-bound steroids dissociate much more rapidly has also facilitated detection of specific receptors.) The specificity of binding has been tested by using different labelled glucocorticoids, by the ability of unlabelled steroids of varying biological potencies to displace bound radioactive steroids (at physiological concentrations) and by correlating the concentration-dependency of a specific physiological response produced by a given steroid with that of its binding to the putative receptor, and so forth[3,16,17].

Using approaches of this sort, four cytosolic receptors for glucocorticoids (and their metabolites) have been found in liver[3]. It appears, however, that only one of these is directly associated with physiological effects such as enzyme induction[17] (see Section 7.3.1.1). Studies by Tomkins and his colleagues have shown that the interaction of glucocorticoids with this receptor fulfils all the requisite temporal characteristics and biological specificities with respect to at least one physiological effect of glucocorticoids, namely induction of the synthesis of tyrosine transaminase[17-22] (see Section 7.3.1.1).

A single class of these receptor species exists within the cytosol and these species exhibit saturation at steroid concentrations which are optimal for induction of the transaminase[18]. (Non-specific binding components do not exhibit saturation kinetics even at high steroid concentrations[12,16,18].) Certain steroids, not known to induce enzymes, exhibit little affinity for the receptor, but others are able to compete with active corticoids for binding. These results strongly support the involvement of these receptors in the physiological action of glucocorticoids since the latter group of steroids (progesterone, in particular) is well-known to competitively inhibit the effects of corticosteroids on liver function[3,20,21]. The receptors appear to be proteins, possibly glycoproteins, on the basis of the sensitivity of *in vitro* binding to enzymatic attack[3,23,24].

The consequence of receptor interaction with a biologically-active corticosteroid appears to be a conformational change in the receptor[3,17,20,21]. It has been proposed (by analogy with the allosteric theory of Monod *et al.*[25]) that receptors exist in two configurations, 'active' or 'inactive' and that the latter predominate in the presence of low concentrations of corticoids or in their absence[17,21]. Upon binding an active glucocorticoid, the receptor is either induced to undergo a conformational shift[26] or else it is stabilised in the active state[25]. Support for this hypothesis is provided by the fact that binding of active glucocorticoids stabilises the receptor to heating, whereas binding of competitive inactive steroids does not[17,21]. The active steroid–receptor complex is then apparently transported to the nucleus where it binds to DNA[22], presumably in the form of chromatin[27]. Inactive steroid–receptor complexes are not found in the nucleus[21,22] nor will they bind to liver DNA *in vitro*[22]. By analogy with progesterone in the oviduct system[28], nuclear acidic proteins may well be involved in the binding of the steroid–receptor complex to specific DNA sites.

These findings demonstrate that the specificity of the biological action of a glucocorticoid is not due simply to its ability to bind to specific receptors. Rather, it is the conformational change in the receptor, which is somehow facilitated by interaction with active glucocorticoids, that determines the overall specificity of the response. Presumably, most of the differences in the biological potencies of various corticosteroids and analogues (other than those due to differences in uptake and metabolism, etc.) can be ascribed to differences in their interaction with the receptor)[17,20]. Clearly, this mechanism provides great flexibility for regulation in that the effects of glucocorticoids can be efficiently and reversibly[3,17-19] inhibited at the first step in the chain of events leading to a biological response.

Although the presence of active steroids in the nucleus argues for a nuclear site of action of glucocorticoids (anti-inducer complexes do not enter the nucleus and adrenalectomy causes depletion of nuclear receptors[17,21] etc.), it neither provides a precise explanation for the mechanism of action of glucocorticoids nor does it rule out effects of the complex in the cytosol[8,29]. There is no reason, *a priori*, to assume that glucocorticoids exert their effects only by virtue of influencing transcriptional processes. There are obviously several steps between transcription of a specific genome and the formation of a finished polypeptide which could be considered as potentially subject to regulation by corticosteroids (e.g. turnover of mRNA, transport of mRNA from the nucleus to the cytoplasm, interaction of mRNA with ribosomal sub-units and initiation factors to form a productive polysome complex, elongation of the growing polypeptide chain and release of the completed protein, etc.).

It is possible that the conformational change in the receptor may be of more significance than its entry into the nucleus. The observation that an analogue of oestradiol with oestrogenic activity exhibits little uptake into uterine nuclei suggests that the latter possibility cannot be eliminated, *a priori*[30]. A similar conclusion is suggested by the report that antiglucocorticoid steroid–receptor complexes can be found in thymus nuclei[31].

It seems most likely, however (as will be seen later), that interaction with chromatin does represent the culmination of the initiation of the biological response to glucocorticoids. Thus, the simplistic view (based in many cases on originally insufficient or incorrect data) that steroids act by influencing the level of specific messenger-RNA species[32] appears to be correct after all. The recent results of O'Malley's and Schimke's groups have demonstrated rather convincingly that this is the case with oestrogenic steroids[33, 34].

7.2.2 Thymocytes

Although the thymus does not appear to concentrate glucocorticoids, thymocytes possess specific receptors for these hormones in the cytosol[3, 31, 35-37]. Isolated thymocytes have proved particularly useful for these studies[3, 35-37]. The kinetics of the uptake and the binding of glucocorticoids are more rapid than in liver and there is a correlation between the ability of various corticoids to compete for binding with their ability to inhibit glucose transport; this appears to be the primary biological effect of glucocorticoids in this tissue[3, 31, 35-37]. (See Section 7.5.)

11-Deoxycortisol has been shown to competitively block the effects of glucocorticoids on thymocytes[31, 35], and it also competes effectively for binding to the specific receptors[31, 35]. Incubation of glucocorticoid–receptor complexes with isolated, purified thymus nuclei at 37°C (but not at 0°C) leads to nuclear uptake of the complex[38] (as has been observed in other tissues with other steroid–receptor complexes[3, 12, 16]). These receptors also appear to be proteins[31, 38].

As pointed out by Munck in the case of the thymus[39] (and by others with other cells[17]), the steroids do not require the receptor in order to enter the nucleus since uncomplexed steroids can be shown to enter the nucleus apparently freely[31, 39]. Conversely, however, it appears likely that (in the liver at least) the steroid represents a necessary vehicle for receptor entry into the nucleus (see above)[17, 21, 22]. The fact that 11-deoxycortisol–receptor complexes apparently gain entry into the nucleus[31, 39] suggests that, in the thymus at least, receptor transport may not depend upon the biological activity of the bound steroid, although initiation of a response clearly does.

Fibroblasts and lymphocytes probably possess similar receptors[40-42]; this would also be predicted to be true for the neural retina[43] (Figure 7.1). As shown in Figure 7.1, there is a clear correlation between the structure and the biological efficacy of a variety of glucocorticoids and other steroids on metabolic processes in liver[5, 20, 44, 45], retina[43], fibroblasts[46], glia[47] and thymocytes[31, 35, 42]. These results suggest that specific glucocorticoid receptors in different target tissues are similar if not identical[17].

It is of interest in this regard that two groups have reported that lymphosarcomas whose growth is resistant to glucocorticoids (see Section 7.5) contain a reduced number of glucocorticoid receptors[48, 49]. Similarly, Levisohn and Thompson[50] have found that clones of HTC hepatoma cells which exhibit

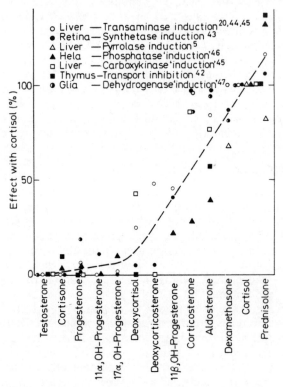

Figure 7.1 Structure–function relationships of various glucocorticoids and other steroids on selected metabolic processes

reduced induction of tyrosine transaminase by corticoids exhibit reduced steroid binding. A fibroblast subline which is resistant to the growth inhibitory effects of glucocorticoids also has a markedly reduced capacity to bind glucocorticoids[40]. Taken together, these results leave little doubt that the receptors under study are specifically involved in the action of glucocorticoids.

7.3 EFFECTS OF GLUCOCORTICOIDS ON THE SYNTHESIS OF SPECIFIC PROTEINS

7.3.1 Liver

In contrast to the general metabolic effects exerted by sex steroids on their target tissues (see Chapters by Liao and O'Malley), glucocorticoids exert a rather selective action on liver metabolism. Although a substantial increase in overall hepatic RNA synthesis is produced within 1–2 h of administration of corticoids to rats[51-53], only a slight increase in liver protein synthesis can be detected[1,54]. However, the activities of a number of liver enzymes have been shown to be increased by the injection of corticosteroids. The following discussion will consider a selected number of these enzymes with special emphasis on cases where it has been established that the increase in activity is due to an increase in the rate of synthesis of the enzyme in question.

7.3.1.1 Tyrosine transaminase: (E.C. 2.6.1.5)

Regulation of tyrosine transaminase by glucocorticoids has been reviewed with increasing regularity[1,4-8,29,55], and although the dramatic effects of glucocorticoids on the activity of this enzyme have been studied intensively for the past 15 years the precise mechanism by which the corticoids act remains elusive. Currently there are two opposing hypotheses to explain the action of glucocorticoids on the transaminase[1,4,7,8,29,56]. Different groups of workers have marshalled impressive bodies of data, which, in certain crucial instances, are totally contradictory; the basis for the disagreement remains inexplicable. These hypotheses have polarised thinking about regulation of the transaminase to an unfortunate extent and little clarification has been provided by recent research. In addition, a series of paradoxical and seemingly unexpected results have been reported during the past few years which have added to the confusion. The present discussion will focus on what appear to be the basic features of the induction process and will not dwell at length on the, so far, poorly understood paradoxes.

Glucocorticoids increase the activity of hepatic tyrosine transaminase in an amazing variety of systems including intact and adrenalectomised rats[1,5,51,53], foetal rat liver in organ culture[57], isolated perfused liver[58], liver slices[59], cultured hepatoma[1,4,7,8,56,60] and normal rat liver cells[61,62]. There is a consistent 45–90 min lag in all these systems before any increase in transaminase activity can be detected, regardless of the dose of steroids, suggesting strongly that the lag period is a fundamental component of glucocorticoid action[1,4-8,51,53,55-62]. This period presumably reflects the time required by processes associated with uptake and interaction of the steroid with specific receptors and the subsequent biological consequences of this interaction (see Section 7.2.1). In this regard, the uptake of labelled dexamethasone into the nucleus of hepatoma cells is at a maximum approximately 30 min after its addition[18].

It has been found that glucocorticoids act by increasing the rate of synthesis of the transaminase[1,29,63]. The earliest increase in the rate of transaminase synthesis (measured radioimmunochemically) can be detected 45–60 min after administration of corticoids with the maximum increase occurring sometime between 1.5 and 3 h[1,29,63]. Although there is considerable controversy over possible changes in the rate of degradation of the enzyme during induction[1,64,65], it is clear that the predominant effect of corticosteroids is on transaminase synthesis[1,4-8,29,55-57,63,66].

In rats, induced transaminase activity reaches a peak at 4–6 h and then begins to return to the basal level after a single injection of cortisol[1,67], but it can be maintained at an elevated level (5–20-fold above the basal level) indefinitely, provided steroid is administered repeatedly[67]. In the various isolated systems, elevated enzyme activity can be maintained for many hours after a single addition of hormone, apparently the result of reduced steroid metabolism[1,4,7,8,29,51,53,56-58,60,61,63]. Upon withdrawal of the steroid, however, transaminase activity also returns rapidly to the basal level after a lag period of 1–2 h[1,56,57,60,63,68-71]. There are claims that the lag is reduced if progesterone, an anti-inducer, is added at the time of corticoid removal[20]. Cycloheximide addition prompts an immediate first-order decay of pre-induced enzyme activity[1,68,71], at least initially (see below).

Actinomycin D added just before, with, or slightly after corticosteroids prevents the increase in transaminase activity in all the liver systems[1,4-8,29,53,55-58]. Addition of low concentrations of the antibiotic (which inhibit RNA synthesis by 80–95%) after pre-induction leads to a decay in enzyme activity after a brief lag period[56,68]. Indeed, the sequence of the decay under these conditions is essentially that seen after withdrawl of steroid[1,56,60,63,68-71]. These data were initially interpreted to suggest that glucocorticoids caused an increase in the rate of synthesis of messenger-RNA for the transaminase[1,51,53]. This seemed especially appealing in view of the increase in RNA synthesis produced in rat liver by cortisol[51,53]. It was suggested that the transaminase template is unstable and that this would account for the decline in elevated enzyme activity after withdrawal of steroid or addition of actinomycin D[1,56]. The lag before initiation of the decline was viewed as suggesting that the half-life of the template was about 1–3 h[56]. Addition of actinomycin D to both basal and steroid-treated hepatoma cells led to a similar rate of decay in the ability of the cells to synthesise the transaminase, as measured radioimmunochemically[1,56,69,72]; this also suggested that the half-life of the template was about 1–3 h.

Perhaps the most convincing support for the concept that glucocorticoid treatment does lead to an increase in the level of transaminase messenger-RNA has been provided by the studies of Peterkofsky and Tomkins[73]. These studies were carried out with HTC hepatoma cells in suspension culture[73]. The cells were incubated with dexamethasone and cycloheximide, either alone or in combination, for 1.5–2.5 h and then the cells were centrifuged, washed thoroughly and re-suspended in fresh medium. Various additions were made as indicated in Figure 7.2(a) and the cells were re-incubated for 3–5 h further. It was found that although little increase in activity took place during the pre-incubation, a rapid but limited rise without any lag occurred upon transfer of cells exposed to dexamethasone and cycloheximide to fresh medium with or without fresh steroid. Pre-incubation with cycloheximide alone, however, did not generate any significant increase during the re-incubation period. If no additions were made, the elevated transaminase activity began to decay after about 2 h, whereas fresh steroid addition sustained the increase well beyond this point.

These results suggested that corticoids cause the formation of an intermediate independent of protein synthesis, which has a half-life of 1.5–3 h. Inclusion of actinomycin D during the pre-incubation suppressed for the most part the subsequent increase in transaminase activity; this suggests that

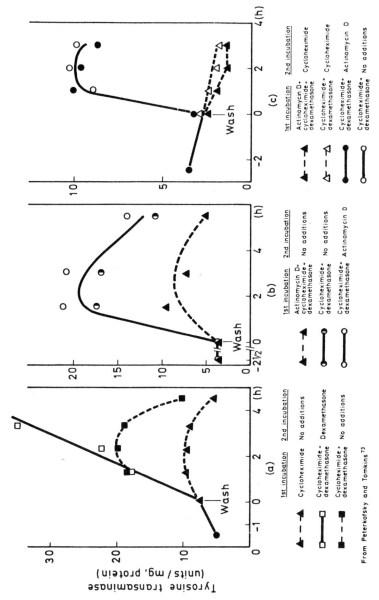

Figure 7.2 Evidence for accumulation of messenger-RNA for tyrosine transaminase during glucocorticoid action in HTC hepatoma cells. (From Peterkofsky[7] and Tomkins[73], by courtesy of National Academy of Science, USA.)

formation of the intermediate is dependent upon RNA synthesis (Figures 7.2(b) and (c)). Addition of actinomycin D in the second incubation period had no effect on the subsequent rise and fall in enzyme activity (Figure 7.2 (c)). On the other hand, cycloheximide added during the second incubation prevented any increase in transaminase activity (Figure 7.2 (c)).

The simplest interpretation of these results is that corticoids stimulate transcription of the structural gene for tyrosine transaminase and that this leads to an increase in translatable messenger-RNA. The elevated messenger-RNA level cannot be expressed in the presence of cycloheximide, but upon its removal an essentially immediate maximal rate of transaminase synthesis can be achieved for a brief period even in the absence of steroid and independently of RNA synthesis. What is not clear from these experiments is whether the steroid actually increases the rate of messenger-RNA synthesis or simply prevents its breakdown. (As pointed out by Tomkins and Martin[8], accumulation of messenger-RNA by an inhibition of degradation produced by glucocorticoid would also require continued RNA synthesis and would, thus, be inhibited by actinomycin D.) This uncertainty lies at the heart of the current controversy over the mechanism of corticoid action.

Unfortunately, several facts came to light which seriously challenged the validity of the simple gene activation hypothesis. One of the first reservations about this model came when it was observed that there was no detectable effect of corticoids on overall RNA synthesis in the isolated culture systems, even though transaminase activity was increased to the same extent[1,57,74]. It had always been difficult to explain the relationship between the magnitude of the increase in RNA synthesis in rat liver (2–3-fold) with the increased synthesis of only a few minor proteins. It became clear that any effect of corticosteroids on RNA synthesis that was related to enzyme synthesis must be very selective and consequently difficult to detect. Tyrosine transaminase and the other hepatic enzymes affected by glucocorticoids constitute a fractional percentage of total liver protein emphasising the fact that detection of an increase in the amount of the relevant messenger-RNAs would be (and still is) a challenging technical feat.

The most serious discrepancy in the simple model came, however, from studies with metabolic inhibitors. It was observed that although high concentrations of actinomycin D also prevented induction of the transaminase the inhibitor either maintained or increased still further the elevated enzyme activity after pre-induction whether steroid was removed or not[1,4,7,8,29,56,70,72]. A similar phenomenon had been observed much earlier with the transaminase in rat liver[75]. The significance of these results was always questioned, however, because of the multiple hormonal regulators of the transaminase[1] and the possibility of enhanced release of a potentially synergistic hormone whose action was not blocked by the antibiotic[69,76]. With the discovery that the same effect could be obtained in cultured cells, however, this explanation was no longer sufficient.

Thompson et al.[60] reported that the elevation in transaminase activity caused by addition of high concentrations of actinomycin D could be blocked by puromycin. These results suggested that the increase depended upon protein synthesis. Later, radioimmunochemical evidence was provided that[66] enzyme synthesis was in fact further enhanced by addition of actinomycin D.

However, the pulse-labelling interval used was greater than one half-life of the enzyme allowing for the possibility that reduced degradation could have contributed to the increased transaminase labelling.

From these and other data (summarised in Refs. 4, 7, 8 and 29), the Tomkins group proposed a model invoking a labile repressor which somehow was required for degradation of transaminase messenger-RNA or which blocked its translation. Although low concentrations of actinomycin blocked transcription of the transaminase structural gene, it was proposed that transcription of the putative regulatory gene was not inhibited until very high concentrations of the antibiotic were added. Under these conditions the labile repressor would decay rapidly, thereby either immortalising transaminase messenger-RNA or rendering its translation constitutive. Since the glucocorticoid was viewed as acting by combining with the repressor, thereby neutralising its effects, there was no requirement for steroid to maintain the superinduced state. The original model[77] has undergone several modifications, and recently it has been proposed[22] that the steroid acts within the nucleus (see Section 7.2.1), suggesting that it either inhibits the synthesis of the regulatory gene product or prevents its action on transaminase messenger-RNA.

At the same time, careful studies by Kenney and his colleagues showed unequivocally that under their conditions high concentrations of actinomycin D drastically and rapidly inhibited the breakdown of the transaminase at the same time that the rate of transaminase synthesis was also inhibited[1,56,72]. Lower concentrations of the antibiotic did not exhibit this effect on breakdown of the enzyme. (As suggested by Lee et al. there is little difference in the extent of inhibition of RNA synthesis with low (0.1–0.2 µg ml^{-1}) and high (2–5 µg ml^{-1}) concentrations of actinomycin D, raising the possibility that superinduction may be unrelated to inhibition of RNA synthesis *per se*[56] [see below].) The increase in transaminase activity during superinduction is explained by these workers as resulting from a rapid inhibition of degradation coupled with an exponential decay in the rate of elevated enzyme synthesis[1,56,72]. Since synthesis continues at an elevated albeit diminishing rate, at the same time that the rate of degradation is greatly reduced transaminase activity would be expected to rise during an interval of time dependent upon the half-life of its messenger-RNA. This explanation appears to be fully consistent with the kinetics of superinduction[1,56,72]. Since continued synthesis of the transaminase would be required for the increase in enzyme activity to take place, the inhibitory effects of puromycin are to be expected[60]. Pulse-labelling studies using much shorter intervals than were employed by Tomkins' group revealed an exponential decay in the rate of enzyme synthesis in both HTC and H35 hepatoma cells with a half-time equivalent to that seen after addition of non-superinducing concentrations of actinomycin D[56]; this suggested that no change in template stability takes place during superinduction.

Tomkins subsequently reported that although the rate of transaminase breakdown is not blocked by actinomycin D under his conditions, considerable inhibition could be achieved if Kenney's 'step-down' conditions were employed[78]. Thus, one is apparently faced with the unappealing prospect of two different explanations for the same phenomenon under two sets of conditions.

The literature is replete with other reports of paradoxical effects of substances such a 5-fluorouracil, 8-azaguanine, mercaptobenzimidazole, cycloheximide and puromycin on the transaminase[1,7,8,29,56,65,71,79-81]. Evidence has been provided which shows that most, if not all, of these agents probably act by inhibiting transaminase breakdown[1,65,79]. It is of interest in this regard that conditions favouring accumulation of aminoacylated transfer-RNA in *E. coli* lead to inhibition of protein breakdown[82]. It has been suggested that an analogous phenomenon could account for many of these effects of inhibitors of RNA and protein synthesis on tyrosine transaminase breakdown[82]. Lee and Kenney have observed that high concentrations of leucine cause both an increase in the rate of transaminase synthesis and a decrease in its rate of degradation in H35 hepatoma cells[83]. An interesting facet of these observations is that one of the three isoaccepting leucyl transfer-RNAs in these cells is apparently uncharged under the usual conditions where the concentration of leucine was moderate, but is charged in the presence of high concentrations of leucine[83]. These results are in accord with the view that unacylated transfer-RNA species may participate in protein degradation[79,82,83]. Alternatively, inhibition of energy metabolism by some of these metabolic inhibitors could also lead to decreased protein breakdown[82].

At the present time it is virtually impossible to decide which of these models, if either, is correct. The fact that the templates for several proteins whose synthesis is (or probably is) increased by glucocorticoids appear to be rather stable (see Sections 7.3.2.1, 7.3.2.2, 7.3.2.3) suggests that the Tomkins model is unlikely to explain all steroid-mediated inductions. The simplest model[56] remains the most likely in this reviewer's opinion; it is to be hoped that more refined techniques will provide information differentiating between either stimulation of the synthesis, or inhibition of degradation, of transaminase messenger-RNA by glucocorticoids.

7.3.1.2 Glutamylalanine transaminase: (E.C. 2.6.1.2.)

In view of the important role of alanine in gluconeogenesis[84], regulation of the activity of hepatic glutamylalanine transaminase is of special interest[85]. Rosen and his collaborators, as well as others, have shown that glucocorticoids produce a slow but large increase in the activity of this enzyme in rat liver[5,6,86,87]; this has been found to be due to an increase in the rate of *de novo* enzyme synthesis[88]. A maximum increase in the rate of enzyme synthesis can be detected at a time when only a marginal increase in enzyme activity has taken place[89]. The stimulation of glutamylalanine transaminase synthesis by corticoids, however, does not appear to be increased as rapidly as that of tyrosine transaminase[90]. These results suggest that the notion of a coordinate effect of steroids on the synthesis of a series of key enzymes may be oversimplified. On the other hand, differential temporal effects could result from differences in rate of transcription of the various genes involved, in transport to the cytosol of the respective template RNAs and initiation of specific protein synthesis. The difference in the rate of increase in alanine transaminase *activity* and that of tyrosine transaminase, however, is primarily attributable to the large differences in the rates of degradation of these two enzymes[1,88].

The report of an increase in the rate of breakdown of alanine transaminase after steroid administration[88] has not been confirmed[91].

Lee and Kenney have demonstrated in studies with H35 hepatoma cells in culture[92] that the effects of corticoids on alanine transaminase result from a direct interaction with liver cells. Addition of cortisol at a concentration of 0.5×10^{-6} mol l^{-1} produced a maximum twofold increase in alanine transaminase activity in these cells[92]. The difference in the magnitude of the increase in alanine transaminase activity *in vivo* and in hepatoma cells may be due in part to the fact that factors other than glucocorticoids appear to contribute to regulation of its synthesis in rats[86, 87]. The similarity with PEP carboxykinase is striking in this regard (see Section 7.3.1.3) and suggests that cyclic-AMP may also be a positive effector of alanine transaminase whereas insulin may exert a repressive influence[93]. Other workers, however, have not observed a significant effect of the polypeptide hormone on the activity of this enzyme[94].

Unfortunately, as a consequence of the slow rise in enzyme activity, it has been difficult to utilise inhibitors of RNA and protein synthesis in animals because of their toxicities. It is to be hoped, however, that through the use of cell culture systems and radioimmunological techniques some information will be forthcoming about the role of RNA synthesis in the effects of corticosteroids on this enzyme.

In contrast to serine dehydratase (see Section 7.3.1.4) there is no evidence that multiple cytosolic forms of alanine transaminase exist[95].

The physiological significance of the regulation of this enzyme is subject to some question because of its slow response to corticosteroids. As pointed out[1] it is difficult to make a case for the participation of induced alanine transaminase in the *initiation* of gluconeogenesis. However, under conditions of extended stress and starvation increases in the activity of this enzyme might play an important role in maintaining the flux of alanine to pyruvate required for continued high rates of glucose synthesis. This may be especially true if extended starvation slows protein catabolism, thus providing smaller amounts of alanine for hepatic utilisation[84].

7.3.1.3 *Phosphoenol pyruvate carboxykinase:* (E.C. 4.1.1.32)

The initial events in the conversion of pyruvate to glucose in rat liver involves a complicated series of reactions which by-pass the physiologically irreversible pyruvate kinase step and which lead ultimately to the synthesis of PEP[94, 96-98]. The last of the enzymes in this by-pass pathway is PEP carboxykinase. This enzyme appears to operate under physiological conditions in the direction of decarboxylation of oxalacetate to PEP (in the presence of ITP or GTP and Mg^{2+} or Mn^{2+} ions)[94, 99, 100]. The enzyme is variably distributed between the cytosol and mitochondrion in mammalian liver[99-102], and the activity of the cytosolic isozyme has been found to be subject to regulation by various hormonal and dietary factors including glucocorticoids[94, 98, 99, 103]. Although the mitochondrial enzyme in sheep liver[101] has been reported to be modified to a slight extent by diabetes, in general the activity of this isozyme remains unchanged.

Shrago et al.[104] were the first to study possible factors involved in the regulation of carboxykinase activity, and found that although adrenalectomy did not significantly depress basal enzyme activity administration of cortisol led to a 2–3-fold increase in its activity. The lack of a decrease in enzyme activity following adrenalectomy implies that basal carboxykinase synthesis does not depend upon circulating glucocorticoids and that resting levels of these hormones do not exert a significant influence on PEP carboxykinase activity (at least in non-diabetic rats). Although alloxan diabetes produces a much greater increase in carboxykinase activity, adrenalectomy reduces this response to about that seen with glucocorticoids alone[104]. It is now known that cyclic-AMP also stimulates the synthesis of the carboxykinase[105-109], and that insulin counteracts this effect[105-107] as well as that of corticoids[104-107]. It seem likely, therefore, that the dramatic increase in the activity of this enzyme in diabetes is the result of the unrestrained effects of *both* adrenal steroids (no longer opposed by insulin) and cyclic-AMP (whose concentration is increased in diabetes[110]), which results in a synergistic increase[111]. Adrenalectomy thus reduces the increase in carboxykinase activity caused by diabetes to that seen with fasting in intact rats[104] ($\sim$ 2–3-fold), which apparently represents the cyclic-AMP component of the response[109,111]. The twofold increase produced by cortisol in fed, adrenalectomised rats presumably largely reflects the steroid component of the response[111,112].

The effects of corticosteroids on carboxykinase activity have been shown to be direct by studies with cultured H35 hepatoma cells[109,111] where addition of cortisol or dexamethasone prompts a 2–3-fold increase in the activity of this enzyme. The concentration-dependency of the response suggests that the regulation of this enzyme can be influenced within the normal fluctuations of glucocorticoid concentrations in rat plasma[113].

Similar temporal relationships have been found in the response of PEP carboxykinase to glucocorticoids as with tyrosine transaminase[111]. The characteristic lag of 1.5–2 h is observed both in rat liver[112,114] and cultured cells[111] after administration of glucocorticoids, in contrast to the almost immediate increase following addition of dibutyryl cyclic-AMP[106,111,115]. The lag in the increase in carboxykinase activity presumably results from the same factors that were discussed for the transaminase (see Section 7.3.1.1). The increase in carboxykinase activity produced by corticoids is prevented both by inhibitors of both RNA[104,111,112,114,115] and protein[104,109,111] synthesis. In addition, immunochemical titration analysis has shown that the increase in catalytic activity generated by glucocorticoids is accompanied by a concomitant increase in immunological reactivity[111]. Although these results suggest that corticoids enhance the *de novo* synthesis of PEP carboxykinase, direct proof of this possibility is still lacking.

A variety of steroids have been tested for their effects on PEP carboxykinase and tyrosine transaminase activities, and although there was a wide range of effects no differences were detected in the relative responses of these two enzymes (see Figure 7.1)[45]. Furthermore, a 20-fold molar excess of progesterone prevented the effects of dexamethasone on both enzymes although the responses to dibutyryl cyclic-AMP were not altered[111]. These results suggest that the same steroid receptor species may be involved in the induction of both enzymes.

Insulin acts more directly on PEP carboxykinase than merely by decreasing the intracellular cyclic-AMP concentration, since the polypeptide hormone prevents the effect of glucocorticoids as well as that of dibutyryl cyclic-AMP on this enzyme in hepatoma cells[111]. The mechanism by which insulin exerts such differential effects on the carboxykinase and tyrosine transaminase[105, 111] is not known at present, but the inhibition of glucocorticoid action appears to be of the non-competitive type[111].

Glucocorticoids seem to influence the synthesis of the carboxykinase and tyrosine transaminase by a similar mechanism. Actinomycin D added together with, or before, dexamethasone suppresses any rise in both enzyme activities[111]. Although pre-incubation of cells with dexamethasone and cycloheximide does not lead to any increases in either enzyme, transfer to fresh medium devoid of additives prompts a dramatic rise in both enzymes during the next 2 h; this rise in activities is not suppressed by actinomycin D[109, 116]. Addition of cycloheximide during the second incubation, however, prevents any increase in the activity of either enzyme. These results suggest that, as with tyrosine transaminase[73] (see Figure 7.2), glucocorticoids somehow give rise to an increase in available template RNA for PEP carboxykinase.

There is one basic difference in the pattern of regulation of these two enzymes in that high concentrations of actinomycin D do not superinduce the carboxykinase[115]. In the absence of any paradoxical effects of inhibitors, it would appear that the simplest interpretation of the available data is that glucocorticoids enhance transcription of the structural gene for PEP carboxykinase.

In striking contrast to the hepatic enzyme, the activity of PEP carboxykinase in adipose tissue is depressed by glucocorticoids and elevated by adrenalectomy[116, 117]. It has been shown by immunochemical analysis that the enzyme in adipose tissue is probably identical with the hepatic cytosol enzyme[118]. The decay in adipose tissue carboxykinase activity generated by triamcinolone was actually reported to be greater than with cycloheximide[116], although only one injection of the antibiotic was administered. These results suggested that the steroid increased the rate of carboxykinase degradation. Recent work, however, suggests that it is more likely that triamcinolone inhibits transcription of the carboxykinase structural gene[117].

This apparent positive–negative influence of glucocorticoids is reminiscent of the inhibition of RNA synthesis in fibroblasts (see Section 7.5) and stimulation of RNA synthesis in liver by glucocorticoids[53]. Clearly the nature of the response of different tissues to corticosteroids is dictated by the specific requirements and state of differentiation of that tissue.

7.3.1.4 Other liver enzymes

(a) *Tryptophan pyrrolase:* (E.C. 1.13.1.12)—The effects of glucocorticoids on tryptophan pyrrolase have been reviewed extensively previously[1, 5-7, 9, 55, 119, 120] and no major new developments have taken place recently. The increase in enzyme activity is due to an increase in new enzyme synthesis with kinetics of response to corticoids very similar to that of tyrosine transaminase[1, 5-7, 9, 55, 75, 119, 120]. Actinomycin D clearly suppresses this effect at low concentrations but also leads to superinduction of the pyrrolase at high concentrations[75], although this has been disputed[121]. In contrast to tyrosine transaminase, however, degradation of this enzyme is not blocked by inhibitors of proteins synthesis[75].

The exact mechanism by which corticoids enhance the synthesis of this enzyme is at present no clearer than is the case with tyrosine transaminase. Progress has been hampered by the curious finding that cultured hepatoma cells appear to contain little, if any, pyrrolase activity. It is clear that steroids act directly on liver cells to induce tryptophan pyrrolase as shown by Goldstein et al. in studies with perfused liver[122].

(b) *Serine dehydratase:* (E.C. 4.2.1.13)—Although serine dehydratase activity is elevated after administration of glucocorticoids[123, 124], the effect is dependent upon the dietary status of the rat. Since cyclic-AMP (and glucagon) also elevate the activity of this enzyme, independent of dietary status, there is reason to question the primacy of glucocorticoids as inducers[106, 125-127]. It seems quite possible that corticoids may act by facilitating the effects of endogenous cyclic-AMP[128]. In the absence of an isolated culture system for study of serine dehydratase regulation, this question may be difficult to answer.

On the other hand, recent reports that serine dehydratase exists as two distinct isozymes (one of whose synthesis appears to be specifically influenced by corticoids and the other by cyclic-AMP (glucagon)) make it difficult to rule out a role for glucocorticoids in regulating the synthesis of this enzyme[129, 130]. The two isozymes may not be different gene products, but could just as easily represent a post-translational modification of the primary gene product whose synthesis is increased by glucocorticoids[130]. In any event, the effect of corticoids on serine dehydratase synthesis are far less substantial than those of cyclic-AMP.

There are other hepatic enzymes which exhibit some degree of response to glucocorticoids, but in most of these cases considerable doubt exists as to the directness of the steroid effect and whether enzyme synthesis is influenced[1, 6].

From the results obtained so far with liver enzymes, it seems clear that the effects of glucocorticoids in each case are qualitatively similar. There is general agreement with the conclusion that the level of specific template RNAs is increased by glucocorticoids, but little agreement on the mechanism involved. It seems certain, however, that an 'induced protein' intermediate, as is the case in the action of oestradiol on uterine metabolism[156], is not involved in the regulation of hepatic enzyme synthesis; the enzymes are themselves induced proteins and the only prerequisite for the response is an effect on RNA metabolism.

There is considerable controversy concerning the regulation of hepatic gluconeogenesis by glucocorticoids and also about the role played by enzyme induction in this process[1, 44, 85, 112, 114, 128, 131-135]. Although the possibility exists that induction of these enzymes is completely unrelated to gluconeogenesis, it seems more probable that they play a supporting role by enhancing metabolite flux over and above that produced by increased delivery of substrates to the liver during extended periods of stress and starvation. The controversy surrounding the directness of the role of corticosteroids in regulating hepatic gluconeogenesis may result in part from the multiplicity of their

ACTION OF GLUCOCORTICOIDS

effects. Thus, glucocorticoids enhance the delivery of glucogenic substrates to the liver[84, 85, 132-134], 'sensitise' the liver to changes in the levels of endogenous cyclic-AMP[128, 135] (see Section 7.4) and stimulate the formation of enzymes whose concentration normally limits the rate of metabolism of various glucogenic intermediates[1, 6, 89, 106, 111, 132]. In this context, it seems more consonant with existing data to propose a supporting, rather than an initiating, role for glucocorticoid-mediated enzyme induction in gluconogenesis[1, 132].

7.3.2 Other tissues

7.3.2.1 Glutamine synthetase—neural retina: (E.C. 6.3.1.2)

Glutamine synthetase is subject to exceedingly complicated regulation in *E. coli*, primarily by covalent modification of the pre-existing enzyme[136]. In the neural retina of chick embryos, as well as in some other tissues[137], this enzyme is also regulated, but in these cases it is by virtue of enhanced synthesis of new enzyme[138]. The activity of this enzyme normally increases spontaneously beginning at about the 16th day of development in chick embryos[139], but it can be induced precociously by glucocorticoids[138-140]. The characteristic lag of about 2 h is also observed in this system[138, 141, 142] as well as inhibition of the responses by cycloheximide and actinomycin D[141, 143]. No gross effects of cortisol on retinal metabolism have been found to occur[141, 142] suggesting that regulation of the synthesis of glutamine synthetase also represents a highly specific action of corticosteroids.

The half-life of the synthetase appears to be very long since it does not decay appreciably in the presence of cycloheximide or after withdrawal of cortisol[140, 141, 143]. The possibility still remains, however, that its half-life is much less than 20 h if (a) the template for this enzyme is quite stable (and thus elevated enzyme synthesis would persist for many hours) and/or (b) cycloheximide blocks the turnover of the enzyme. In any case, the temporal increase in synthetase activity is consistent with an enzyme half-life of at least 10–15 h[140, 143].

Incubation of retinas with cortisol and cycloheximide, followed by transfer to medium without cycloheximide, led to a nearly normal increase in enzyme activity even in the presence of actinomycin D suggesting, once again, the involvement of RNA synthesis in the initial action of glucocorticoids[141]. Paradoxical effects of inhibitors of protein and RNA[141, 143, 144] synthesis have also been observed with this enzyme, illustrating further the parallelism between glutamine synthetase and tyrosine transaminase. A major difference exists, however, in that the enzyme and template for the synthetase appear to be stable[141] and thus inhibition of degradation of the enzyme (or template) is not likely to explain the resistance to actinomycin D[143]. It seems more likely that the template RNA for glutamine synthetase may be more stable than the average mRNA[141] and is able to compete very efficiently for inoperative ribosomal sub-units. This would result in an enhanced frequency of initiation of translation of the messenger for glutamine synthetase to an extent which more than compensates for the slow turnover of template[145].

Recent work has provided suggestive evidence that cortisol does stimulate the formation of template RNA for glutamine synthetase[142, 146]. Sarkar and Moscona observed that cortisol increased the incorporation of [^{14}C]aspartate (the synthetase is especially rich in this amino acid) into nascent polypeptides associated with polysomes of the requisite size to be engaged in the synthesis of synthetase sub-units[146]. Pulse-labelling with [^{3}H]uridine after cortisol pretreatment, followed by incubation with actinomycin D for 2 h to allow decay of short-lived templates[141], revealed an increase in rapidly-labelled RNA species associated with the polysomes putatively engaged in synthetase synthesis[146].

Similar conclusions were reached by Schwartz using refined techniques including radioimmunochemical analysis of enzyme synthesised on polysomes *in vitro* after steroid treatment *in vivo*[142]. Although no effect of cortisol on overall RNA (or protein) synthesis was detectable, the synthesis of a high molecular weight RNA species associated with polysomes of the appropriate size was found to be enhanced[142]. The molecular weight of this RNA species is of sufficient size to code for synthetase sub-units suggesting that it could be the messenger-RNA for this protein. In the absence of evidence demonstrating that this RNA species is capable of directing the synthesis of glutamine synthetase (see Refs. 33, 34), these results must be interpreted as being only strongly suggestive. These data do, however, provide the only direct evidence that glucocorticoids act by stimulating formation of messenger-RNA rather than by inhibiting its degradation. It is of interest that the steroid also seems to enhance the formation of a low molecular weight RNA species (5–7S) whose function is totally unknown.

7.3.2.2 Glycerol phosphate dehydrogenase—glial cells:
(E.C. 1.1.1.8)

It has been known for many years that changes in the level of circulating glucocorticoids can influence brain function[11]. Little is known, however, as to either the basis for this influence or whether it represents a direct action on brain cells. Recently, a direct effect of glucocorticoids has been reported on what appears to be specific protein synthesis in cultured glial cells. This opens up a new area of potentially exciting research relative to hormonal control of brain function.

DeVellis and Inglish have reported that the development of glycerol phosphate dehydrogenase in rat brain can be stimulated precociously by cortisol administration[147]. Subsequently, it was found that this phenomenon can be reproduced in cultured glial cells but not in other cell types[47, 148]. There appears to be a lag of a few hours followed by a maximum (*ca.* twofold increase) in 24 h[47, 148]. A secondary increase (50–75%) in enzyme activity is reported to occur between 72–96 h of continuous exposure to cortisol.

Removal of the glucocorticoid prompts a very slow decay in enzyme activity after a considerable lag period[47]. Cycloheximide blocks any increase in dehydrogenase activity if given together with cortisol, but if added after maximum induction it also prompts only a very slight decrease in enzyme activity over the next 48 h. Since only a slight decrease in enzyme activity was noted after addition of cycloheximide to basal cultures for 48 h, it appears that the half-life of the dehydrogenase is very long[47]. These data suggest that *de novo* enzyme synthesis accounts for the elevation of dehydrogenase activity produced by corticoids. This awaits direct verification.

Actinomycin D prevents the increase in dehydrogenase activity if added during the first few hours of exposure to cortisol, but if added after pre-induction addition of the antibiotic leads to only a slow decay or to a degree of superinduction[47]. The fact that dehydrogenase activity decays very slowly after withdrawal of cortisol suggests the possibility that the messenger-RNA for this enzyme is also quite stable. The lack of decay in basal dehydrogenase activity after 24 h exposure to actinomycin D is consistent with this possibility. If it is assumed that cortisol increases the level of messenger-RNA for the dehydrogenase, then the effects of actinomycin D added after pre-induction may be explained, as above (see Section 7.3.2.1), by more favourable competition of the relatively stable dehydrogenase template for ribosomal sub-units in the initiation of protein synthesis.

Overall RNA and DNA synthesis was actually depressed to some extent by cortisol in the glial cultures although total protein synthesis was unaffected[47, 148]. These results once again underscore the selectivity of the effects of corticosteroids on specific proteins.

Although only incomplete data are available, the pattern of regulation of glycerol phosphate dehydrogenase is similar to that of the induction of other enzymes in suggesting an effect of the glucocorticoids at the level of transcription.

7.3.2.3 *Alkaline phosphatase—HeLa cells:* (E.C.3.1.3.1)

The activity of this enzyme can be increased by glucocorticoids in HeLa cells[149], human leucocytes[150] and during development in amphibian, chick and mouse intestinal epithelial cells[151]. Most of the work pertinent to the scope of this review has been performed in cultured HeLa cells. A characteristic lag period invariably precedes any increase in phosphatase activity, of about 12 h length in asynchronous cultures but which can be reduced to 6–8 h in partially synchronised cells[46, 152, 153]. Maximum increases of 5–10-fold can be achieved within 48–60 h[152, 153], and withdrawal of hormone, after an initial lag period[152], leads to a decay in enzyme activity approximately equal to the half-life of the enzyme (16–20 h determined radioimmunochemically[153]).

The increase in phosphatase activity can be prevented by simultaneous addition of inhibitors of protein or RNA synthesis[152, 153]. Actinomycin D does not exhibit any inhibitory effect for at least 10–12 h however if added 8 h after addition of corticosteroids[152]. Although these results are essentially identical with those obtained for other induced enzymes, careful immunochemical analysis has shown that apparently *no* increase in enzyme synthesis occurs after exposure to glucocorticoids[46, 153, 154]. The same quantity of antiserum precipitates the alkaline phosphatase from equivalent numbers of cells even though there is a 5–6-fold difference in specific activity of the enzyme between control and steroid-treated cultures[46, 153, 154]. Radioimmunochemical analysis did not reveal any detectable increase in the incorporation of labelled leucine into alkaline phosphatase nor did glucocorticoids have any apparent effect on the rate of degradation of this enzyme in pulse-chase experiments[153].

Since the response of the phosphatase to corticoids requires protein synthesis, it has been suggested that the steroids stimulate the synthesis of another protein, which in turn leads to an increase in the catalytic efficiency of pre-existing or newly synthesised phosphatase[153]. There is a distinct possbility that this might involve covalent modification of the phosphatase since extensive purification does not lead to inactivation of the enzyme from steroid-treated cultures[46,153]; such might have been expected if the hypothetical modifier protein must be associated with the enzyme for activation to occur. In addition, there are at least two isozymes of alkaline phosphatase[155]; this also is compatible with the hypothesis of covalent modification.

The enzyme from corticoid-treated cells has been found to be more heat-labile[154] and to exhibit different kinetics of inhibition by EDTA[153] consistent with an altered conformation, presumably due to the action of the modifier. Although these results represent a deviation from the usual pattern of glucocorticoid action, the data are none the less consistent with the hypothesis that RNA synthesis (but not protein synthesis) is initially required for the response. Thus, pre-incubation of HeLa cells with prednisolone and cycloheximide for 16 h, followed by transfer to medium devoid of additions for 6 h, provokes the full increase in phosphatase activity normally seen at this time in the absence of the antibiotic[153]. Fresh steroid is not required, and actinomycin D added during the second incubation does not block the response although cycloheximide completely suppresses the increase[153]. These data are fully compatible with results obtained in other systems (see Sections 7.3.1.1, 7.3.1.3. 7.3.2.1, etc.) which indicate that an intermediate dependent on RNA synthesis (mRNA?) but not on protein synthesis accumulates during the early stages of glucocorticoid action and which can then be expressed for some time (dependent upon its half-life, etc.) in the total absence of further RNA synthesis.

The proposed mechanism of the induction of alkaline phosphatase appears to be formally analogous to that of the 'induced protein' hypothesis for the action of oestrogen in the uterus[156]. Whereas corticoids generally appear to act directly on the synthesis of the proteins in question, in this case apparently an intervening step occurs.

7.4 GLUCOCORTICOIDS, CYCLIC-AMP AND THE PERMISSIVE EFFECT

7.4.1 General considerations

The so-called permissive effect of glucocorticioids has been known for years[157]. The term results from the observation that many effects of catecholamines or glucagon, in particular, are reduced, or in some cases apparently abolished, by adrenalectomy[128,135,157-160]. In some instances the effects of cyclic-AMP itself are also diminished[128,135,160]. The precise basis for the apparent requirement for glucocorticoids is not clear and some contradictory results have been reported[76, 111, 128, 161, 162]. In any event, the effect of adrenalectomy is not due to inhibition of cyclic-AMP formation by glucagon or catecholamines[128, 135], as would also be predicted from the reduced effects of exogenous cyclic-AMP. Increasing concentrations of cyclic-AMP can overcome the defect due to corticoid deficiency and the results suggest a relative, and not an absolute, dependency upon glucocorticoids[128, 135].

There are several possibilities which could be advanced to explain the permissive effects of glucocorticoids:

(a) Protein kinase is not as readily activated by cyclic-AMP in the absence of corticoids. Although this possibility has not been examined carefully, there is indirect evidence[163-165], as well as preliminary direct evidence, against it[135].

(b) The amount of protein kinase is reduced by adrenalectomy, There has been at least one report of an increase in total protein kinase activity after steroid treatment[162], but the very weak effect of cyclic-AMP on kinase activity and the discrepancy between the dose–response relationship for steroid effects on the kinase and the cellular process (tyrosine transaminase induction) influenced by cyclic-AMP makes these results of questionable significance. Furthermore, transaminase induction by cyclic-AMP has been shown not to require corticoids[76, 106, 111] (see Section 7.3.1.1).

(c) The amount of the protein inhibitor of the cyclic-AMP-dependent protein kinase is negatively influenced by glucocorticoids. The fact that f_1 histone phosphorylation in perfused liver from adrenalectomised rats is normally stimulated by cyclic-AMP argues strongly against this possibility as well as possibility (a)[135].

(d) The rate of cyclic-AMP turnover is increased by corticoid deficiency. Although the results of Schaeffer et al.[160] and De Wulf and Hers[164] suggest that this is not likely, other workers have provided some evidence for it[166, 167].

(e) The level of the respective protein substrates for protein kinase is diminished after adrenalectomy. This possibility seems unlikely in the case of glycogen synthetase[159] and phosphorylase b kinase[135].

(f) Cyclic-AMP and steroids both influence the same process and, in the absence of potentially synergistic corticoids, cyclic-AMP exerts a reduced effect. Although this explanation is appealing in the case of enzyme induction[106, 111] (and possibly gluconeogenesis), the opposing effects of cyclic-AMP and glucocorticoids on glycogen synthetase[159, 163, 164, 168] argue against it as a universal mechanism.

(g) Steroids influence some extrahepatic process which is crucial for the effect of cyclic-AMP on liver. This hypothesis has received support in relation to gluconeogenesis[134, 135, 169, 170]. but it seems unlikely to account for deficiencies in glycogen metabolism.

(h) Some undefined factor(s), possibly related to ion flux, is regulated by corticoids and is required for proper 'sensitivity' to cyclic-AMP[128, 135]. It is conceivable, for example, that the affinity of a protein substrate for protein kinase could be influenced by some unknown factor which in turn was subject to regulation by glucocorticoids.

Unfortunately, insufficient data are currently available to allow any definitive statements to be made relative to which of these possibilities is most likely. It may well turn out that each regulatory process will have to be considered separately and that no unifying hypotheses will be possible.

7.4.2 Evidence for or against glucocorticoid requirement for the action of cyclic-AMP and mediation of the effects of glucocorticoids by cyclic-AMP

A series of observations have been made which indicate rather strongly that cyclic-AMP can produce its biological effects in the absence of glucocorticoids. In addition, there is evidence against cyclic-AMP as a mediator of the effects of glucocorticoids. There are, however, limited data which have suggested that this may not always be true. Fain and his colleagues have observed that dexamethasone in the presence of growth hormone increases the levels of cyclic-AMP in adipose tissue, albeit only after a lag period[158]; this increase is inhibited by cycloheximide, puromycin and actinomycin D[158]. It has been found that growth hormone is responsible for the increase in cyclic-AMP, however, and the requirement for steroid may be to sensitise adipose tissue to the effects of cyclic-AMP[135, 171].

Braun and Hechter have found that ACTH does not exert its lipolytic effects in adrenalectomised rats[172]. Administration of glucocorticoids restores the sensitivity to ACTH by repairing some defect in a membrane-bound receptor for ACTH which is associated with adenylate cyclase. This repair is blocked both by actinomycin D and cycloheximide suggesting that the corticoid increases the synthesis of a proteinaceous component(s) of the receptor complex. This explanation of a permissive effect of glucocorticoids, however, does not seem to hold for other permissive effects[135].

The induction of tyrosine transaminase by dibutyryl cyclic-AMP has been reported to require glucocorticoids in hepatoma cells[162], but other workers have not found such a requirement[76, 106, 111].

Finally, it should be stated that cyclic-AMP has been found to reproduce some of the effects of the sex hormones as well as those of vitamin D[173, 174]. However, there are no conclusive data which demonstrate that cyclic-AMP *mediates* the action of steroids. In fact, Gorski has reported that the putative primary event in the oestrogenised rat uterus, namely synthesis of the 'induced protein', is not stimulated by even heroic concentrations of dibutyryl cyclic-AMP[175]. That steroid hormones may cause an increase in cyclic-AMP levels in certain tissues is not at issue here, only the concept that cyclic-AMP is responsible for all the effects of steroids. For a further discussion of this issue from a different viewpoint consult Ref. 176.

The following list provides, in this reviewer's opinion, additional compelling evidence that cyclic-AMP neither mediates the actions of glucocorticoids nor always requires steroids for its effects on metabolic processes.

(a) Glucocorticoids do not elevate hepatic cyclic-AMP concentrations[161, 168].

(b) Dibutyryl cyclic-AMP induces tyrosine transaminase, PEP carboxykinase and serine dehydratase in intact and adrenalectomised rat liver, but tryptophan pyrrolase shows no response[106, 127].

(c) Dibutyryl cyclic-AMP induces tyrosine transaminase and PEP carboxykinase in cultured foetal liver, but only the transaminase responds to glucocorticoids[105].

(d) Dexamethasone induces both tyrosine transaminase and PEP carboxykinase in MH_1C_1 hepatoma cells, whereas only the carboxykinase responds appreciably to dibutyryl cyclic-AMP[45].

(e) Dexamethasone induces tyrosine transaminase in HTC cells under conditions where dibutyryl cyclic-AMP is ineffective[177] (although Stellwagen has recently found conditions in which massive amounts of the cyclic nucleotide are effective, but not to the extent of corticoids[178]). Addition of dibutyryl cyclic-AMP together with the steroid did not produce any greater effect than did dexamethasone alone.

(f) Glycogen synthetase is inactivated by cyclic-AMP and activated by glucocorticoids[159, 163, 164].

(g) The effects of glucocorticoids and cyclic-AMP on enzyme induction are additive or synergistic[105, 106, 111].

(h) Dibutyryl cyclic-AMP produces a maximum increase in the rate of tyrosine transaminase synthesis prior to any increase produced by corticosteroids[1, 45, 106].

(i) The transient release of renin produced by ACTH and cyclic-AMP can be suppressed by dexamethasone[179]. It is of interest that the ability of dexamethasone to block renin release is prevented by actinomycin D administered 1 h prior to the steroid.

(j) Dexamethasone suppresses the release of ACTH produced by CRF (corticosteroid releasing factor) or dibutyryl cyclic-AMP in isolated rat pituitaries as well as in whole rats[180, 181]. Actinomycin D given prior to the steroid prevented its inhibition of ACTH release but not if administered afterwards[180]. (CRF is presumed to act via cyclic-AMP[181].)

(k) Whereas glucocorticoids cause involution of the thymus (see Section 7.5), cyclic-AMP has been found to stimulate thymocyte proliferation[182].

In conclusion, it seems clear that although glucocorticoids influence many processes (both positively and negatively) also affected by cyclic-AMP, the steroids in all probability act by independent means. The actions of corticoids and cyclic-AMP do appear to be inextricably interwoven in a complex regulatory matrix, and it is difficult to separate the two when considering any metabolic process. Nevertheless, there are no data supporting the conclusion that *any* effect of glucocorticoids is directly mediated by cyclic-AMP. It is clear, however, that in some instances for reasons which remain obscure glucocorticoids are required for maximum responsiveness to cyclic-AMP[128, 135].

7.5 REGULATION OF THE GROWTH AND FUNCTION OF LYMPHOID TISSUES BY GLUCOCORTICOIDS

It has been known for almost 30 years that glucocorticoids exert an inhibitory effect on the metabolic functions of lymphoid tissues[39, 42, 48, 183-185]. This action is manifested by inhibition of mitosis and cell growth, pycnosis of cells, karyorrhexis of nuclei and involution of lymph nodes, spleen and thymus after treatment of mice, rats and rabbits with glucocorticoids[9, 34, 42, 48, 183-185]. (Skin and adipose tissue also exhibit to some extent a similar response to glucocorticoids[39].) Studies at early intervals after steroid administration revealed inhibition of respiration[186], transport processes[39, 42, 187, 188], as well as a decrease in the content of and incorporation of isotopic precursors into RNA[39, 42, 185, 189, 190], DNA[39, 42, 185, 188] and protein[39, 42, 185, 189, 191]. Analysis of the mechanism by which corticosteroids influence these processes has been greatly facilitated by the use of suspended cells and cultured normal and malignant cells[39, 42, 185, 188, 189]. The potencies of various corticoids on lymphoid involution have been found to be similar to those on other metabolic processes (see Figure 7.1) regulated by these hormones[39, 42, 185]. In addition, progesterone and 11-deoxycortisol, although essentially inactive alone, competitively inhibit the effects of active glucocorticoids on lymphoid tissues[39, 42, 192]. These two steroids are also competitive inhibitors of glucocorticoid action on hepatic enzyme induction[20].

A variety of contradictory results has been obtained with regard to possible direct catabolic effects of glucocorticoids in lymphoid involution[42, 189, 193, 194]. It should be pointed out that the interpretation of many of these results has been complicated by effects of corticoids, whether direct or indirect, on transport of precursors and on the size of various intracellular metabolite pools[39, 42, 185, 188, 193]. Thus one report of increased catabolism of protein in lymphoma cells generated by corticosteroids was shown to be actually due to decreased re-utilisation of radioactive amino acids liberated by degradation[185]. These results point out some of the pitfalls associated with dissection of the precise nature of the effects of glucocorticoids on lymphoid tissue. Additional complications have arisen from the use of diverse tissue and cell preparations. Many of the conflicting results obtained to date concerning the effects of corticoids on nucleic acid and protein synthesis may derive from these considerations[188]. There seems to be little doubt that macromolecular synthesis is inhibited to some degree by corticosteroids[39, 42, 185, 188, 191, 195, 196], but this inhibition of macromolecular synthesis does not seem to be the primary effect of the steroids here[39, 187, 191, 196].

The most persuasive (and also probably the oldest[197]) theory put forward to explain the initial effects of glucocorticoids on lymphoid tissue metabolism is that they inhibit glucose transport[39]. Munck and his colleagues have been the major proponents of this theory and have gathered a considerable amount of data, primarily in thymocytes, to support it[39, 187, 191, 192]. It has been observed that a significant inhibition of glucose uptake occurs as early as 15-20 min after addition of cortisol to isolated thymocytes[187]. (The effect of corticoids on this process does not appear to involve antagonism of insulin since the latter hormone does not stimulate glucose transport in the thymus[39]. This rapid onset of glucocorticoid action stands in contrast to the pronounced (albeit variable) lag in other tissues (see Section 7.3). Studies of steroid-receptor interaction, however, have also shown that much less time is required for complex formation and subsequent transport to the nucleus in thymocytes, the process being complete in less than 10 min[39, 192].

Through the use of inhibitors and temperature changes, Munck and co-workers have identified four discrete stages between steroid-receptor association and inhibition of glucose transport[39, 187]. The initial phase has been referred to as the 'irreversible' step (between 0-5 min) after which time steroid is no longer required for the effect to occur[39, 42, 192]. The nature of this 'irreversible' step is not clear but it could relate to formation of a long-lived nuclear steroid-receptor complex from which the active steroid is not easily displaced, or to a stable intermediate which results from the action of the steroid (see below).

The second step has been identified in terms of actinomycin D sensitivity[39,192]. Addition of the antibiotic at any point prior to 5 min after the addition of corticoids prevents the inhibition of glucose transport, but has no effect if added thereafter. These results suggest that glucocorticoids stimulate the synthesis of a stable RNA species which then mediates the steroid effect. There appear to be two additional steps, one identified on the basis of its temperature-sensitivity and the other by its sensitivity to inhibition by cycloheximide[39,192]. Addition of cycloheximide at any time up to 15 min but not later, blocks the steroid effect. These results clearly suggest the formation of a stable intermediary protein which then, putatively, inhibits glucose transport by some unknown mechanism. Thus, as with alkaline phosphatase (see Section 7.3.2.3) and oestradiol in the uterus[175], a key 'induced protein' is apparently formed by the action of glucocorticoids in lymphoid tissue. The exact nature of this hypothetical protein, however, remains elusive for the present.

It has been proposed that most, if not all, of the effects of corticoids on lymphoid tissue flow from the initial inhibition of glucose transport and, consequently, reduced ATP concentration[39,191,192]. It seems clear that many of the effects on nucleic acid and protein synthesis can be ascribed to this effect. For example, in the absence of glucose, corticosteroids have no effect on DNA, RNA and protein synthesis in lymphoid cells[39,42,190,191]. Elimination of glucose from the medium, in fact, mimics the effects of glucocorticoids[42,198]. Deficiency in ATP production would clearly inhibit macromolecular synthesis and could lead to enhanced protein turnover. A reduction in the rate of aminoacylation of transfer-RNA, by analogy with $E.$ $coli$[82], would be expected to stimulate protein degradation. It is possible that a similar phenomenon may hold true for RNA breakdown, since conditions which enhance protein breakdown in $E.$ $coli$ are associated with a depression of ribosomal RNA synthesis[82]. It is of special interest, therefore, to note that ribosomal RNA synthesis appears to be preferentially inhibited in thymocytes early after corticoid administration[199].

On the other hand, the decreases in RNA polymerase activity[195,196] and protein synthesis in isolated microsomes[200] are not as readily explained by the glucose transport hypothesis. If protein synthesis is reduced and breakdown enhanced, then labile components associated with RNA or protein synthesis could become limiting with the passage of time. Thus, a defect might persist even in the presence of excess ATP and other co-factors. An alternative possibility is that the induced protein exerts multivalent effects or that several different proteins are induced. At present it is not possible to make a decision between these possibilities.

It seems unlikely that inhibition of the breakdown of a pre-existing template for the 'induced protein' could account for the effects of glucocorticoids on lymphoid cells. The template for this protein appears to be quite stable (based on lack of requirement for steroid and resistance to actinomycin D after 5 min[39,192]), and thus, any rapid increase in its concentration would most likely have to result from an increased rate of transcription.

As pointed out by Munck[39], the effect of glucocorticoids on lymphoid tissue is actually anabolic at first and only after synthesis of the 'induced protein' do the anti-anabolic and catabolic effects become apparent. That not all the effects of glucocorticoids in lymphoid tissue are anti-anabolic is also suggested by the findings of Rosen and his colleagues that alanine transaminase activity is increased in the thymus and steroid-sensitive lymphosarcomas after administration of corticoids[5]. In steroid-resistant tumors, however, there is little change in transaminase activity after treatment with glucocorticoids. As mentioned earlier, steroid-resistant lymphosarcomas appear to have a reduced ability to bind active glucocorticoids. It is of interest that glucose transport is also not markedly influenced by corticoids in these resistant tumors[39, 188]. It seems unlikely, however, that reduced binding will invariably be the cause of resistance to steroids. Some alteration in the effects of the steroid–receptor complex on the synthesis of the 'induced protein' or a defect in the protein itself could also endow a tumour with resistance to glucocorticoids.

In spite of differences in the rate of onset of, and apparent (relative?) irreversibility of, the steroid effect, the pattern of glucocorticoid action on lymphoid tissues is essentially the same as in the regulation of specific protein synthesis discussed previously.

7.6 CONCLUDING REMARKS

From the studies discussed in this review, an amazingly similar pattern of the action of glucocorticoids emerges regardless of which tissue or response is considered. Upon entry of the steroid into the cell, a variable period of time is required for interaction with a specific receptor and transport of the resulting complex to the nucleus before any cellular response can be detected. This 'processing' of the steroid presumably explains the universal lag in metabolic responses to glucocorticoids. Within the nucleus, the steroid–receptor complex apparently binds to chromatin in the course of its action. At present there is no way of knowing for certain whether this interaction turns on transcription of specific structural genes or turns off the transcription of specific repressor genes, or even whether it has *any* biological consequences.

Additional, and indeed the earliest, support for the concept that steroids do influence specific RNA synthesis is provided by the phenomenon of puffing of *chironomus* salivary gland chromosomes which is generated by the insect steroid hormone, ecdysone[32]. It seems quite likely that steroids do have a more or less common mechanism of action; this does not preclude a wide diversity of metabolic responses. The diversity is, therefore, considered not to be a reflection of dissimilar mechanisms of steroid action but rather of the state of differentiation of the tissue in question.

The recent successes with cell-free synthesis of specific proteins directed by polysomal RNAs (see Chapter by O'Malley) suggest that, eventually, quantitation of template RNAs coding for even proteins synthesised in very small amounts will be feasible[33, 34]. Isolation of specific messenger-RNAs by the selective immunoprecipitation of polysomes holds great promise as a method for achieving this goal[201]. It is to be hoped that through appropriate pulse and pulse-chase labelling procedures in the presence of glucocorticoids, followed by the isolation of specific polysomes, it will be possible to determine whether the steroids increase the rate of synthesis or decrease the rate of degradation of the messenger-RNA in question. This problem still clearly provides an enticing challenge to the industrious and resourceful scientist, and the next few years should be filled with exciting new discoveries in this area.

Note added in proof

Schutz, Beato and Feigleson (1973; *Proc. Nat. Acad. Sci., USA*, **70**, 1218) have recently demonstrated directly that cortisol treatment enhances the level of translatable mRNA for tryptophan pyrrolase in rat liver. Their results, however, do not distinguish between an effect of the steroid on the rate of the synthesis or degradation of the specific mRNA.

References

1. Kenney, F. T. (1970). *Mammalian Protein Metabolism*, Vol. IV, 131 (H. N. Munro, editor) (New York: Academic Press)
2. Ashmore, J. and Morgan, D. (1967). *The Adrenal Cortex*, 249 (A. B. Eisenstein, editor) (London: Churchill)
3. Litwack, G. and Singer, S. (1972). *Biochemical Actions of Hormones*, 113 (G. Litwack, editor) (New York: Academic Press)
4. Tomkins, G. M. and Gelehrter, T. D. (1972). *Biochemical Actions of Hormones*, 1 (G. Litwack, editor) (New York: Academic Press)
5. Rosen, F. and Nichol, C. A. (1963). *Vitam. Horm.*, **21**, 135
6. Rosen, F. and Milholland, R. J. (1971). *Enzyme Synthesis and Degradation in Mammalian Systems*, 77 (M. Rechcigl, editor) (Baltimore: University Park Press)
7. Gelehrter, T. (1971). *Enzyme Synthesis and Degradation in Mammalian Systems*, 165 (M. Rechcigl, editor) (Baltimore: University Park Press)
8. Tomkins, G. M. and Martin, D. W. (1970). *Ann. Rev. Genetics*, **4**, 91
9. Schimke, R. T. (1966). *Bull. Soc. Chim. Biol.*, **48**, 1009
10. Feigelson, P. and Feigelson, M. (1964). *Mechanisms of Hormone Action*, 246 (P. Karlson, editor) (New York: Academic Press)
11. Sayers, G. and Travis, R. H. (1970). *The Pharmacological Basis of Therapeutics*, 1604 (L. S. Goodman and A. Gilman, editors) (London: Macmillan)
12. Baulieu, E. E., Alberga, A., Jung, I., Lebeau, M. C., Mercier-Bodard, C., Milgrom, E., Raynaud, J. P., Raynaud-Jammet, C., Rochefort, H., Truong, H. and Robel, P. (1971). *Rec. Progr. Horm. Res.*, **27**, 351
13. Robison, G. A., Butcher, R. W. and Sutherland, E. W. (1971). *Cyclic AMP*, 22 (New York: Academic Press)
14. Sandberg, A. A., Slaunwhite, W. R. and Antoniades, H. N. (1957). *Rec. Prog. Horm. Res.*, **13**, 209
15. Litwack, G., Sears, M. L. and Diamondstone, T. I. (1963). *J. Biol. Chem.*, **238**, 302
16. Jensen, E. V., Numata, M., Brecher, P. I. and DeSombre, E. R. (1971). *The Biochemistry of Steroid Hormone Action*, 133 (R. M. S. Smellie, editor) (London: Academic Press)
17. Rousseau, G. G., Baxter, J. D. and Tomkins, G. M. (1972). *J. Mol. Biol.*, **67**, 99
18. Baxter, J. D. and Tomkins, G. M. (1970). *Proc. Nat. Acad. Sci. USA*, **65**, 709
19. Baxter, J. D. and Tomkins, G. M. (1970). *Proc. Nat. Acad. Sci. USA*, **68**, 932
20. Samuels, H. H. and Tomkins, G. M. (1970). *J. Mol. Biol.*, **52**, 57
21. Baxter, J. D., Rousseau, G. G. and Tomkins, G. M. (1971). *Fed. Proc.* **30**, 1048 abs.
22. Rousseau, G. G., Baxter, J. D. and Tomkins, G. M. (1972). *IV Int. Congr. Endocrinology*, 146, abstracts (Washington: Excerpta Medica)
23. Beato, M., Bieswig, D., Braendle, W. and Sekeris, C. E. (1969). *Biochim. Biophys. Acta*, **192**, 494
24. Beato, M., Braendle, W., Bieswig, D. and Sekeris, C. E. (1970). *Biochim. Biophys. Acta*, **208**, 125
25. Monod, J., Wyman, J. and Changeux, J. P. (1965). *J. Mol. Biol.*, **12**, 88

26. Koshland, D. E. (1963). *Cold Spring Harbor Symp. Quant. Biol.*, **28**, 473
27. Spelsberg, T. C., Steggles, A. W. and O'Malley, B. W. (1971). *J. Biol. Chem.*, **246**, 4186
28. O'Malley, B. W., Spelsberg, T. C., Schrader, W. T., Chytil, F. and Steggles, A. W. (1972). *Nature (London)*, **235**, 141
29. Tomkins, G. M., Gelehrter, T. D., Granner, D., Martin, D. M., Samuels, H. H. and Thompson, E, G. (1969). *Science*, **166**, 1474
30. Muldoon, T. G. (1971). *Biochemistry*, **10**, 3780
31. Munck, A. and Wira, G. R. (1972). *Advances in the Biosciences*, Vol. 7 (G. Raspe, editor) (Oxford: Pergamon Press), in the press
32. Karlson, P. (1963). *Perspect. Biol. Med.*, **6**, 203
33. Rosenfeld, G. C., Comstock, J. P., Means, A. R. and O'Malley, B. W. (1972). *Biochem. Biophys. Res. Commun.*, **46**, 1965
34. Rhoads, R. E., McKnight, G. S. and Schimke, R. T. (1971). *J. Biol. Chem.*, **246**, 7407
35. Munck, A. and Brinck-Johnson, T. (1968). *J. Biol. Chem.*, **243**, 5556
36. Brunkhorst, W. K. and Hess, E. L. (1966). *Biochim. Biophys. Acta*, **82**, 385
37. DeVenuto, F. and Chader, G. (1966). *Biochim. Biophys. Acta*, **121**, 151
38. Wira, C. and Munck, A. (1970). *J. Biol. Chem.*, **245**, 3436
39. Munck, A. (1971). *Perspect. Biol. Med.*, **14**, 265
40. Hackney, J. F., Gross, S. R., Aronow, L. and Pratt, W. B. (1970). *Mol. Pharmacol.*, **6**, 500
41. Hackney, J. F. and Pratt, W. B. (1971). *Biochemistry*, **10**, 3002
42. Makman, M. H. Nakagawa, S. and White, A. (1967). *Rec. Prog. Horm. Res.*, **23**, 195
43. Moscona, A. A. and Piddington, R. (1967). *Science*, **158**, 496
44. Kenney, F. T. and Lee, K.-L. (1971). *Proc. III int. Congr. Horm. Steroids*, 472 (V. H. T. James and L. Martini, editors) (Amsterdam: Excerpta Medica)
45. Barnett, C. A., McKibbin, J. B. and Wicks, W. D., unpublished results
46. Cox, R. P. (1971). *Ann. N.Y. Acad. Sci.*, **179**, 596
47. DeVellis, J., Inglish, D., Cole, R. and Molson, J. (1971). *Proc. Int. Soc. Psychoneuroendocrinology*, 25 (Karger: Basel)
48. Baxter, J. D., Harris, A. W., Tomkins, G. M. and Cohn, M. (1971). *Science*, **171**, 189
49. Hollander, N. and Chiu, Y. W. (1966). *Biochem. Biophys. Res. Commun.*, **25**, 191
50. Levisohn, S. R. and Thompson, E. B. (1972). *Nature (London), (New Biol.)*, **235**, 102
51. Kenney, F. T. and Kull, F. J. (1963). *Proc. Nat. Acad. Sci. USA*, **50**, 493
52. Feigelson, M., Gross, P. R. and Feigelson, P. (1962). *Biochim. Biophys. Acta*, **55**, 495
53. Kenney, F. T., Wicks, W. D. and Greenman, D. L. (1965). *J. Cell. Comp. Physiol.*, **66**, Suppl. 1, 125
54. Silber, R. H. and Porter, C. C. (1953). *Endocrinology*, **52**, 518
55. Rechcigl, M. (1971). *Enzyme Synthesis and Degradation in Mammalian Systems*, 236 (M. Rechcigl, editor) (Baltimore: University Park Press)
56. Lee, K.-L., Reel, J. R. and Kenney, F. T. (1970). *J. Biol. Chem.*, **245**, 5806
57. Wicks, W. D. (1968). *J. Biol. Chem.*, **243**, 900
58. Hager, C. B. and Kenney, F. T. (1968). *J. Biol. Chem.*, **243**, 3296
59. Liberti, J. P., DuVall, C. H. and Wood, D. M. (1971). *Can. J. Biochem.*, **49**, 1357
60. Thompson, E. B., Tomkins, F. M. and Curran, J. F. (1966). *Proc. Nat. Acad. Sci. USA*, **56**, 296
61. Gerschenson, L. E., Anderson, M., Molson, J. and Okigaki, T. (1970). *Science*, **170**, 859
62. Haung, Y. L. and Ebner, K. E. (1969). *Biochim. Biophys. Acta*, **191**, 161
63. Granner, D. K., Thompson, E. B. and Tomkins, G. M. (1970). *J. Biol. Chem.*, **245**, 1472
64. Boctor, A. and Grossman, A. (1970). *J. Biol. Chem.*, **245**, 6337
65. Levitan, I. B. and Webb, T. E. (1970). *J. Mol. Biol.*, **48**, 339
66. Granner, D. K., Hayashi, S. I., Thompson, E. B. and Tomkins, G. M. (1968). *J. Mol. Biol.*, **35**, 291
67. Grossman, A. and Boctor, A. (1972). *Proc. Nat. Acad. Sci. USA*, **69**, 1161
68. Wicks, W. D. (1971). *J. Biol. Chem.*, **246**, 217
69. Reel, J. R., Lee, K.-L. and Kenney, F. T. (1970). *J. Biol. Chem.*, **245**, 5800
70. Thompson, E. B., Granner, D. K. and Tomkins, G. M. (1970). *J. Mol. Biol.*, **54**, 159
71. Levinson, B. B., Tomkins, G. M. and Stellwagen, R. H. (1971). *J. Biol. Chem.*, **246**, 6297
72. Reel, J. R. and Kenney, F. T. (1968). *Proc. Nat. Acad. Sci. USA*, **61**, 200

73. Peterkofsky, B. and Tomkins, G. M. (1968). *Proc. Nat. Acad. Sci. USA*, **60**, 222
74. Gelehrter, T. and Tomkins, G. M. (1967). *J. Mol. Biol.*, **29**, 59
75. Garren, L. D., Howell, R. R., Tomkins, G. M. and Crocco, R. M. (1964). *Proc. Nat. Acad. Sci. USA*, **52**, 1121
76. Butcher, F. T., Becker, J. E. and Potter, V. R. (1971). *Exp. Cell. Res.*, **66**, 231
77. Tomkins, G. M., Thompson, E. B., Hayashi, S., Gelehrter, T., Granner, D. K. and Peterkofsky, B. (1976). *Cold Spring Harbor. Symp. Quant. Biol.*, v**31**, 349
78. Aurricchio, F., Martin, D. and Tomkins, G. M. (1969). *Nature (London)*, **224**, 806
79. Barker, K. L., Lee, K.-L. and Kenney, F. T. (1971). *Biochem. Biophys. Res. Commun.*, **43**, 1132
80. Magus, R. D., King, S. W. and Harrison, J. D. (1971). *Biochem. Pharmacol.*, **20**, 2239
81. Levitan, I. B. and Webb, T. E. (1969). *J. Biol. Chem.*, **244**, 4684
82. Goldberg, A. (1971). *Proc. Nat. Acad. Sci. USA*, **68**, 362
83. Lee, K.-L. and Kenney, F. T. (1971). *J. Biol. Chem.*, **246**, 7595
84. Marliss, E., Aoki, T. T., Felig, P., Pozefsky and Cahill, G. F. (1970). *Advan. Enzyme Regul.*, **8**, 3
85. Eisenstein, A. B. (1965). *Advan. Enzyme Regul.*, **3**, 121
86. Nichol, C. A. and Rosen, F. (1969). *J. Biol. Chem.*, **234**, 476
87. Nichol, C. A. and Rosen, F. (1963). *Advan. Enzyme Regul.*, **1**, 341
88. Segal, H. L. and Kim, Y. S. (1963). *Prov. Nat. Acad. Sci. USA*, **50**, 912
89. Segal, H. L. and Kim, Y. S. (1965). *J. Cell. Comp. Physiol.*, **66**, Suppl. 1, 11
90. Kim, Y. S. (1968). *Mol. Pharmacol.*, **4**, 168
91. Kim, Y. S. (1969). *Mol. Pharmacol.*, **5**, 105
92. Lee, K.-L. and Kenney, F. T. (1970). *Biochem. Biophys. Res. Commun.*, **40**, 469
93. Rosen, F., Harding, H. R., Milholland, R. J. and Nichol, C. A. (1963). *J. Biol. Chem.*, **238**, 3725
94. Lardy, H. A., Foster, D. O., Young, J. W., Shrago, E. and Ray, P. D. (1965). *J. Cell. Comp. Physiol.*, **66**, suppl. 1, 39
95. Gatehouse, P. W., Hopper, S., Schatz, L. and Segal, H. L. (1967). *J. Biol. Chem.*, **242**, 2319
96. Utter, M. F. (1963). *Iowa State J. Sci.*, **38**, 97
97. Keech, D. B. and Utter, M. F. (1963). *J. Biol. Chem.*, **238**, 2609
98. Lardy, H. A., Foster, D. O., Shrago, E. and Ray, P. D. (1964). *Advan. Enzyme Regul.*, **2**, 39
99. Nordlie, R. C. and Lardy, H. A. (1963). *J. Biol. Chem.*, **238**, 2259
100. Holten, D. D. and Nordlie, R. C. (1965). *Biochemistry*, **4**, 723
101. Taylor, P. H., Wallace, J. C. and Keech, D. B. (1971). *Biochim. Biophys. Acta*, **237**, 179
102. Diesterhaft, M., Shrago, E. and Sallach, H. J. (1971). *Biochem. Med.*, **5**, 297
103. Nordlie, R. C., Varrichio, F. E. and Holten, D. D. (1965). *Biochim. Biophys. Acta*, **97**, 214
104. Shrago, E., Lardy, H. A., Nordlie, R. C. and Foster, D. O. (1963). *J. Biol. Chem.*, **238**, 3188
105. Wicks, W. D. (1969). *J. Biol. Chem.*, **244**, 3941
106. Wicks, W. D., Kenney, F. T. and Lee, K.-L. (1969). *J. Biol. Chem.*, **244**, 6008
107. Yeung, D. and Oliver, I. T. (1968). *Biochemistry*, **7**, 3231
108. Reshef, L. and Hanson, R. W. (1972). *Biochem. J.*, **127**, 809
109. Wicks, W. D., Van Wijk, R. and McKibbin, J. B., (1973). *Advan. Enzyme Regul.*, **11**, 117
110. Jefferson, L. S., Exton, J. H. Butcher, R. W., Sutherland, E. W. and Park, C. R. (1968). *J. Biol. Chem.*, **243**, 1031
111. Barnett, C. A. and Wicks, W. D. (1971). *J. Biol. Chem.*, **246**, 7201
112. Foster, D. O., Ray, P. D. and Lardy, H. A. (1966). *Biochemistry*, **5**, 555
113. Ader, R. and Friedman, S. B. (1968). *Neuroendocrinology*, **3**, 378
114. Ray, P. D., Foster, D. O. and Lardy, H. A. (1964)). *J. Biol. Chem.*, **239**, 396
115. Wicks, W. D. and McKibbin, J. B. (1972). *Biochem. Biophys. Res. Commun.*, **48**, 205
116. Reshef, L., Ballard, F. J. and Hansen, R. W. (1969). *J. Biol. Chem.*, **244**, 557
117. Reshef, L. and Hanson, R. W. (1972). *Biochem J.*, **127**, 809
118. Ballard, F. J. and Hanson, R. W. (1969). *J. Biol. Chem.*, **244**, 5625
119. Schmike, R. T. and Doyle, D. (1970). *Ann. Rev. Biochem.*, **39**, 929
120. Schimke, R. T. (1970). *Mammalian Protein Metabolism*, Vol. IV, 177 (H. N. Munro, editor) (New York: Academic Press)

121. Csyani, V., Greengard, O. and Knox, W. D. (1967). *J. Biol. Chem.*, **242**, 2688
122. Goldstein, L., Stella, E. J. and Knox, W. E. (1962). *J. Biol. Chem.*, **237**, 1723
123. Pitot, H. C., Cho, Y. S., Lamar, C. and Peraino, C. (1965). *J. Cell. Comp. Physiol.*, **66**, Suppl. 1, 163
124. Rosen, F. and Nichol, C. A. (1964). *Advan. Enzyme Regul.*, **2**, 115
125. Peraino, C. and Pitot, H. C. (1964). *J. Biol. Chem.*, **239**, 4308
126. Pitot, H. C. and Jost, J.-P. (1967). *Nat. Can. Inst. Monograph*, **26**, 145
127. Jost, J.-P., Hsie, A., Hughes, S. D. and Ryan, L. (1970). *J. Biol. Chem.*, **245**, 351
128. Exton, J. H., Mallette, L. E., Jefferson, L. S., Wong, E. H. A., Friedmann, N., Miller, T. B. and Park, C. R. (1970). *Rec. Progr. Horm. Res.*, **26**, 411
129. Inoue, H. and Pitot, H. C. (1970). *Advan. Enzyme Regul.*, **8**, 289
130. Inoue, H., Kasper, C. B. and Pitot, H. C. (1971). *J. Biol. Chem.*, **246**, 2626
131. Greengard, O., Weber, G. and Singhal, M. (1963). *Science*, **141**, 160
132. Ray, P. D. (1968). *Functions of the Adrenal Cortex*, Vol. 2, 1138 (K. W. McKerns, editor) (New York: Appleton-Century-Crofts)
133. Dunn, A., Chenoweth, M. and Hemington, J. G. (1971). *Biochim. Biophys. Acta*, **237**, 192
134. Williamson, J. R., Thurman, R. G. and Browning, E. T. (1971). *The Action of Hormones*, 143 (P. Foa, editor) (Springfield: C. C. Thomas)
135. Exton, J. H., Friedmann, N., Wong, E. H. A., Brineaux, J. P., Corbin, J. D. and Park, C. R. (1972). *J. Biol. Chem.*, **247**, 3579
136. Stadtman, E. R., Shapiro, B. M., Ginsburg, A., Kingdon, H. S. and Denton, M. D. (1968). *Brookhaven Symp. Biol.*, **21**, 378
137. Raina, P. N. and Rosen, F. (1968). *Biochim. Biophys. Acta*, **165**, 470
138. Alescio, T. and Moscana, A. A. (1969). *Biochem. Biophys. Res. Commun.*, **34**, 176
139. Piddington, R. and Moscana, A. A. (1965). *J. Cell. Biol.*, **27**, 247
140. Reif-Lehrer, L. and Amos, H. (1968). *Biochem. J.*, **106**, 425
141. Moscana, A. A., Moscana, M. H. and Saenz, N. (1968). *Proc. Nat. Acad. Sci. USA*, **61**, 160
142. Schwartz, R. J. (1972). *Nature (New Biol.)*, **237**, 121
143. Reif-Lehrer, L. (1971). *J. Cell. Biol.*, **51**, 303
144. Moscana, A. A., Moscana, N. H. and Jones, R. E. (1970). *Biochem. Biophys. Res. Commun.*, **39**, 943
145. Palmiter, R. D., Oka, T. and Schmike, R. T. (1971). *J. Biol. Chem.*, **246**, 724
146. Sarkar, P. K. and Moscana, A. A. (1971). *Proc. Nat. Acad. Sci. USA*, **68**, 2308
147. DeVellis, J. and Inglish, D. (1968). *J. Neurochem.*, **15**, 106
148. DeVellis, J., Inglish, D. and Galey, F. (1971). *Cellular Aspects of Neural Growth and Differentiation*, 23 (D. Pease, editor) (Los Angeles: University of California Press)
149. Cox, R. P. and MacLeod, C. M. (1961). *Nature (London)*, **190**, 85
150. Valentine, W. N., Follette, J. H., Hardin, E. B., Beck, W. S. and Lawrence, J. S. (1954). *J. Lab. Clin. Med.*, **44**, 219
151. Moog, F. (1971). *Enzyme Synthesis and Degradation in Mammalian Systems*, 47 (M. Rechcigl, editor) (Baltimore: University Park Press)
152. Griffin, M. J. and Cox, R. P. (1966). *J. Cell. Biol.*, **29**, 1
153. Cox, R. P., Elson, N. A., Tu, S.-H., and Griffin, M. J. (1971). *J. Mol. Biol.*, **58**, 197
154. Cox, R. P. and Griffin, M. J. (1966). *Proc. Nat. Acad. Sci. USA*, **56**, 946
155. Cox, R. P. and Griffin, M. J. (1967). *Arch. Biochem. Biophys.*, **122**, 552
156. Notides, A. and Gorski, J. (1966). *Proc. Nat. Acad. Sci. USA*, **56**, 230
157. Ingle, D. (1952). *J. Endocrinol.*, **8**, XXIII
158. Fain, J. N., Dodd, A. and Novak, L. (1971). *Metabolism*, **20**, 109
159. Mersmann, H. J. and Segal, H. L. (1969). *J. Biol. Chem.*, **244**, 1701
160. Schaeffer, L. D., Chenoweth, M. and Dunn, A. (1969). *Biochim. Biophys. Acta*, **192**, 292
161. Rinard, G. A., Okuno, G. and Haynes, R. C. (1969). *Endocrinology*, **84**, 622
162. Sahib, M. K., Jost, Y.-C., Jost, J.-P. (1971). *J. Biol. Chem.*, **246**, 4539
163. DeWulf, H. and Hers, H. G. (1968). *Eur. J. Biochem.*, **6**, 552
164. DeWulf, H. and Hers, H. G. (1968). *Eur. J. Biochem.*, **6**, 558
165. Fain, J. N. (1968). *Endocrinology*, **82**, 825
166. Stoff, J. S., Handler, J. S. and Orloff, J. (1972). *Proc. Nat. Acad. Sci. USA*, **69**, 805
167. Senft, G., Schultz, G. S., Munske, K. and Hoffman, M. (1968). *Diabetologia*, **4**, 330

168. Van den Berghe, G., De Wulf, H. and Hers, H. G. (1970). *Eur. J. Biochem.*, **16,** 358
169. Eisenstein, A. B., Spencer, S. and Brodsky, A. (1966). *Endocrinology*, **79,** 182
170. Friedmann, B., Goodman, E. H. and Weinhouse, S. (1965). *J. Biol. Chem.*, **240,** 3729
171. Honeyman, T. W. and Goodman, H. N. (1972). *IV Int. Congr. Endocrinology*, 159, abstracts (Washington: Excerpta Medica)
172. Braun, T. and Hechter, O. (1970). *Proc. Nat. Acad. Sci. USA*, **66,** 995
173. Singhal, R. L. and LaFreniere, R. T. (1972). *J. Pharm. Expt. Ther.*, **180,** 86
174. Neville, E. and Holdsworth, E. S. (1969). *FEBS Lett.*, **2,** 313
175. Katzenellenbogzen, B. S. and Gorski, J. (1972). *J. Biol. Chem.*, **247,** 1299
176. Hechter, O. and Soifer, D. (1971). *Basic Actions of Sex Steroids on Target Organs*, 93 (P. O. Hubinot, F. Leroy and P. Gland, editors) (Basel: Karger)
177. Granner, D. K., Chase, L. R., Auerbach, G. D. and Tomkins, G. M. (1968). *Science*, **162,** 1018
178. Stellwagen, R. H. (1972). *Biochem. Biophys. Res Commun.*, **47,** 1144
179. Hauger-Klevene, J.-H. (1970). *Acta Physiol. Latinoam*, **20,** 373
180. Arimura, A., Bowers, C. Y., Schally, A. V., Saito, M. and Miller, M. C. (1969). *Endocrinology*, **85,** 300
181. Fleischer, N., Donald, R. A. and Butcher, R. W. (1969). *Amer. J. Physiol.*, **217,** 1287
182. MacManus, J. P., Whitfield, J. F. and Youdale, T. (1971). *J. Cell. Physiol.*, **77,** 103
183. Dougherty, T. F. (1952). *Physiol. Rev.*, **32,** 379
184. Blecher, M. and White, A. (1959). *Rec. Progr. Horm. Res.*, **15,** 391
185. Gabourel, J. and Aronow L. (1962). *J. Pharm. Exp. Ther.*, **136,** 213
186. Blecher, M. and White, A. (1958). *J. Biol. Chem.*, **233,** 1161
187. Munck, A. (1968). *J. Biol. Chem.*, **243,** 1039
188. Rosen, J. M., Fina, J. J., Milholland, R. J. and Rosen, F. (1970). *J. Biol. Chem.*, **245,** 2074
189. Pratt, W. B., Edelmans, S. and Aronow, L. (1967). *Mol. Pharmacol.*, **3,** 219
190. Drews, J. (1969). *Eur. J. Biochem.*, **7,** 200
191. Young, D. A. (1969). *J. Biol. Chem.*, **244,** 2210
192. Mosher, K. M., Young, D. A. and Munck, A. (1971). *J. Biol. Chem.*, **245,** 2074
193. Haynes, R. C. and Sutherland, E. W. (1967). *Endocrinology*, **80,** 297
194. Werthamer, S., Pachter, B. and Amaral, L. (1971). *Life Sci.*, **10,** 1039
195. Nakagawa, S. and White, A. (1966). *Proc. Nat. Acad. Sci. USA*, **55,** 900
196. Gomez, J., Kemper, B. W., Pratt, W. B. and Aronow, L. (1970). *Biochem. Pharmacol.*, **19,** 1471
197. Ingle, D. (1941). *Endocrinology*, **31,** 419
198. Ariyoshi, Y. and Plager, J. E. (1970). *Endocrinology*, **86,** 996
199. Drews, J. and Wagner, L. (1970). *Eur. J. Biochem.*, **13,** 231
200. Pena, A., Dvorkin, B. and White, A. (1966). *J. Biol. Chem.*, **241,** 2144
201. Palmiter, R. D., Palacois, R. and Schimke, R. T. (1972). *J. Biol. Chem.*, **247,** 3296

8
Steroid Hormonal Analogues

H. B. ANSTALL
University of Utah College of Medicine

8.1	INTRODUCTION		244
8.2	ANDROGENIC AND ANABOLIC STEROID ANALOGUES		245
	8.2.1 *Metabolic considerations*		245
		8.2.1.1 *Ketonic and hydroxyl substituents*	245
		8.2.1.2 *Hydroxylation and related processes*	248
		8.2.1.3 *Aromatisation of the A-ring*	249
		8.2.1.4 *Integrity of the 4,5-double bond*	250
		8.2.1.5 *Miscellaneous metabolic effects*	251
		8.2.1.6 *Effects of structural modification on metabolism*	251
		(a) *Protection of the 17β-hydroxyl group*	251
		(b) *Protection of the Δ^4-3-one system*	251
		(c) *Unsaturation at C-1*	251
		(d) *17α-Methylation*	251
		(e) *Protection against A-ring aromatisation*	251
		(f) *Modifications promoting oral activity*	252
	8.2.2 *The assessment of activity*		252
		8.2.2.1 *The anabolic/androgenic ratio*	252
		8.2.2.2 *Functional comparison of anabolic v. androgenic effects*	254
	8.2.3 *Structure–activity relationship*		258
		8.2.3.1 *Modification in ring size*	258
		8.2.3.2 *Effects of substituent groups*	258
	8.2.4 *Dissociation of effect*		260
8.3	PROGESTATIONAL AGENTS AND ORAL CONTRACEPTIVES		260
	8.3.1 *Chemical and metabolic considerations*		261
		8.3.1.1 *Compounds available*	261
		(a) *Progesterone derivatives*	261
		(b) *Testosterone and 19-nortestosterone derivatives*	261
	8.3.2 *The assessment of activity*		264

8.4	OESTROGENS AND OESTROGENIC ANALOGUES	265
	8.4.1 Chemistry and metabolic considerations	265
	8.4.2 Assessment of activity	271
8.5	CORTICOID ANALOGUES	271
	8.5.1 Chemical and metabolic considerations	271
	8.5.2 Structure–activity relationship	273
	8.5.2 Principal available corticoid analogues	273
	8.5.3.1 Monosubstituted derivatives of major corticoids	275
	8.5.3.2 Disubstituted derivatives of major corticoids	276
	8.5.3.3 Trisubstituted derivatives of major corticoids	276
8.6	CONCLUSION	276
	ADDENDUM A	277
	ADDENDUM B	278

8.1 INTRODUCTION

Following the momentous advances, earlier in the century, in our understanding of the functions of the adrenal cortex, and of the physiology of the testis and ovary, and the inter-relationships of all of these to each other, rapid progress in the knowledge of steroid chemistry and biochemistry was inevitable. Much of this fascinating field has been discussed in detail in earlier chapters. These have been largely concerned with naturally-occurring steroidal hormones and their respective mechanisms of action. However, knowledge of the relationships between molecular structure and biological activity, laboriously worked out from both experimental evidence and theoretical considerations, led to the development of large numbers of synthetic or semi-synthetic steroids in an effort to produce specific substances of highly-predictable biological activity. These, it was hoped, would enable a clinician to select a compound of highly-specific physiological effect for the remedy of some aspects of a disease process—'tailor-made', as it were, to fulfil a particular function in the human organism.

This dream is of similar substance to the one which led to the synthesis of many antibiotics, and like that other great series of adventures in biochemical exploration, it has been partially successful. The clinical uses to which steroid analogues may be put are chiefly to reproduce the physiological functions of the corresponding natural hormones, but in many cases for more general purposes other than replacement of an acquired deficiency. These other uses include (a) utilisation of an anabolic effect, and (b) utilisation of more general effects, such as anti-inflammatory activity, for the management of a variety of diseases of varied specificity. From this, it will be apparent that the ability to select a compound whose action was largely limited to one or other of these functions, but excluded the rest, would be a singular advantage in many cases. A hormone possessing the anabolic attributes of testosterone but without its androgenic properties, for example, would be of obvious superiority in the management of senile osteoporosis in a female over the naturally-occurring androgen, in avoiding virilisation. Likewise, the use of a corticoid analogue with marked anti-inflammatory propensities but with little or no sodium-retaining properties, has obvious major advantages over cortisol, in the management of certain 'collagen' diseases or neoplastic conditions.

What is attempted is a concise review of four biological classes of steroids-androgens, progestogens, oestrogens, and 'corticoid' analogues, in terms of their degree of biological activity, as related to their structure, and thus to see to what extent the above goals have been achieved.

8.2 ANDROGENIC AND ANABOLIC STEROID ANALOGUES

Androgens may be defined as biologically-active steroids which promote the various male secondary sexual characteristics[1]. Naturally-occurring androgens are steroids with 19 carbon atoms (C-19 compounds) having an oxygen function (ketonic or hydroxyl) at C-3 and C-17, and partial unsaturation of the A-ring. This definition must be variously modified for some of the analogous synthetic compounds possessing varied androgenic v. anabolic activity. Anabolic steroids are, essentially, those which promote a positive nitrogen balance in the intact organism, i.e. nitrogen retention. Androgenic and anabolic properties are usually combined to a varying degree in the same compound such as in testosterone. The capability to produce compounds in which these two activities can be completely dissociated would be of obvious clinical advantage. So far, this goal has been only partially attained.

Chemical modifications of testosterone have led to some compounds which display a fairly high degree of dissociation of androgenic and anabolic effect. Some of these are capable of oral administration; a great advantage from the clinical standpoint, over compounds requiring injection[2,3].

8.2.1 Metabolic considerations

It is necessary to be able to answer certain questions about a given steroid in order to decide its likely biological efficacy and activity. These centre around the possible metabolic fate of the compound when presented to the body. First, is the drug orally effective, or is its structure such as to preclude absorption? Secondly, can it be transported in the body sufficiently rapidly to the sites where its action is desired? Thirdly, what is likely to be its metabolic fate, and can its structure be modified, if required, to permit effective biological action for a reasonable length of time? Fourthly, will the natural receptors recognise an acceptable structure, not significantly altered by metabolic effects within the body? Many of these questions can now be answered with a reasonable degree of accuracy. We may now examine some of the known metabolic effects upon various parts of the steroid structure in order to see what these answers are.

8.2.1.1 Ketonic and hydroxyl substituents

All of these are liable to be the subject of redox transformations in living organisms, in which such reactions are extremely frequent. Hydroxyl groups are also likely to be the subject of conjugation, for example, with sulphate or glucuronide.

Redox transformations are very important at certain key groups. Oxidation of the 17β-hydroxyl group of testosterone, for example, results in the loss of much of the androgenic potency of the compound. It is thus desirable, for the preservation of androgenic activity in analogues, to prevent the metabolic oxidation of this 17β-hydroxyl group.

It was formerly believed that, in the conversion of testosterone to other metabolites, oxidation of this 17β-hydroxyl was a necessary prerequisite.

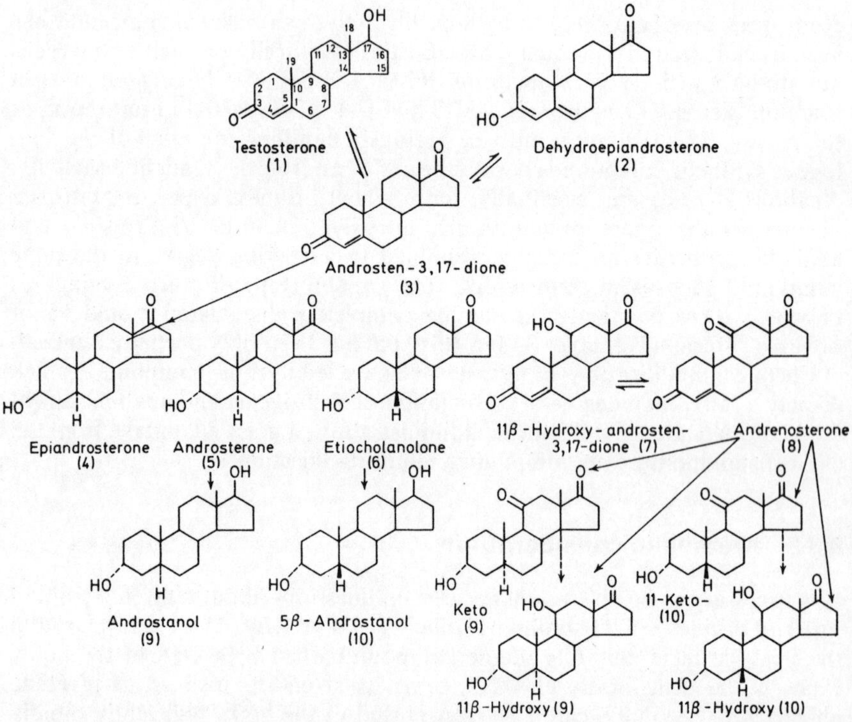

Figure 8.1 Some of the metabolites of the 'keto' pathways of natural androgen degradation. Note that reductions, notably at C-3 and C-11, occur in a number of derivatives

This would imply that androstenedione was the common precursor of other metabolites. (Figure 8.1). This 'ketonic' pathway, in which the 17β-hydroxyl is first oxidised to a 17-keto group is now known to be only one of several routes for the degradation of testosterones and, presumably, similar compounds. It has been shown that neither androsterone nor 5β-androsterone (Figure 8.2) is derived uniquely from a pool in equilibrium with androstenedione[4-6]. A 'hydroxyl' pathway thus also exists whereby testosterone and, presumably, related compounds can be converted to metabolites such as androsterol and 5β-androstanol (Figure 8.2) which have negligible biological activity.

For steroids with the Δ^4-3-keto system, it seems that reduction of the 4,5 double bond is followed by the reduction of the 3-keto-group. This results in the formation of an asymmetric centre at C-3, such that 3a- and 3β-hydroxy stereoisomers become possible. Both isomers may be produced *in vivo*, the 3a-hydroxy-compounds generally predominating in males. Interestingly, 3a-hydroxy-compounds tend to possess more androgenic activity than 3β-compounds but there are occasional exceptions.

In steroids possessing the Δ^4-3-keto system (enone system), saturation of the 4,5 double bond proceeds initially as the rate-limiting step[8-9]. After this, reduction to one or other of the hydroxy-compounds proceeds apparently independently of the redox potential. This is probably so because the hydroxylated end-products are continuously removed by the formation of

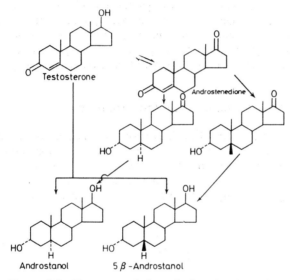

Figure 8.2 The proposed direct 'hydroxy' pathway from testosterone to its reduced 3-hydroxyl-derivatives, and its relationship to the 'keto' pathway

'conjugates', principally by the esterification of the newly-formed hydroxyl with glucuronic acid or sulphate. Thus, by removal of the newly-formed hydroxylated compounds from the ketone–hydroxyl equilibrium, the forward reaction is favoured[10]. Glucuronide conjugation may represent, by and large, a virtually irreversible inactivation[11]. Formation of the sulphate conjugate does not seem to have the same degree of finality as that of the glucuronide. Sulphates are known to be eliminated more slowly by the kidney, but a relatively greater degree of reversibility for sulphate conjugation appears to exist[12-13].

Further metabolism of conjugates appears possible. Thus dehydroepiandrosterone glucuronide may be partially converted, albeit to a small extent only, to androsterone; similar conversions may be presumed to occur in the case of certain analogous conjugates[14]. Transconjugation, i.e. the interconversion of one conjugate ester to another, also takes place to varying degrees. Thus, Baulieu et al.[11] showed that intravenously administered dehydroepiandrosterone and dehydroepiandrosterone glucuronide were both partially converted to dehydroepiandrosterone sulphate, but that more of the conversion was from the free steroid rather than from its glucuronide. However, it should be noted that the free steroids in general have a greater tendency to undergo further degradative metabolism than to undergo sulphate conjugation.

It has been supposed previously that this conjugation occurs in the liver as a function of hepatic cells. Naturally occurring steroids were supposedly synthesised as free compounds by the cells of various endocrine organs and underwent metabolic inactivation by partial degradation and esterification (conjugation) in non-endocrine tissues, chiefly, the liver. It is now known, however, that endocrine tissues may synthesise conjugates directly, and that at least sulphate conjugates, may be primary products of adrenal cortical cells[15, 16]. It must be remembered, however, that such sulphates, although themselves biologically inactive, are probably largely reconverted into free steroids with their characteristic degree of physiological activity.

8.2.1.2 Hydroxylation and related processes

Hydroxylation of the steroid nucleus may take place at any position which may be considered 'activated', for example, adjacent to a carbonyl grouping or at an allylic methylene position. In general, so far as androgenic and anabolic functions are concerned, this usually results in a compound of reduced biological activity. Hydroxylation can be regarded as a further mode

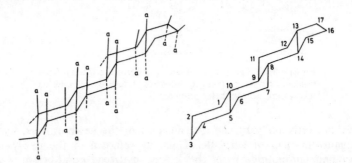

Figure 8.3 The structure of androstane, showing principal substituent bond-positions (a) = axial bonds. Basic steroid skeleton shows conventional numbering

of metabolic attack upon the functional steroid which must be taken into consideration in attempting to predict the biological utility of a given compound.

An interesting hypothesis for the mechanism of hydroxylation of activated positions was proposed by Ringold[17]. Such hydroxylations may occur through enol intermediates. The enolic form present at the site of hydroxylation would provide a high electron density, so that, assuming attack by an electrophilic species, a marked acceleration of reaction rate might be anticipated. As an axial approach of the hydroxylating species would provide a maximum overlap of π-electrons, then potential hydroxylations at positions 2β, 6β, 10β, and 17α of the steroid nucleus, which are axial, are favoured over the corresponding equatorial orientations. (Figure 3.8). This is of some physiological importance for many natural steroids, since, in a number of them, as will be seen later, the 17α-hydroxy group is necessary for adequate biological activity.

8.2.1.3 Aromatisation of the A-ring

The aromatisation of the A-ring of natural androgens to oestrogens has been known for many years. Indeed, it is a normal step in the synthesis of oestrogens[18]. This reaction involves an intermediate hydroxylation step at C-19[19], followed by conversion to the oestrogenic steroid with a totally phenolic A-ring: this conversion is mediated by the enzyme 10,19-desmolase. (Figure 8.4) In the case of androgenic analogues, those lacking the angular methyl group at C-19 (as in 19-nortestosterone and its derivatives) resist

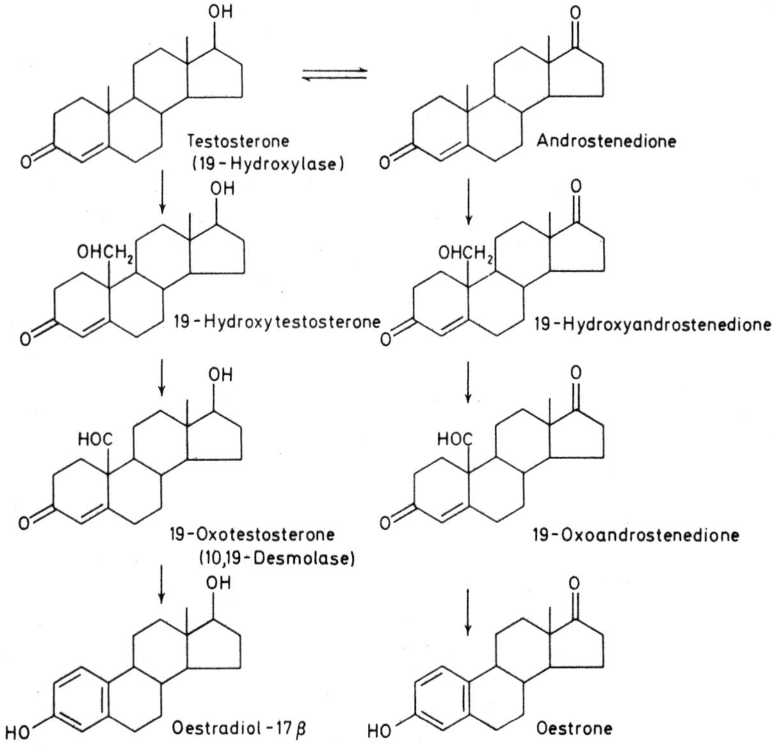

Figure 8.4 Natural aromatisation sequences leading to the synthesis of oestradiol-17-β and oestrone

aromatisation strongly, since the usual mechanism via C-19 hydroxylation is not possible here. Since the normal 'desmolisation' step requires first the hydroxylation of C-19, then double bond formation at C1–2 with final removal of C-19 this is hardly surprising. Thus absence of the C-19 angular methyl group affords protection against aromatisation, which would lead to the loss of androgenic properties. The risk of aromatisation is also lessened by substitutions at positions 11, 16 and 17. Some 17α-derivatives, e.g. 17α-hydroxytestosterone, can be converted to oestrogens; esterification may not prevent this aromatisation[20]. Alkylation of the 17α-hydroxy group is much more although not completely, effective in preventing aromatisation. Thus 17α-methyltestosterone shows *ca.* 44% activity.

Methylation at the C-1 position markedly protects against aromatisation, presumably by interfering with the initial binding to the hydroxylating enzyme thus inhibiting the first step in aromatisation[21]. Chlorine substitution at C-4, as in 4-chloro-17β-hydroxyandrost-4-en-3-one 17β-p-chlorphenoxyacetate, administration of which leads to no increase in oestrogen excretion, appears to protect against aromatisation.[22]

8.2.1.4 Integrity of the 4,5-double bond

Many active androgens and anabolic steroids have a double bond at 4:5. Reduction of this double bond in general results in a marked degree of inactivation, and, as has been mentioned above, probably represents the rate-limiting step in metabolic inactivation in many cases[8,9]. Reduction of this bond results in C-5 becoming asymmetrical so that two stereoisomers, one *cis* with respect to the A/B ring junction (5β-), the other *trans*-(5α-), are possible. *In vitro* catalytic reduction of Δ^4-compounds results in a mixture of isomeric forms with general predominance of the 5β-product. However, the course of such reactions is largely affected by the prevailing environment in which the reaction takes place, so that in neutral environment, 5β-isomers often prevail, while in an acidic medium 5α-products prevail[23]. Moreover, this relative preponderance of one or other isomer is subject to direction by neighbouring substituent groups. Thus the presence of an 11β-hydroxyl group leads to the exclusive formation of the 5α-product, whereas an 11α-hydroxyl substituent leads to formation of the 5β-product[24,25].

The above considerations are based on the study of catalytic hydrogenations and other *in vitro* procedures. However, enzymatic reductions behave in a very similar manner. In general, metabolic reduction of a Δ^4-3-ketone system gives a mixture of 5α- and 5β-products. Their relative proportion, in the presence of a reasonably constant and controlled environment with respect to pH, is thus determined by the number and nature of other substituent groups in the molecule. Under these conditions, an 11β-hydroxyl function leads to great predominance of 5α-products[26], while in the absence of an oxygen function at C-11, 5β-and 5α-compounds are formed to about the same extent.

The 5β-dihydro-compounds have little or no anabolic or androgenic activity. The greater preponderance of the 5β-product, the greater the degree of inactivation.

8.2.1.5 Miscellaneous metabolic effects

In vivo experiments with differentially-labelled testosterone and testosterone glucuronide [^{3}H]testosterone and [^{14}C]testosterone glucuronide) indicate that not all of the metabolites derived from testosterone glucuronide can be accounted for by the prior conversion of the glucuronide into free testosterone which then would presumably undergo the normal conversion to 5α-derivatives (epiandrosterone, androsterone, androstanol) and 5β-derivatives- (5β-androstanol, 5β-androsterone). [Figure 8.1] The [^{14}C]labelled glucuronide yielded almost exclusively [^{14}C]-labelled 5β-derivatives, suggesting a new, more direct pathway from the glucuronide conjugate to 5β-metabolites[28, 29].

8.2.1.6 Effects of structural modification on metabolism

From what has been discussed above, it will be apparent that certain structural modifications of the steroid molecule will tend to protect certain groups necessary for biological activity from metabolic attack. Other modifications will facilitate uptake of the steroid from the alimentary tract. The more important of these are now given

(a) *Protection of 17β-hydroxyl group*—Substitution of methyl groups at C-1, C-2 and C-6 in synthetic steroids stabilises the 17β-OH group, protecting it from NAD-dependent 17β-hydroxy (testosterone) oxidoreductase attack[30]. Alkylation at the 17α-position[31], converting the 17β-hydroxyl into a tertiary hydroxyl group, also protects this group against enzymatic oxidation. Other structural modifications, including 4-chloro substitution[32], also have a retarding effect.

(b) *Protection of the* Δ^4-*3-one-system*—2α-methylation, and 6α-methylation significantly retard the rate of reduction of the 4,5 double bond in the A-ring, thus impairing a further route of metabolic degradation[33]. Introduction of such groups into some steroid analogues, which include not only anabolic and androgenic steroids but also corticoids, may be a valuable means of preserving activity.

(c) *Unsaturation at C-1*—This affects metabolic degradation. The analogue 17β-hydroxyl-1-methyl- Δ^1-androstene-3-one is more resistant to metabolic attack than is testosterone, large amounts of the compound remaining unchanged, some 1-methyl- Δ^1-androstanedione and 1α-methyl androstanedione are formed, but little further change is apparent[34]. Thus etiocholane derivatives are not produced (Figure 8.1) as might be expected.

(d) *17α-Methylation*—17α-Methylation in testosterone derivatives increases biological activity, both with respect to anabolic (nitrogen-retaining) and androgenic effects. This is even more apparent in the 19-nortestosterone series[35]. It is presumed that this substituent group protects the 19-nor configuration, ensuring the persistence of the potent steroid.

(e) *Protection against aromatisation of the A-ring*—This would lead to oestrogen formation and is afforded partially by removal of the angular methyl group at C-10, as in the 19-nortestosterone series, since the normal hydroxylation at C-19, which precedes the oxidation of the A-ring is not possible. Partial protection is also afforded by substitution at C-11, C-16 and C-17, and by alkylation of the 17α-hydroxy-group. Methylation at the C-1 position affords a substantial measure of protection against aromatisation, probably by steric hindrance of the attachment of the substrate to the hydroxylating enzyme which catalyses the hydroxylation of C-19. 4-Chloro-substitution is also partially effective in inhibiting aromatisation. The above statements are summarised in the composite Figure 8.5.

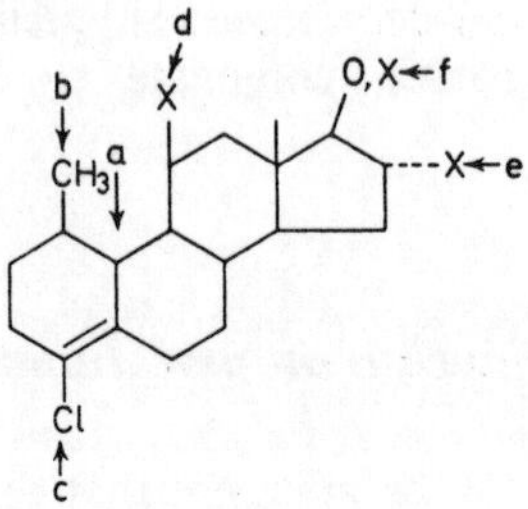

Figure 8.5 Schema of steroid nucleus showing positional changes tending to protect against aromatisation of A-ring. a = removal of C-19 methyl group, b = 1-methylation, c = 4-chloro-substitution, d = 11-substitution, e = 16α-substitution, f = esterification of 17-hydroxy group

(f) *Modifications promoting oral activity*—These include esterification of the 17β-hydroxyl group (e.g. with acetate, propionate, decanoate, etc.), which both protects the group from oxidative attack and prolongs the biological effect, probably by retardation of transport of the analogue within the body[36]. 17α-Alkylation also appears to prolong the biological effect by retarding excretion of the compound as a 17-ketosteroid; the alkyl group produces a protective effect. 17α-Alkyl substitution, may, however, be at least partially responsible for some of the hepatotoxic effects of these compounds, which include primarily cholestasis and varying, but generally slight, intrahepatic biliary obstruction and are reflected in mildly abnormal liver function tests[37, 38].

8.2.2 The assessment of activity

8.2.2.1 The anabolic/androgenic ratio

So far, we have been concerned with certain general principles relating the structure of androgenic and anabolic steroid analogues to their probable metabolic fate. We must now turn our attention to those aspects of their molecular structure which relate to their basic biological activities—androgenic and anabolic—and determine what molecular modifications are likely to bring about a predominance of one or other of these functions. First, however, it is necessary to say something about how these respective activities may be assessed, and what kind of quantitative criterion may be used to record differences in the two types of activity.

It has long been known that castration of male animals produces a gradual reduction in total muscle mass, which can be reversed by the administration of natural androgen[39] or of an anabolic hormone analogue. This myotrophic effect, dependent upon the promotion of nitrogen retention by the steroids, differs in its magnitude with different species of test animal. Moreover, the effect may be further modified by other factors including age, sex, dosage and whether or not the adrenal cortex and pituitary are intact[40]. In general, castration results in decreases of muscle mass, with loss of protein, water and mineral elements in proportion in which they occur in the normal muscle. Administration of anabolic steroids or natural androgens reverses these changes, restoring the *status quo*[41].

The question as to whether the anabolic effects, as measured by changes in muscle mass, differ with natural steroids and synthetic anabolic analogues is not yet settled. There is some experimental evidence to suggest that some differences may exist. Certain anabolic steroids appear to promote a greater myotrophic effect on some types of muscle rather than on others[42]. Moreover, striking time differences with respect to the occurrence of maximum biological effect between the various analogues.

Various experimental techniques were developed making use of the myotrophic effect upon skeletal musculature, as well as of other androgenic effects (e.g. weight gain of prostate and seminal vesicles) shown to be exerted by anabolic steroids, in order to quantitate anabolic *v.* strictly androgenic activities of given compounds. Assays are based on the performance of certain standard androgenic steroids as points of reference; their activity, for the purposes of the tests, could be regarded as 100%. They included testosterone, *a*-methyl testosterone, and testosterone propionate. Experiments of Wainman and Shipounoff[43] demonstrated that the atrophic musculature of the perineum (levator ani, ischiocavernosus, bulbospongiosus) of castrated male rats respond by rapid weight gain to administered androgen. Further experiments[44] showed that administration of anterior pituitary extracts rich in growth hormone produced a similar effect. That such resulted from stimulation by substances other than androgens was taken as evidence that such changes really represented an anabolic non-androgenic effect. That such changes could also be brought about by steroids having androgenic properties was regarded as evidence that they also possessed *anabolic* properties, distinct from their androgenic functions.

Various experimental techniques are now in use. These are admirably reviewed by Kürskemper[45], and will not be described here. Basically the anabolic/androgenic index as determined by these techniques, is expressed as the ratio of the weight of the levator ani muscle to the prostate or seminal vesicle after administration of a test steroid in carefully standardised dosage for a standard time-period. Comparison can then be made with the activity of the standard reference steroids mentioned above, whose effects, with regard to both activities, are taken to represent 100%. Stimulation of weight-gain of the prostate or seminal vesicle is taken to represent an expression of *androgenic* activity, as distinct from anabolic. However, since certain androgenic analogues promote the weight-gain of either the prostate *or* the seminal vesicle in disproportionate amounts, care must be taken to calculate the anabolic/androgenic ratio by reference to one or other androgenic reference organ[46]. Methods dependent upon the ratio (*Wt. gain levator ani*/Wt. gain prostate (or vesicle)) have been criticised also because different activity ratios may be found for different dosages of steroid. Furthermore, the concept that levator ani weight-gain represents a pure anabolic effect may well be erroneous, since, in castrated rats on a protein free diet, administration of androgen causes a marked myotrophic effect, whereas wasting of other muscles is not greatly retarded. Under these circumstances the differential growth of the levator ani may represent an *androgenic* effect[47, 48].

Administration of [^{14}C]glycine to castrated male rats treated with androgen leads to highest incorporation into the seminal vesicles, and levator ani[49]. There was little incorporation elsewhere, with the exception of the skin, where marked incorporation into newly-generated collagen occurred[50]. It is thus very hard to dissociate the extent to which androgenic, as distinct from anabolic, effect is represented by levator ani weight gain. However, some criterion for bioassay is required, and an unequivocal method is not yet available. Other criteria of androgenic or anabolic activity, based on nitrogen and mineral balance determinations are of little value as screening procedures for potential therapeutic agents, since they involve multiple and complex parameters which are difficult to assess.

8.2.2.2 Functional comparison of anabolic v. androgenic effects

Based on various modifications of the procedures discussed, and by careful comparison of the observed quantitative effects of large numbers of anabolic and androgenic steroid analogues with those of 'reference' steroids (testosterone, testosterone propionate, 17a-methyl testosterone,) much significant information about many compounds has been obtained. In Tables 1–3 a selected list of examples of such analogues is presented. Many of these compounds have been selected because they have been found useful for clinical purposes. In clinical use, where androgenic deficiency has to be replaced, a steroid displaying a combination of androgenic and anabolic properties is generally valuable. Where an anabolic effect alone is desired, and especially in the management of female patients, maximal dissociation of anabolic from androgenic action is obviously required. The tables classify steroids primarily into the following, purely arbitrary, groups, based on their assessment by bioassay procedures; those showing increase in both androgenic and anabolic effect; those showing decreased androgenic effect, and those showing increased anabolic activity with reduced or relatively unchanged androgenic effect. This classification, although imperfect in that the various categories overlap to some extent depending on the techniques used to assay the types of function, is convenient from the clinical point of view.

In the tables, anabolic activity is expressed as a percentage of the activity of the 'standard' reference steroid (identified in the appropriate column), in increasing the weight of the levator ani muscle of the castrated male rat. Androgenic activity is expressed as the percentage activity of each substance, compared to the appropriate reference standard, in increasing the weight of the seminal vesicle or ventral prostate of the rat. Reference standards in each case are presumed to have 100% activity.

Table 8.1 Anologues showing increased androgenic effect combined with increased or dominant anabolic effect

Structure	Chemical name	Sem Ves.	Pros.	L.A.	Standard	Ref.
	9α-Fluoro-11β,17β-hydroxy-17α-methyl testosterone (Fluoxymesterone)	750	120	1745	17α-MT	51
	7α,17α-Dimethyl testosterone (Bolasterone)	660	224	1140	Test	52
	17α-Methyl-17β-hydroxy oestra-4,9,11-trien-3-one.	500	600	120	17α-MT	53
	7α-Methyl-17β-acetoxy-19 norandrost-4-en-3-one.	580	290	350	Test	54
	17α-Methyl-17β-hydroxy androstane-1-one	190	180	830	T.P.	55
	Androstane-17β-ol-3-one (1-methoxy) cyclopentyl ether	270	122	308	17α-MT	56
	17β-Hydroxyandrost-1-en-3-one 17-cyclopent-1′-enyl ether.	240	—	400	17α-MT	57
	13β,17α-diethyl-17β-hydroxygona-4,9(10)-dien-3-one	—	860	1450	17α-ethyl-nortest	58

Abbreviations: Sem. Ves.,=Seminal vesicle, Pros=Ventral prostrate, L.A.=Levator ani. Test=testosterone, T.P.=Testosterone propionate, 17α-MT=17α-methyltestosterone.

Table 8.2 Analogues showing significant anabolic activity with decreased androgenic activity

Structure	Chemical name	Sem. Ves.	Pros.	L.A.	Std.	Ref.
	19-Nortestosterone	36	39	390	Test.	59
	19-Nortestosterone propionate	41	—	2.59	T.P.	60
	17α-Ethyl-17β-hydroxy-19-nor-androst-4-en-3-one (norethandrolon)	29	37	120	Test.	61
	4-Hydroxy-19-nortestosterone-17-cyclopentyl propionate (oxabalon)	—	—	—	—	62
	19-nortestosterone 17-n-decanoate (Deca-Durabolin)	—	—	—	—	63
	19-Nortestosterone cyclohexylpropionate (Nortestosterone cypionate) (Depostestosterone)	—	—	—	—	—
	4-Chloro-17β-acetoxy-androst-4-en-3-one (chlortestosterone acetate)	48.3	55	126	T.P.	64
	17α-Methyl-17β-hydroxyandrost-1,4-dien-3-one (Methandienone, Methandrostenolon)	60 6	60 12	210 30	17α-MT 17α-MT	65 66
	2α-Methyl-17β-hydroxy-androstan-3-one. (Drostanolone)	10	10	100	Test	67

Table 8.3 Analogues showing marked augmentation of anabolic effect with diminished androgenic effect

Structure	Chemical name	Sem Ves.	Pros.	L.A.	Std.	Ref.
	17α-Methyl-17β-hydroxy-2-oxa-5α-methyl androstan-3-one (Oxandrolone)	24	24	322	17α-MT	68
	1-Methyl-17β-hydroxyandrost-1-en-3-one (Methenolone)	57	44	88	Test	69
	17α-Ethyl-17β-hydroxy-19-norandrost-4-ene-3-one (Ethylestrenol)	40	40	200		70
	17α-Methyl-17β-hydroxyandrostan-[3,2,-c] pyrazol (Winstrol, Stanazol)	100	—	183		71
	17α-Methyl-17β-hydroxy-androstan-[3,2,-c] isoxazole	—	22	155	17α-MT	72
	17α,13β-Diethyl-17β-hydroxygon-4-en-3-one. (Norbolethone)					

8.2.3 Structure–activity relationship

It seems that the basic 5α-androstane configuration itself possesses some androgenic activity[74], a property which may be shared by certain other polycyclic compounds, such as those illustrated in Figure 8.5[75,76]. Oxygen functions at C-3 and C-17 are usually associated with enhanced androgenic activity. The effects of other major modifications are briefly summarised in the following discussion.

8.2.3.1 Modifications in ring size

These have various effects. Ring *contraction* decreases androgenic and anabolic activity. Ring *expansion* as in the homosteroids in which one of the rings A, B or D contains an additional carbon atom, are generally biologically active. Activities tend to be of the same order of magnitude or somewhat

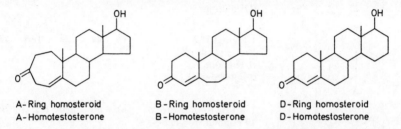

A-Ring homosteroid   B-Ring homosteroid   D-Ring homosteroid
A-Homotestosterone   B-Homotestosterone   D-Homotestosterone

Figure 8.6 Configuration of A-, B-, D-type homosteroids

lower than that of testosterone. In D-homosteroids, in which C and D represent two fused 6-membered rings, the *trans*-junction of the rings is relatively stable, so that the normal stabilising influence of the C-18 methyl group is probably not required. D-homosteroids, likewise, are biologically active (Figure 8.6).

8.2.3.2 Effects of substituent groups

Considerations of the mechanism of action of androgenic steroids suggests the importance of attachment of the steroids to receptors by the α-face of the A-ring at the C-1 position. 1α-Substitution thus decreases androgenic activity. Anabolic effects are not reduced so markedly so that 1α-substitution may help to dissociate anabolic from androgenic effect. 1β-Substitution tends to produce some decrease in both activities, that of the androgenic being usually the greater. Vida[77] points out that this should not be regarded as supportive evidence for β-face receptor attachment since 1β-substituents lie in the plane of the molecule. This is construed as evidence for a three-dimensional attachment of the steroid to receptor sites.

2α-Substituents lie in the plane of the molecule. They generally decrease the androgenic activity, sometimes substantially and decrease the anabolic effect to a lesser extent, so that again a useful degree of dissociation may be effected by such substitution. Planar substitution at C-2 in Δ^1 or Δ^2-unsaturated steroids also produce this effect. 2β-Substitutions, where they occur, generally diminish both types of activity thought not necessarily to the same degree.

3α-Substitution, particularly 3α-hydroxy-substitution, usually leads to enhanced androgenic activity. 3β-Substitution produces variable and unpredictable results suggesting the probable non-involvement of this position in attachment to receptors. Substitution of the 3-keto group by nitrogen or by a methylene group increases both bond lengths and bond angles considerably. This is the rationale for introducing heterocyclic rings across the C-2—C-3 positions. (See Table 8.3). The effect is usually to produce dissociation of anabolic and androgenic functions, producing a favourable anabolic/androgenic ratio. Most 4α- and 4β-substitutions decreased the anabolic effect slightly and the androgenic effect markedly. The anabolic–androgenic ratio thus becomes favourable. Di-substitution, as in 4,4-dimethyl-3-keto- Δ^5, may abolish all biological activity[78].

5α-Substituted steroids generally show decreased biological activity, suggesting again the probable importance of α-face attachment at this position (see 1α-substitution, above).

6β-Substitution reduces overall biological activity, emphasising the importance of β-face receptor attachment at C-6.

7α-Substitution markedly increases both anabolic and androgenic activity. (See Table 8.1, compounds 2 and 4.) This implies non-involvement of the α-face at the C-7 position in receptor attachment. The increase in activity is more likely related to the protection of other configurations such as the 4,5 double bond. 7β-Substitution interferes with both androgenic and anabolic activity. The 7β-substituent again lies within the plane of the molecule, and its interference with biological activity affords further support for the three-dimensional concept of steroid–receptor interaction. 8β-Substitution behaves similarly[78].

9-Substitutions have somewhat variable results; this suggests little involvement of the C-9 position in receptor attachment. Some heavy substituents diminish activity while fluorine tends to increase it. The substitution of bulky groups in the C-10 position tends to diminish both activities.

11α-Substitutions result in increase of both types of activity. 11β-Substituents generally produce diminished activities, probably due to severe steric interference with the C-19 methyl group. This would be expected to impede β-face attachment to a receptor at the C-19 position. In the absence of the 19-methyl group, as in 11β-substituted nor-testosterones, there is no such steric hindrance. In such cases the 11β-substituent may itself become the vehicle of attachment to the receptor, and so increase biological activity.

13β-Substitution generally augments both types of activity. Such substitution has been studied only in 19-nor compounds.

14 and 15 Substitution has been incompletely studied; Segaloff and Gabbard[79] demonstrated that Δ-^{14}-5α-dihydro-testosterone had greater biological activity (androgenic) than did Δ^{14}-testosterone.

Both 16α- and 16β-substitution decrease overall biological activity. This is interesting, since in the corticoid analogues 16α-substitution may increase anti-inflammatory activity (see later).

The introduction of a 17β-hydroxy-substituent into the 5α-androstane skeleton significantly enhances biological activity. This substituent is found in almost all effective anabolic agents and androgens. No other substituent at the 17-position is as effective in enhancing biological activity as the 17β-hydroxyl group. Esterification, ether formation, acetal formation, etc. do not interfere with biological activity and contribute to its preservation in the body. It is not entirely clear whether the active species is the ester in such cases, or whether hydrolysis to the free hydroxyl is required for attachment to receptors. This is suggested by the observations of Van der Vies[80]. The importance of the 17β-oxygen atom in receptor attachment is apparent. 17α-alkyl substitution is limited since the steroid-receptor site cannot easily accommodate alkyl substituents larger than ethyl groups. Larger groups result in a significant reduction of activity.

The stereochemical significance of many of these considerations cannot be discussed in detail here. For further details the reader is referred to excellent discussions by Vida[81], Bush[82] Klimstra et al.[83] and to other chapters in this book. Vida's emphasis is on the total steroid molecule being a three-dimensional structure, enabling steroid-receptor attachment not only through axially oriented bonds at C-1, C-3 and C-5 on the α-face of the A-ring, C-6, C-10 and C-13 through axially orientated bonds on the β-face rings, B, C, D, but also through planar linkage to planes perpendicular to the α-and β-faces.

8.2.4 Dissociation of effect

It will be apparent from what has been said that, so far, total dissociation of anabolic from androgenic activity has not been achieved, although a significant degree of separation has contributed many useful therapeutic compounds to the clinical armamentarium. Whether total separation will eventually be achieved cannot be answered; greater knowledge of the mechanisms of action of these compounds must be obtained before this problem is resolved.

8.3 PROGESTATIONAL AGENTS AND ORAL CONTRACEPTIVES

Following the original discovery of progesterone in the corpus luteum of the pig[84], a relatively long period of time elapsed before clinically and biologically effective analogues were forthcoming. This occurred primarily in the decade 1950–1960 when new synthetic progestational agents with prolonged and augmented action became available. Steroids having progestational activity are structurally diverse, and a number are now in regular clinical use—both for their primary progestational effect and, combined with oestrogens, as contraceptive agents.

The progestational effect in a substance reflects its ability to induce secretory changes in the endometrium and reproductive tract of a female animal pre-conditioned by oestrogen. It also implies the maintenance of pregnancy

STEROID HORMONAL ANALOGUES 261

in an ovariectomised animal. Progestational steroids act as antifertility compounds by suppression of oestrus or prevention of ovulation, via a feedback effect upon the secretion of gonadotrophins by the anterior pituitary.

8.3.1 Chemical and metabolic considerations

Many of the general principles which applied to our detailed examination of the structure–activity aspects and metabolic stability of androgenic and anabolic agents apply here also. They do not require further elaboration. Just as we made a physiological division between 'anabolic' and 'androgenic' activities, so progestogens can be regarded as possessing, to a greater or lesser degree, 'progestational' and 'anti-fertility' (i.e. antigonadotrophic) activity. Again, total dissociation of the two effects has not yet been achieved[85]. In general, it has been possible to enhance progestational activity while retaining a limited antifertility effect, but the reverse has not been possible to date.

8.3.1.1 Compounds available

Compounds available belong to two basic series of steroids, progesterone derivatives and 19-nortestosterone derivatives. Those in common use are illustrated in Table 8.4.

(a) *Progesterone derivatives*—Hydroxylation at C-17 in the α-position to 17α-hydroxyprogesterone causes loss of activity, which is restored by esterification. Such compounds (Table 8.4), tend to have a marked antifertility activity[86]. Esterification with longer-chain fatty acids leads to a reduction of this effect[87].

6α-Methylation generally leads to augmentation of the antifertility effect, and unsaturation at the 6,7-position augments this effect even further[88, 89]. Chlorine substitution at the C-6 position, as in pregna-4,6-dien-6-chloro-17α-acetoxy-3,20-dione (Chlormadinone) also enhances the ability to suppress ovulation.

(b) *Testosterone and 19-nortestosterone derivatives*—Until 1950 Ethisterone (17α-ethynyl testosterone) was the only synthetic progestogen available, and had the advantage of oral activity, possessing significant progestational effect with inconspicuous anti-fertility effect[90]. Subsequent introduction of alkyl groups produced various results, most noticeably the marked enhancement of progestational effect resulting from 6α-methyl, or 6α- and 21-dimethyl, substitution. The antifertility effect, however, was not enhanced, so that the compound 4-androsten-6α,21-dimethyl-17α-ethynyl-17β-ol-3-one (Dimethisterone) is a poor suppressor of ovulation, requiring very high doses[89].

19-Norprogesterone possesses enhanced progestational activity, an observation which led to the synthesis of a large number of 19-nortestosterone derivatives. Substitution at the 17α-position gave rise to a number of compounds, some of which possessed varying degrees of anabolic or androgenic activity in addition to progestational effects, e.g. 17α-ethyl nortestosterone (norethandrolone), which was used more for its anabolic action than its progestational activity. Norethisterone (Table 8.4). 17α-Ethynyl-19-nortestosterone, is a potent progestogen but retains some androgenic and anabolic effect. Shift of the double bond to the 5(10) position produces a substance with augmented anti-fertility effect (Norethynodrel), one of the earliest oral contraceptives still in extensive use. Esterification of the 3-oxygen function of norethisterone (17α-Ethynyl-19-nortestosterone) gives 4-oestren-17α-ethynyl-3β,17β-diol diacetate (Ethynodiol diacetate), a compound of marked progestational activity.

Removal of the angular methyl group at C-10 gives rise to the oestrenols; their biological activity depends upon the nature of the substituent groups at C-17. 17α-Ethynyl substitution, as in 4-oestren-17α-ethynyl-17β-ol (lynestrenol) leads to significant progestational activity, with limited anabolic or androgenic effect, noticeable only in very high dosage[91]. The alkyl substituent

Table 8.4 Important progestational analogues. (a) Progesterone derivatives (b) Testosterone and 19-nortestosterone derivatives

Structure	Chemical name	Trivial name
(a)	Δ⁴-Pregnen-17α-caproyloxy-3,20-dione	17α-hydroxyprogesterone acetate
	4-Pregnen-6α-methyl-17α-acetoxy-3,20-dione	Medroxy-progesterone acetate
	4,6-Pregnadien-6-methyl-17α-acetoxy-3,20-dione	Megestrol acetate
	4,6-Pregnadien-6-chloro-17α-acetoxy-3,20-dione	Chlormadinone acetate
	6α-Methyl-17α-acetoxy-4-pregnen-20-one	Anagesterone acetate

STEROID HORMONAL ANALOGUES

Structure	Chemical name	Trivial name
(b)	4-Oestren-17α-ethynyl-17β-ol-3-one (Ethynyl-19-nortestosterone)	Norlutin, Norethisterone
	17α-Ethynyl-17β-hydroxyoestra-5(10)-ene-3-one	Norethynodrel
	4-Oestren-17α-ethynyl-17β-acetoxy-3-one	Norethisterone acetate
	4-Oestren-17α-ethynyl-3β,17β-diol diacetate	Ethynodiol diacetate
	4-Oestren-17α-enthynyl 17β-ol	Lynestrenol
	4-Gonen-13β-ethyl-17α-ethynyl-17β-ol-3-one	Norgestrel
	4-Androstan-6α-21-dimethyl-17α-ethynyl-17β-ol-3-one	Dimethisterone

has moderate progestational effect, and the ethyl substituent has a much more marked androgenic/anabolic effect with progestational side-effects.

Other progestational agents have been derived from the gonene nucleus. One such compound, 4-Gonen-13β-ethyl-17α-ethynyl-17β-ol-3-one, is a potent progestogen. A singularly powerful progestogen is 6-chloro-6-dehydro-17a-acetoxy-1,2a-methylene progesterone (cyproterone acetate). This substance is of great interest because, in addition to its remarkable potency as a progestational agent, approaching 1000 times that of progesterone in standardised tests in rabbits[92], it also has striking antiandrogen activity. Treatment of pregnant rats with doses of 1.0–10.0 mg daily resulted in feminisation of the male foetuses, with rudimentary development of the male genitalia (vestigial prostate, tiny penis resembling a clitoris, small, undescended testes.

In mature male animals, administration of cyproterone acetate produces atrophy of the prostate, seminal vesicles and perineal musculature, together with atrophic changes in the skin and sebaceous glands—all structures variously dependent upon the anabolic and androgenic effects of natural androgens. Cellular changes occur in the anterior pituitary reminiscent of those following castration[93]. The effects of androgens administered to castrated male animals are inhibited, apparently in a dose-related fashion, large doses produce almost total inhibition[94]. In human trials of cyproterone acetate marked loss of libido and suppression of the sexual urge[95] were observed.

8.3.2 The assessment of activity

Many of the new synthetic steroid agents display a spectrum of activities, often having not an exclusive oestrogenic or progesterone-like effect, but generally having a predominance of one or other of these activities. Thus some progestational agents showing some degree of androgenic and anabolic effect, may undergo aromatisation of the A-ring to produce substances with oestrogenic activity. Substances useful for clinical purposes are selected for their overwhelming preponderance of the desired activity. Steroids with a multiplicity of activities tend to be of little value, for their use leads to undesirable side-effects resulting from the varied expression of all their potential physiological activities. Some minor degrees of overlap, however, may be unavoidable.

Several bioassay procedures are in use. These include:

(a) The Corner and Allen[96] technique (as modified by Clauberg). In this procedure, immature female rabbits are treated with standard doses of oestrogen; then the substance to be assayed is administered for 5 days and the degree of progestational (secretory) change in the endometrium is evaluated histologically and compared to that of standard doses of progesterone as reference. The need for *immature* animals is obvious to eliminate interference by endogenous progesterone production following ovulation and the formation of a corpus luteum.

(b) The Miyake and Pincus procedure[97] is similar but assays the carbonic anhydrase activity of the endometrium, which is significantly increased by progestogens.

(c) The Hooker–Forbes procedure[98] requires the instillation of the test substance into the lumen of an isolated segment of mouse uterus, and the evaluation is made on nuclear changes induced in the cells of the endometrial stroma. This method has the advantage of a high degree of sensitivity but suffers from the defect of low specificity since a positive response can be obtained with non-progestational materials in some cases.

(d) The McGinty procedure[99] is similar, but the isolated segment of rabbit uterus is used after pretreatment with oestrogen as in the Clauberg procedure, and the assay is considerably more sensitive.

Various other techniques have been used. These include assessment of the development of the decidual reaction in rat endometrium after a standardised oestrogenic pretreatment; and the ability to maintain pregnancy in the rabbit, mouse or rat after oöphorectomy of the pregnant animal. Suppression of ovulation in the rabbit[100] is used as a criterion of progestational activity. This is a convenient and simple technique since the rabbit will ovulate *ca.* 10 h after copulation with a high degree of predictability.

Unfortunately, none of these tests are directly comparable as a measure of progestational activity. Results with one procedure may differ significantly from those of another. First, not all measure the same parameters. Secondly, species differences between the test animals cause major difficulties due to varying degrees of response, even with the same substance. This is a major criticism of the use of biological methods. A substance, markedly progestational by one parameter in the rabbit may not be so in the mouse. The effect on a human being is even more difficult to predict until clinical trial is carried out.

For a full discussion of the varied and multiple biochemical effects of progestational preparations, the reader is referred to the excellent reviews by Briggs *et al.*[101].

8.4 OESTROGENS AND OESTROGENIC ANALOGUES

Oestrogenic activity is defined on the basis of its ability to promote the development of the female secondary sexual characters. Naturally-occurring oestrogens are steroids with 18 carbon atoms and a phenolic A-ring. However, a wide range of substances of dissimilar molecular structure may have varying degrees of oestrogenic activity. A variety of non-steroidal, phenolic substances from plants have significant oestrogenic potency. For example, the fertility of sheep in Australia has been shown to be affected adversely by allowing them to graze on the 'subterranean clover'. Grazing resulted in the consumption of *genistein*, an oestrogenic isoflavone occurring in this plant[102]. However, all of these plant substances are significantly less potent than the oestrogens of the mammalian ovary, adrenal cortex and placenta.

8.4.1 Chemistry and metabolic considerations

The natural oestrogens from animal sources are all steroids fitting the general description given above. The three main oestrogens of man are oestradiol-17β, oestriol and oestrone. Principal pathways of oestrogenic synthesis are shown in Figure 8.7.

Oestradiol-17β is by far the most potent of these. It is the major secretory product of the ovary and readily undergoes oxidation to oestrone. Quantitatively, the most important site of this conversion is the liver where a free interconversion between these two steroids appears to take place[103]. Hydroxylation of oestradiol to oestriol, in which a 16α-hydroxyl function is added, also takes place. All three are excreted as conjugates, principally of glucuronide or sulphate. A variety of metabolic degradation products are formed by

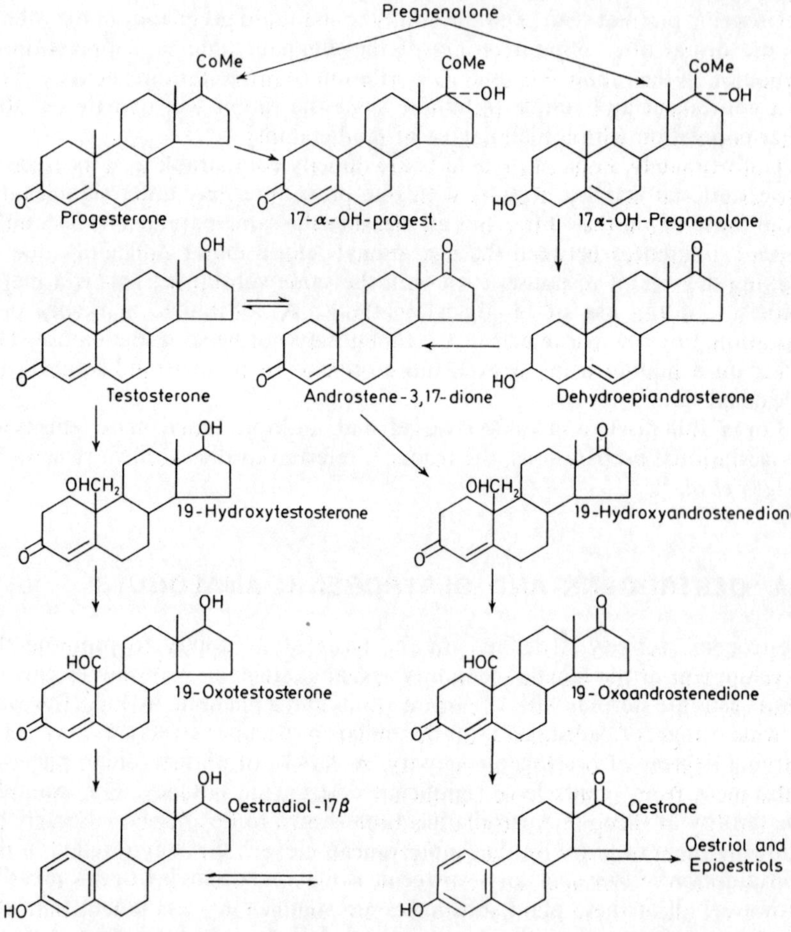

Figure 8.7 Principal pathways of oestrogen formation in man, beginning with pregnenolone

various metabolic degradative steps. The most significant are summarised in Figure 8.8. Large quantities of oestrogens are produced in pregnancy by the placenta. Both sexes of the family *equidae* are able to synthesise oestrogens in remarkable amounts. Thus stallions produce prodigious quantities of oestrogen; the equine oestrogens, *equilin*, and *equilenin* and their derivatives, show additional unsaturation of the B-ring.

Chemical modification of some of the natural oestrogens may lead to

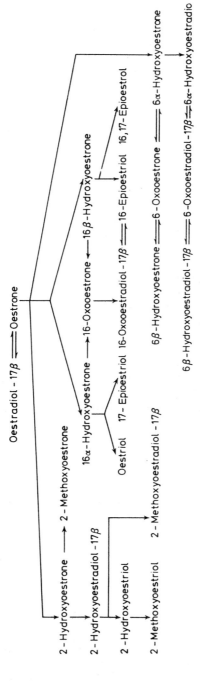

Figure 8.8 Principal routes of oestrogen degradation in man

enhancement of activity; the phenolic A-ring is required for oestrogenic activity. Thus 17α-ethynyloestradiol-17β- obtained synthetically by treatment of (+)oestrone with potassium acetylide in liquid ammonia[104], is not only the most potent oestrogen known but is also orally active. The addition of this 17α-ethynyl group appears to be protective of the 17β-hydroxyl group, which also favours biological activity (see androgens and anabolic steroids, Section 8.2). The importance to biological activity of the 17β-hydroxy group is further emphasised by the conspicuously reduced activity of 17α:hydroxy compounds. Thus oestradiol-17α is 1/40th as active as oestradiol-17β. In this substance, the 17α-OH group may be regarded as axial in orientation, whereas the 17β-OH is equatorial. Oestradiol-17α is much more readily dehydrated than oestradiol-17β. Small amounts of oestradiol-17α occur naturally[105].

Esterification of the 17β-hydroxyl group of oestradiol-17β leads to protection of the group from metabolic attack and to prolongation of effect. Thus oestradiol benzoate, administered by injection, has a much more sustained biological action than the free hormone.

Protection of the oxygen function at C-3 may be paralled by protection of biological activity, as in *quinestrol* the 3-cyclopentyl ether of ethynyl oestradiol. This substance is capable of a very sustained biological effect; a single dose of 5 mg gives detectable effects on vaginal epithelium for *ca.* 3 months. The prolonged activity is probably the result of slower transport within the organism, and to an enhanced solubility in depot fat from which it is slowly released.

Non-steroidal synthetic oestrogens have been prominent in the clinical armamentarium for a long period. The first of these compounds to be introduced were a series of stilbene-4,4'-diols with alkyl substituents on the a,a' carbons[106]. Stilbene-4,4' diol (stilboestrol) is the parent compound of the series and alkylstilboestrols are substituted in the manner shown in Figure 8.9. The

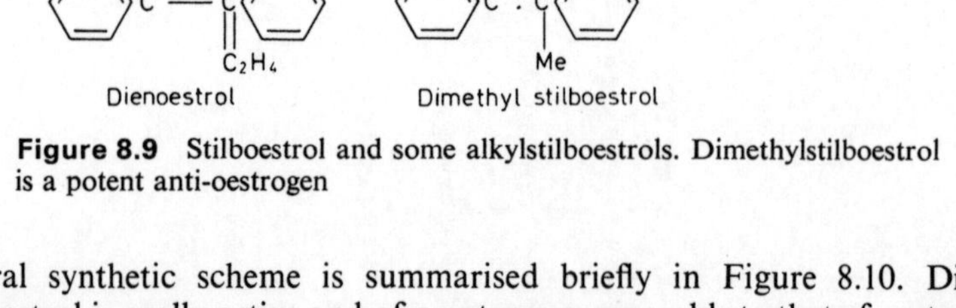

Figure 8.9 Stilboestrol and some alkylstilboestrols. Dimethylstilboestrol is a potent anti-oestrogen

general synthetic scheme is summarised briefly in Figure 8.10. Diethyl stilboestrol is orally active and of a potency comparable to that of oestradiol-17β. Hexestrol and dienestrol are very similar and have much the same oestrogenic properties. They tend, however, to be a little less active if administered by mouth.

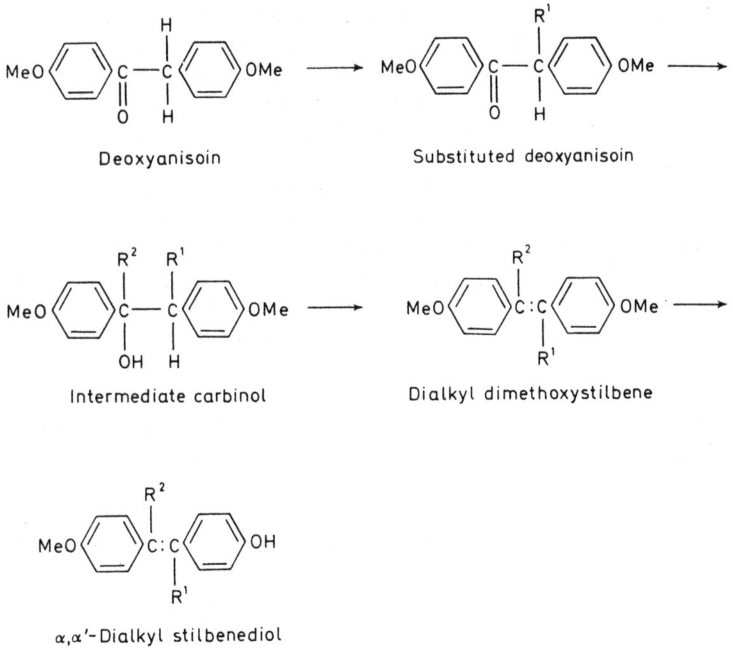

Figure 8.10 Synthesis of a,a'-diakyl stilbenes from deoxyanisoin

The three substances described appear to be oestrogenic in their own right. Other members of the stilbene-4,4'-diol series, however, are pro-oestrogens (i.e. substances which are themselves inactive but are converted to oestrogens by *in vivo* metabolism) while others are potent anti-oestrogens, that is, physiological antagonists of oestrogenic activity[107-109].

It is interesting to speculate as to what molecular conformation in this series is responsible for oestrogenic activity. It is certainly possible to rearrange the structural formulae of some of them into a configuration resembling that of oestradiol. (Figure 8.11). In this context, the behaviour of three closely-related members of the series is remarkable, for one is a potent oestrogen, the second shows maximal pro-oestrogenic potency, and the third maximal anti-oestrogenic potency. These are diethylstilboestrol, methylethyl stilboestrol, and dimethyl stilboestrol respectively. The reasons for these

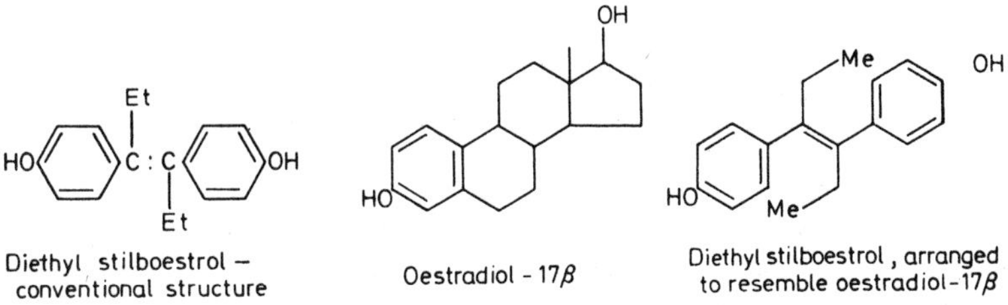

Figure 8.11 Possible arrangement of molecule of diethyl stilboestrol to resemble oestradiol-17β

remarkable differences are not yet clear but presumably involve such factors as reversibility of binding to receptor sites, differences in metabolic fate and duration of effect at target organs. Structural similarities between anti-oestrogens and oestrogens in this series, and the ability of anti-oestrogens to inhibit oestrogenic responses at low dosage, without significant toxicity, suggest that they are competitive inhibitors.

Tri-*p*-anisylchloroethylene (TACE) is an interesting example of a tricyclic pro-oestrogen closely resembling the stilboestrol series in basic structure. It is orally active and undergoes conversion into weakly active metabolites with oestrogenic properties. With respect to some activities, it also appears to have mild anti-oestrogenic properties[110]; for example, it fails to produce the enlargement of the pituitary which is consistently observed with oestradiol in rats. Simultaneously-administered TACE inhibits this effect of oestradiol, at least partially. The fact that pro-oestrogenic substances usually possess varying degrees of anti-oestrogenic activity has been recognised for some time[109], although it is not universally true. Complete dissociation of activities, however, is not yet achieved.

Mention should be made of two other substances structurally very similar to TACE. These are ethamoxytriphetol and clomiphene (Figure 8.12). Both

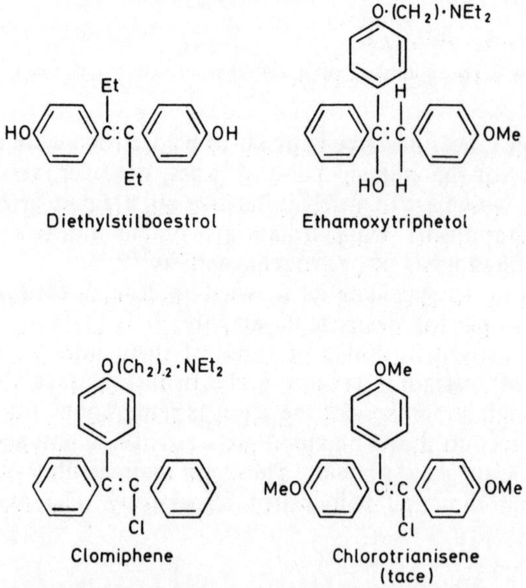

Figure 8.12 Comparison of the structure of diethylstilboestrol with the anti-oestrogens ethamoxytriphetol and clomiphene, and the pro-oestrogen chlorotrianisene

are fully oestrogenic but also show significant anti-oestrogenic activity. Notably, clomiphene in animals is a powerful inhibitor of the release of pituitary gonadotrophin, in addition to its anti-oestragenic effect. In humans, however, this substance induces marked enlargement of the ovaries, which is reversible after cessation of administration. These effects in man, as distinct from those in test animals, indicate that clomiphene in the human organism *promotes* the release of pituitary gonadotrophin and thereby induces the ovarian changes. This is presumably the result of its anti-oestrogenic effect whereby the normal feed-back inhibition of the anterior pituitary by oestrogen is blocked. Evidence suggests the augmentation of secretion of both FSH and LH. The compound is thus a valuable adjunct to the induction of ovulation in the Stein–Levinthal syndrome and other forms of infertility[111].

8.4.2 Assessment of activity

A number of methods for assaying oestrogenic activity have been devised. Many variations of the procedure of Allen and Doisy[112], in which oestrogen-induced cornification of the vaginal epithelium in spayed rats or mice is used as the basis of quantitation. Spaying is generally performed on immature females soon after weaning. The animals are not used for the assay for *ca.* 2 weeks; during this time a priming dose of a short-acting, potent oestrogen is given to maintain sensitivity of tissues to oestrogen and to assure a uniform response. Test substances are later administered by subcutaneous injection in oil, and the cornification response is estimated by cytologic examination of vaginal smears. A detailed discussion of the methodology of this procedure and its modifications is given by Emmens[113].

Local application of test substances to the vaginal epithelium of sprayed animals is another sensitive technique for assessment of oestrogenic activity[114]. Certain potential oestrogens cannot be measured effectively by this means, since they require prior metabolic activation in the body before they become effectively oestrogenic.

In other techniques the increase in weight of the uterus in spayed female rodents is measured. These, however, tend to be less specific than those dependent upon vaginal cornification.

Yet other methods depend upon the reduction of tetrazolium salts to a stable insoluble formazan by the oxidoreductases of vaginal epithelial cells; this activity is sharply increased by oestrogen pre-treatment[115,116]. This method correlates well with other methods dependent upon an early vaginal response. Other procedures depend upon the oestrogen-induced increase of the uptake of [5-^{3}H] uridine by the uterus or vagina. Because of its greater response, the uterine uptake is measured preferably. It is a relatively sensitive method.[117-119].

Villee[120,121] demonstrated the stimulation of NAD-dependent isocitrate dehydrogenase activity in human placental extracts by oestradiol-17β. Reduction of NAD to NADH$^+$ is stimulated both by oestradiol and oestrone, but not by other steroids. The method is thus severely limited in its applicability.

8.5 CORTICOID ANALOGUES

8.5.1 Chemical and metabolic considerations

The adrenal cortex synthesises a large variety of steroids from cholesterol. These are the adrenal androgens and oestrogens, and a third major class of compounds containing 21 carbon atoms. The physiological activities of this group of compounds are of two major types; first, there is an effect on sodium retention as measured by reduction of sodium excretion by the kidney of adrenalectomised animals. Secondly, there are broad metabolic effects upon glycogen deposition in liver, blood sugar levels, an anti-inflammatory effect and an involutional effect upon lymphoid tissue. Different steroids, and in particular, synthetic analogues, show combinations of these two types of

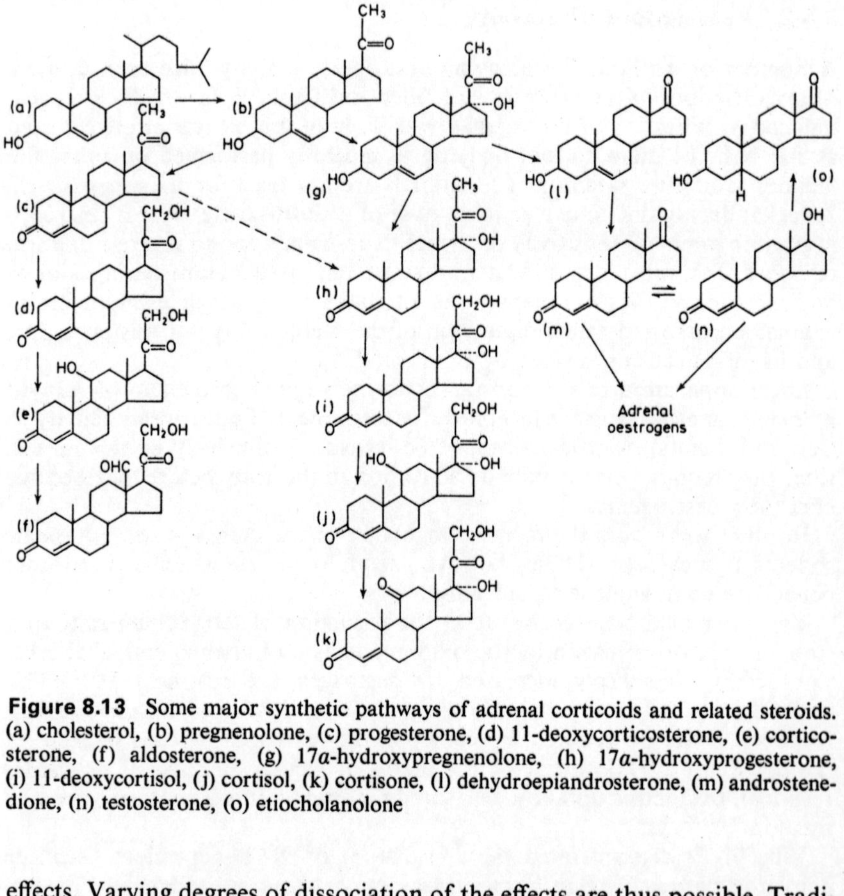

Figure 8.13 Some major synthetic pathways of adrenal corticoids and related steroids. (a) cholesterol, (b) pregnenolone, (c) progesterone, (d) 11-deoxycorticosterone, (e) corticosterone, (f) aldosterone, (g) 17α-hydroxypregnenolone, (h) 17α-hydroxyprogesterone, (i) 11-deoxycortisol, (j) cortisol, (k) cortisone, (l) dehydroepiandrosterone, (m) androstenedione, (n) testosterone, (o) etiocholanolone

effects. Varying degrees of dissociation of the effects are thus possible. Traditionally, corticoids in which the sodium-retaining potency predominates markedly (as in deoxycorticosterone and aldosterone) have been called *mineralocorticoids*; those in which metabolic and anti-inflammatory properties predominate have been called *glucocorticoids*. Overlap of activity between the two categories, however, is frequent. One of the primary requirements for the production of clinically-useful corticoid analogues, therefore, is the attainment of as complete as possible a dissociation of these two types of activity. Some major pathways of natural corticoid synthesis are shown in Figure 8.13.

8.5.2 Structure–activity relationship

This is conveniently expressed in terms of the ratio between anti-inflammatory and sodium-retaining activities. So far, in available compounds studied, total dissociation of effect has not been possible, although augmentation of one or other effect and significant reduction of the other is now possible. In general categories of molecular, modification discussed under Androgens and Anabolic steroids in Section 8.2 apply in a parallel fashion to corticoid analogues. (see also Fried and Borman[122]). Some of the major factors in molecular structure and activity are as follows:

(a) A-ring modifications—The Δ^4-3-one configuration is required for significant biological activity. Thus cortisol is potent, while tetrahydrocortisol is inactive. 1,2-unsaturation, as in prednisone (Δ^1 cortisone) and prednisolone (Δ^1-cortisol), greatly enhances anti-inflammatory and 'glucocorticoid' activities with little effect on sodium retention.

(b) B-ring modifications—6α-substitution has variable effects. Thus 6α-methylation of cortisol increases both the anti-inflammatory and sodium retaining activities and promotes a nitrogen-wasting (protein-catabolic) effect. In 6α-methyl prednisolone, however, there is greater dissociation, the anti-inflammatory effect is augmented to a greater degree than that of sodium-retention. 9α-substitution with fluorine as in 9α-fluorocortisol, augments both categories of activity, thought the effect upon sodium retention is generally much greater, unless modified by other substituents elsewhere. Thus 16α-substitution tends to offset the augmentation of 'mineralocorticoid' effect promoted by 9α-fluorination, as in triamcinolone (9α-fluoro-16α-hydroxyprednisolone), dexamethasone (9α-fluoro-16α-methylprednisolone) and betamethasone (9α-fluoro-16β-methylprednisolone). All of these compounds show marked augmentation of anti-inflammatory effect, with no augmentation of sodium-retention.

(c) C-ring modifications—An oxygen function at C-11 is required for anti-inflammatory activity and effects upon protein and carbohydrate metabolism. This effect is most pronounced with 11β-hydroxylation. Oxidation of this hydroxyl to a keto-group results in significant reduction of activity[123] (see C-11 substitution in androgens and anabolic steroids Section 8.2).

(d) D-ring modifications—16-hydroxylation or methylation, as noted above, greatly reduces sodium-retaining activity but has little other effect. However, 16α-substitution may increase anti-inflammatory activity. 17α-hydroxylation appears to be a prerequisite for anti-inflammatory activity of major significance. C-21 hydroxylation appears to be required for sodium-retaining activity, and augmentation of anti-inflammatory effects.

In steroids having a 20-ketone group, reduction of this ketone to a hydroxy-group is generally associated with marked reduction of biological activity.

8.5.3 Principal available corticoid analogues

The basic molecular structure, relative potencies, and trivial names of some of the more important of these analogues is summarised in Table 8.5. (See also Refs. 124 and 125). An annotated list of the major analogues found by partial and total synthesis is given below. Selected references are given.

Table 8.5 Relative glucocorticoid and mineralocorticoid potencies of the major corticoid analogues compared with those of cortisol

Structure	Trivial name	Anti-inflamatory effect (v. Cortisol)	Relative Na^+ retaining activity
	Cortisol	1	1
	Cortisone	0.8	0.8
	Prednisolone	3.5	0.8
	Prednisone	5	0.8
	Triamcinolone	5	0

Structure	Trivial name	Anti-inflamatory effect (v. Cortisol)	Relative Na⁺ retaining activity
	Paramethasone (6α-Fluoro-16α-methylprednisolone)	10	0
	Betamethasone (9α-Fluoro-16β-methyl-prednisolone)	25	0
	6-Methylprednisolone. (9α-Fluoro-6α-methyl-prednisolone)	5	0
	Dexamethasone (9α-Fluoro-16α-methylprednisolone)	25	0
	9α-Fluorohydrocortisone	10	125

8.5.3.1 Monosubstituted derivatives of major corticoids

(a) Δ^4-Pregnen-17α-21-diol-3,11,20-trione (cortisone) derivatives include substituents as follows: 2α-Me, 6α-Me, 6α-F, 6α-Cl, 6β-Cl, 7α- and 7β-Me, 9α-F, 16α-Me, 16β-Me, 16β-Cl, 16β-Br, 17α-F, 17α-Cl, 7α-OH, 14α-OH, and 16α-OH and 16α-OHAc (Refs. 126–132).

(b) Δ^4-Pregnen-11β,17α,21-triol-3,20-dione (cortisol) derivatives include:

2a-Me, 2a-Et, 4-Me, 6a-Me, 7a-Me, 7β-Me, 9a-Me, 11a-Me, 16a-Me, 16β-Me, 6a-F, 6a-Cl, 6β-F, 9a-F, 9a-Cl, 9a-Br, 9a-I, Δ^6-9a-F, 12a-F, 12a-Cl, 16β-F, 16a-OH, 16a-OAc (Refs. 133–135).

(c) Δ^1-Cortisone (prednisone) derivatives include: 6a-Me, 7a-Me, 7β-Me, 16a-Me, 16β-Me, 6a-F, 6a-Cl, 6β-Cl, 9a-F, 9a-Cl, 7a-OH, 16a-OAc (Refs. 136–138).

(d) Δ^1-Cortisol (predisolone) derivatives include: 6a-Me, 6a-F, 6a-Cl, 9a-F, 9a-Cl, 9a-Br, 16a-F, 16β-F, 12β-OH, 16a-OH, 16a-OAc. (Refs. 139–140).

8.5.3.2 Disubstituted derivatives of major corticoids

A very large number of substances have been synthesised in this category. They include Me/halogen, Me/hydroxy, dihalogen, and dihydroxy substituents of cortisol, cortisone, prednisone and prednisolone. Included in this category are 9a-fluoro-16a-hydroxyprednisolone (Triamcinolone)[141], 6a-fluoro-16a-methyl-prednisolone (paramethasone)[142], 9a-fluoro-16a-methyl-prednisolone[142] (Dexamethasone). For a detailed summary of other compounds see Ref. 145.

8.5.3.3 Trisubstituted derivatives of major corticoids[140-145]

(a) Cortisol derivatives include: (i) 2a-Me, 9a-F, 16a-OH; (ii) 6a-Me, 9a-F, 21-F; (iii) 6a-F, 9a-F, 16a-Me; (iv) 6a-F, 9a-F, 16a-OH; (v) 9a-F, 16a-OH, 21-Cl.

(b) Prednisolone derivatives include (i) 2a-Me, 9a-F, 16a-OH, (ii) 6a-Me, 9a-F; 21-F; (iii) 6a-F, 9a-F, 16a-OH; and (iv) 6a-Cl, 9a-F, 16a-OH; and others

Some of the major corticoid analogues selected from the above list have found important clinical uses. These are illustrated in Table 8.5 and their relative potencies with regard to mineralocorticoid and glucocorticoid activities are compared. The points made in the preceding paragraphs regarding the effect of ring unsaturation and substituent group upon these two different activities are well illustrated by the examples given. In particular the remarkable modifying effect of a-methyl substitution at the C-6 or C-16 positions on the Na^+ retaining influences of 9a-fluoro-substitution is conspicuous.

8.6 CONCLUSION

The principal objective of this article was to discuss in some detail the major aspects of the relationship between molecular structure and hormonal activity of some of the more important steroid or quasi-steroid hormonal analogues, with emphasis on compounds that have found a valuable place in clinical usage. At best, this article can be only a précis of the subject, for the field has beome too extensive to permit comprehensive treatment in a limited space. Discussion of mechanisms of physiological actions have been deliberately avoided here since they are discussed admirably elsewhere in this book. Again, restrictions of space have precluded discussion of pharmaceutical preparations and of clinical indications for the use of the substances described. A selected, if extensive, bibliography is included which indicates excellent source material for the further pursuit of these subjects. One may express the hope that the matter presented suffices as a review of the major principles of this fascinating aspect of applied steroid chemistry, and that the reader will find here a balanced digest of the subject.

ADDENDUM 1

On steroid receptor purification: major progress in biospecific affinity chromatography

The ultimate decisive progress in the elucidation of the mechanism of action of sex steroid hormone awaits receptor purification. The very small amount of the receptor protein is the greatest obstacle to purification. Affinity chromatography[146] utilises a ligand of great affinity, the hormone or a derivative or an analogue, which is linked through an appropriate 'spacer' arm to a solid polymeric matrix. Ideally it becomes possible with such a biospecific adsorbent, to select the highest affinity proteins among the vast majority of non-binding or non-specific (low affinity) binding components. There is no reason for not obtaining at least a 10^3-fold purification in one chromatographic step with a successful technique, if one recalls that the K_A of the steroid receptors is of the order of 10^{10} M^{-1}, while the K_A of non-specific proteins is less than 10^7 M^1.

Much work has been done already with the calf uterus oestradiol receptor[147,148]. From past and present experiments, it appears very important to select some position(s) on the hormone to which can be grafted the spacer without losing high affinity for the receptor. Moreover, the hormone and/or the hormone spacer derivative should not be released when exposed to biological extracts during chromatography, whenever it has been absorbed onto the polymer during chemical synthesis, or if it has been secondarily released during chromatography because of some instability of the chemical link[147-149]. In such cases, the receptor will bind the released compound but will not be selected by the chromatography. These difficulties, not apparent during early attempts[150,151], have been described quantitatively, evaluated and overcome[147,148].

Moreover, methods have been devised for transfering radioactive hormone onto the receptor and then labelling it for further purifications, in exchange for the biospecific chromatographic adsorbent[147,148]. With such improvements, evaluation of different absorbents is rapidly progressing, including their capacities, their stability and their specificity. Most materials show non-suitable adsorption of the receptor hydrophobic and/or ionic interactions with many proteins, which result in a decrease of the purification.

These conceptual and methodological advances have led to several hundred fold purification[147,148] of one of the molecular forms of the uterus oestradiol receptor ('4S–Calcium':7), chosen because it displays little aggregation compared to others. The results indicate that now there is a real possibility of purifying the steroid hormone receptor by affinity chromatography. The instability of the purified molecule in minute amounts, and the still lower recovery give at present much concern.

ADDENDUM 2

On the cytosol → nucleus 'translocation' of the receptor: the hormone dependent, temperature and ionic strength accelerated 'activation' of the receptor

The apparent transfer of the hormone-cytosol receptor complex to nuclei in the target cell has been fully reported. Very recent experiments, using *in vitro* 'reconstituted' systems (cytosol + isolated nuclei) have thrown some light on some of the underlying mechanisms.

The cytosol receptor, when it has been incubated at low temperature and low ionic strength with the hormone, which then binds to it, does not transfer to the nuclei or only very slowly. A temporary exposure to higher temperature and/or ionic strength gives to the hormone receptor complex the ability to bind to nuclei, even if this is subsequently tested at low temperature and in low salt medium[153-155]. Such a new property of the receptor indicates than an 'activation' step has taken place. The previous binding of the hormone is a prerequisite to this activation: the receptor exposed to a rise in temperature or ionic strength in the absence of hormone and secondarily incubated with the steroid does not bind to nuclei[155]. In fact, the activated rat liver glucocorticosteroid receptor is now able to bind not only to liver nuclei but also to various polyanions: various DNAs, RNAs or even to substituted Sephadex[155].

Therefore, it appears that the receptor can be present under two different conformations which possibly differ by the presence (activated form) or the absence (inactivated form) of positively charged groups at the surface of the molecule, thus changing its affinity for polyanions. The utmost role of the steroid would be to regulate the concentration of the activated receptor in the cell[155]. Since the activated receptor acquires the ability to bind to any polyanion, the problem of the experimental identification of the physiological and specific nuclear acceptor becomes even more difficult.

References

1. Kochakian, C. D. and Murlin, J. R. (1935), *J. Nutrit.*, **10**, 437
2. Krüskemper, H.-L. (1968). *Anabolic Steroids*, 119 (New York: Academic Press)
3. Camerino, B. and Scala, G. (1960). *Progr. Drug. Res.*, **2**, 71
4. Korenman, S. G. and Wilson, H. (1966). *Steroids*, **8**, 729
5. Baulieu, E. E. and Mauvais-Jarvis, P. (1964). *J. Biol. Chem.*, **239**, 1569, 1578
6. Lipsett, M. B., Korenman, S. G., Wilson, H. and Bardin, C. W. (1966). *Steroid Dynamics*, 117, (G. P. Pincus, J. F. Tatit, and T. Nakao, editors) (New York: Academic Press)
7. Dorfman, R. I. (1966). *Recent Progr. Hormone Res.*, **22**, 272
8. Tomkins, G. M. (1957). *J. Biol. Chem.*, **225**, 13
9. Tomkins, G. M. (1956). *Recent Progr. Hormone Res.*, **12**, 125

10. Isselbacher, K. J. (1956). *Recent Progr. Hormone Res.*, **12**, 234
11. Baulieu, E. E. (1967). *Hormone Steroids*, **132**, 37 (L. Martini, F. Fraschini and M. Motta, editors) (Amsterdam: Excerpta Medica Foundation)
12. Bush, I. E. (1962). *Pharmacol. Rev.*, **14**, 317
13. Tait, J. F. and Burstein, S. (1964). *Hormones*, **5**, 551
14. Baulieu, E. E., Corpechot, C., Dray, F., Emiliozzi, R., Lebeau, M. C., Mauvais-Jarvis, P. and Robel, P. (1965). *Recent Progr. Hormone Res.*, **14**, 317
15. Baulieu, E. E. (1960). *C. R. Acad. Sci. Fr.*, **251**, 1421
16. Gurpide, E., MacDonald, R. C., Vande Wiele, R. L. and Lieberman, S. (1963). *J. Clin. Endocrinol. Metab.*, **23**, 346
17. Ringold, H. J., quoted by Hayano, M. (1972). *Oxygenases*, 227 (O. Hayasishi, editor) (New York: Academic Press)
18. Bagett, B., Engel, L. L. Savard, K. and Dorfman, R. I. (1955). *Fed. Proc. (Fed. Amer. Soc. Exp. Biol.)*, **14**, 175
19. Meyer, A. S. (1955). *Biochem. Biophys. Acta*, **17**, 440
20. Longchamp. J. E., Hayano, M., Ehrenstein, M. and Dorfman, R. I. (1960). *Endocrinology*, **67**, 843
21. Kuji, N. (1961). *Acta Endocrinol.*, **37**, 71
22. Dörner, G., Stahl, F. and Zabel, R. (1963). *Endokrinologie*, **45**, 121
23. Shoppee, C. W. (1958). *Chemistry of the Steroids*, 76 (New York: Academic Press)
24. Pataki, J., Rosenkranz, G. and Djerassi, C. (1952). *J. Biol. Chem.*, **195**, 791
25. Mancera, O., Ringold, H. J., Djerassi, C. Rosenkranz, G. and Sondheimer, F. (1953). *J. Amer. Chem. Soc.*, **75**, 1286
26. Bradlow, H. L. and Gallagher, T. F. (1959). *J. Clin. Endocrinol. Metabol.*, **19**, 1575
27. Dorfman, R. I (1964). *Recent Progr. Hormone Res.*, **9**, 5
28. Baulieu, E. E. (1964). *Research on Steroids* Vol. 1, 1 (C. Cassano, editor) (Rome: Il Pensiero Sci. Publ. Co)
29. Baulieu, E. E., Corpechot, C., Dray, F., Emiliozzi, R., Lebeau, M. C., Mauvais-Jarvis, P. and Robel, P. (1965). *Recent Progr. Hormone Res.*, **21**, 411
30. Kürskempker, H.-L. (1968). *Anabolic Steroids*, 24 (New York: Academic Press)
31. Kürskemper, H.-L. and Breuer, H. (1961). *Verhandl. Dent. Ges. Inn. Med*, **67**, 387
32. Sala, G. and Castegnaro, (1958). *Lolia, Endocrinol. (Pisa)*, **11**, 348
33. Bush, I. E., and Makesh, V. B. (1959). *Biochem. J.*, **71**, 718
34. Langecker, H., (1962). *Arzneimittle Forsch*, **12**, 213
35. Segaloff, A. (1963). *Steroids*, **1**, 299
36. Gilbert, E. F., Da Silva, A. P. and Queen, D. M. (1963). *J. Amer. Med. Assoc.*, **185**, 538
37. Hogarth, W. J. (1961). *J. Can. Med. Assoc.*, **88**, 368
38. Overbeck, G. A., Bonta, I. L., de Vos, C. L., de Visser, J. and Van der Vies, J. (1967). *Hormonal Steroids*, **132**, 68 (Amsterdam: Excerpta Medica Foundation)
39. Kochakian, C. D., Bartlett, M. N., and Gongora, J. (1948). *Amer. J. Physiol.*, **153**, 210
40. Korner, A. and Young, F. G. (1955). *J. Endocrinol.*, **13**, 78
41. Kochakian, C. D., Tilotson, C., Austin, J., Daughtery, E., Haag, V. and Coolson, R. (1956). *J. Endocrinol.*, **58**, 315
42. Kochakian, C. D. and Tilotson, C. (1957). *Endocrinology*, **60**, 607
43. Wainman, P. and Shipounoff, (1941). *Endocrinology*, **29**, 975
44. Eisenberg, E. and Gordon, G. S. (1950). *J. Pharmacol. Exp. Thes.*, **99**, 38
45. Kürskemper, H.-L. (1968). *Anabolic Steroids*, 185 (New York: Academic Press)
46. Saunders, F. J. (1962). *Excerpta Med. Congr. Ser.* **51**, 207
47. Nimni, M. E. and Geiger, E (1952). *Proc. Soc. Expt. Biol. (N.Y.)*, **76**, 406
48. Geiger, E. and Rawi, I. E. (1952). *Metabolism*, **1**, 145
49. Nimni, M. E. and Bavetta, L. A. (1961). *Proc. Soc. Exp. Biol. (N.Y.)*, **106**, 738
50. Boucek, R. J., Noble, N. L. and Woessner, F. (1961). *Ann. N.Y. Acad. Sci.*, **72**, 1016
51. Dorfman, R. I. and Kinel, F. A. (1963). *Endocrinology*, **72**, 259
52. Segaloff, A. (1963). *Steroids*, **1**, 299
53. Feyel-Cabanes, T. (1963). *Compt. Rend. Soc. Biol.*, **157**, 1428
54. Campbell, J. A., Lyster, S. C., Duncan, F. W. and Babecock, J. C. (1963). *Steroids*, **1**, 317
55. Nutting, E. F., Klimstra, P. D., Counsell, R. E. (1966). *Ecta Endocrinol.*, **53**, 627
56. Newmann, F. and Wiechert, R. (1965). *Arzneimittel Forsch.*, **15**, 1168
57. Encoli, A., Gardi, R. and Vitali, R. (1962). *Chem. and Ind. (London)*, 1984

58. Edgren, A., Peterson, D. L., Jones, R. C., Nagra, C. L., Smith, H. and Hughes, G. A. (1966). *Recent Progr. Hormone Res.*, **22**, 305
59. Desaulles, P. A. and Kruhenbuhl, C. (1960). *Acta Endocrinol.*, **40**, 217
60. Barnes, L. E., Stafford, R. O., Guild, M. E., Thole, L. C. and Olson, K. J. (1954). *Endocrinology*, **55**, 77
61. Perrine, J. W. (1961). *Acta Endocrinol.*, **37**, 376
62. Sala, G. and Baldratti, G. (1957). *Endocrinology*, **72**, 494
63. Rapala, R. T., Kraay, R. J. and Gerzan, K. (1965). *J. Med. Chem.*, **8**, 580
64. Sala, G. and Baldratti, G. (1957). *Proc. Soc. Exp. Biol. Med.*, **95**, 22
65. Camerino, B. and Sala, G. (1969). *Progr. Drug Res.*, **2**, 71
66. Overbeek, G. A., Delver, A. and de Visse, J. (1961). *Acta Endocrinol.* Suppl. **63**, 7
67. Kincl, F. A. (1965). *Methods Hormone Res.*, **4**, 21
68. Lennon, H. D. and Saunders, F. J. (1964). *Steroids*, **4**, 689
69. Cekan, Z. and Pelc, B. (1966). *Steroids*, **8**, 209
70. Arnold, A., Potts, G. O. and Beyler, A. L. (1963). *Acta Endocrinol.*, **40**, 217
71. Potts, G. O., Reyler, A. L. and Burnham, D. F. (1960). *Proc. Soc. Exp. Biol. Med.*, **103**, 383
72. Arnold, A., Potts, G. O. and Reyler, A. L. (1963). *Endocrinology*, **72**, 408
73. Strike, D. P., Herbot, D. and Smith, H. (1967). *J. Med. Chem.*, **10**, 446
74. Segaloff, A. and Gabbard, R. B. (1960). *Endocrinology*, **67**, 887
75. Dorfman, R. I., (1960). *Science*, **131**, 1096
76. Mamlock, L., Horeau, A. and Jacques, J. (1965). *Bull. Soc. Chem. Fr.*, 2359
77. Vida, J. A. (1969). *Androgens and Anabolic Steroids, Chemistry and Pharmacology*, 46 (New York: Academic Press)
78. Nagata, W. (1967). *Proc. Internat. Sympos. Drug Res.* 1967, 134, (Montreal: Chemical Institute of Canada)
79. Segaloff, A. and Gabbard, R. B. (1963). *Steroids*, **1**, 77
80. van der Vies, J. (1965). *Acta Endocrinol.*, **49**, 271
81. Vida, J. A. (1969). *Androgens and Anabolic Agents, Chemistry and Pharmacology* 37, 73 (New York: Academic Press)
82. Bush, I. E. (1962). *Pharmacol. Rev.* **14**, 317
83. Klimstra, P. D., Nutting, E. F and Counsell, R. E. (1966). *J. Med. Chem.*, **9**, 924
84. Corner, G. W. and Allen, W. M. (1929). *Amer. J. Physiol.*, **88**, 326
85. Falconi, G. and Bruni, G. (1962). *J. Endocrinol.*, **25**, 169
86. Pincus, G. and Merrill, A. P. (1961). *Control of Ovulation*, 37 (Oxford: Pergamon Press)
87. Bryan, H. S. (1960). *Proc. Soc. Exp. Biol. Med.*, **105**, 23
88. Falconi, G., Gardi, R., Bruni, G. and Ecoli, A. (1961). *Endocrinology*, **69**, 638
89. Hartley, F. (1962). *J. Endocrinol.*, **26**, 24
90. Siechter, R. F., Chang, M. C. and Pincus, G. (1954). *Fertil. Steril.*, **5**, 282
91. Overbeek, G. A., Madjerek, Z. and de Visser, J. (1962). *Acta. Endocrinol.*, **41**, 351
92. Hamada, H., Neumann, F. and Junkmann, K. (1963). *Acta Endocrinol.*, **44**, 380
93. Neumann, H. (1966). *Acta Endocrinol.*, **53**, 53
94. Neumann, H., Berswordt-Wallrabe, R. von; Elger, W., Steinbeck, H., Hahn, J. and Kramer, M. (1970). *Recent Progr. Hormone Res.*, **26**, 337
95. Laschet, U., Laschet, L. Felzner, H.-R., Glaesel, H.-U., Mall, G. and Naab, M. (1967). *Acta Endocrinol.*, **50**, Suppl. 119, 55
96. Corner, G. W. and Allen, W. M. (1929). *Amer. J. Physiol.*, **88**, 326
97. Mujake, T. and Pincus, G. (1958). *Endocrinology*, **52**, 287
98. Hooker, C. W. and Forbes, T. R. (1947). *Endocrinology*, **41**, 158
99. McGintz, D. A., Anderson, L. P. and McCullough, N. B. (1939). *Endocrinology*, **24**, 829
100. Jackson, H. (1966). *Antifertility Compounds in the Male and Female*, 135 (Springfield, III: Charles C. Thomas)
101. Briggs, M. H., Pitchford, A. G., Sterniford, M., Barker, H. M., and Taylor, D. (1970) *Advances in Steroid Biochemistry and Pharmacology*, **2**, 112 (M. H. Briggs, editor (London and New York: Academic Press)
102. Curnow, D. H. (1954). *Biochem. J.*, **58**, 283
103. Ryan, K. J. and Engel, L. L. (1953). *Endocrinology*, **52**, 287
104. Inhoffen, H. H., Logemann, W., Hohlweg, W. and Serini, A. (1938). *Ber.*, **71**, 1024

105. Klyne, W. and Wright, A. A. (1957). *Biochem. J.*, **66,** 92
106. Dodds, E. C., Goldberg, L., Lawson, W. and Robinson, R. (1939). *Proc. Roy. Soc. (London),* **B127,** 140
107. Emmens, C. W. and Cox, R. I. (1958). *J. Endocrinol.*, **17,** 265
108. Emmens, C. W., Cox, R. I. and Martin, L. (1962). *J. Endocrinol.*, **20,** 198
109. Emmens, C. W., Cox, R. I. and Martin, L. (1962). *Recent Progr. Hormones Res.*, **18,** 415
110. Segal, S. J., and Thompson, C. R. (1956). *Proc. Soc. Exp. Biol. Med.*, **91,** 623
111. MacGregor, A. H., Johnson, J. E. and Bunde, C. A. (1968). *Fert. Steril.*, **19,** 616
112. Allen, E. and Doisy, E. A. (1923). *J. Amer. Med. Assoc.*, **81,** 819
113. Emmens, C. W. (1969). *Methods in Hormone Research*, **2A,** 67
114. Emmens, C. W. (1941). *J. Endocrinol.*, **2,** 444
115. Martin, L. (1960). *J. Endocrinol.*, **20,** 187
116. Martin, L. (1964). *J. Endocrinol.*, **30,** 21
117. Miller, B. G. and Emmens, C. W. (1967). *J. Endocrinol.*, **39,** 473
118. Miller, B. G., Owen, W. H. and Emmens, C. W. (1968). *J. Endocrinol.*, **41,** 189
119. Miller, B. G., Owen, W. H. and Emmens, C. W. (1968). *J. Endocrinol.*, **42,** 351
120. Villee, C. A. (1955). *J. Biol. Chem.*, **216,** 171
121. Villee, C. A. and Gordon, E. E. (1955). *J. Biol. Chem.*, **216,** 203
122. Fried, J. and Borman, A. (1963). *Vitamins Hormones*, **16,** 303
123. Sweat, M. L. and Bryson, M. J. (1960). *Biochem. Biophys. Acta*, **44,** 217
124. Doughterty, T. F., Brown, H. E. and Berliner, D. L. (1900). *Endocrinology*, **62,** 455
125. Berliner, D. L. and Doughterty, T. F. (1961). *Pharmacol. Rev.*, **3,** 329
126. Hogg, J. A., Lincoln, F. H., Jackson, R. W. and Schneider, W. P. (1955). *J. Amer. Chem. Soc.*, **77,** 6401
127. Spero, G. B., Thompson, J. L., Mageslein, B. J., Hanze, A. R., Murray, H. C., Sebek, O. K. and Hogg, J. A. (1956). *J. Amer. Chem. Soc.*, **78,** 6213
128. Ringold, H. J., Mancera, O., Djerassi, C., Bowers, A., Batres E., Martinez, H., Necoechea, E., Edwards, J., Velasco, M., Campillo, C. C. and Dorfman, R. I. (1958). *J. Amer. Chem. Soc.* **80,** 6464
129. Cambell, J. A. and Babcock, J. C. (1959). *J. Amer. Chem., Soc.*, **81,** 4069
130. Fried, J. and Sabo, E. F. (1954). *J. Amer. Chem. Soc.*, **76,** 1455
131. Taub, D., Hoffsommer, R. D., Slates, H. L. and Wendler, N. L. (1958). *J. Amer. Chem. Soc.*, **80,** 4635
132. Herzog, H. L., Gentles, M. J., Marshall, H. M. and Heubberg, E. B. (1960). *J. Amer. Soc.*, **82,** 3691
133. Bowers, A. and Ringold, H. J. (1958). *J. Amer. Chem. Soc.*, **80,** 4423
134. Hogg, J. A. (1958). *6th Med. Chem. Symp. Amer. Chem. Soc.* (Madison)
135. Ayer, D. E. and Schneider, W. P. (1960). *J. Amer. Chem. Soc.*, **82,** 1249
136. Beyler, R. E., Oberster, A. E., Hoffmann, F., and Sarett, L. H. (1960). *J. Amer. Chem. Soc.*, **82,** 170
137. Fried, J. Florey, K., Sabo, E. A., Hery, J. E. Restivo, A. R., Bowman, A. and Linger, F. M. (1955). *J. Amer. Chem. Soc.*, **77,** 4181
138. Nussbaum, A. L., Brabayon, G., Popper, T. F. and Oliveto, E. P. (1958). *J. Amer. Chem. Soc.*, **80,** 2722
139. Zderic, J. A., Carpio, H. and Djerassi, C. (1960). *J. Amer. Chem. Soc.* **82,** 446
140. Bernstein, S., Lenhard, R. H., Allen, W. S., Heller, M., Litteld, R., Stolar, S. M., Feldman, L. I. and Blank, R. H. (1959). *J. Amer. Chem. Soc.*, **81,** 1689
141. Arth, G. E., Fried, J., Johnston, D. B. R., Hoff, D. R., Sarett, L. H., Sieber, R., Stoerk, H. and Winter, C. (1958). *J. Amer. Chem. Soc.*, **80,** 3161
142. Edwards, J. A., Ringold, H. J. and Djerassi, C. (1960). *J. Amer. Chem. Soc.*, **82,** 2318
143. Oliveto, E. P., Rausser, R., Herzog, H. L., Hershberg, E. B., Tolksdorf, S., Eisler, M., Perlman, P. L. and Pechet, M. M. (1958). *J. Amer. Chem. Soc.*, **80,** 6687
144. Oliveto, E. P., Rausser, R., Weber, L., Nussbaum, A. L., Gebert, W., Coniglio, C. T., Hershberg, E. B., Tolksdorf, S., Eister, M., Perlman, P. E. and Pechet, M. M. (1958). *J. Amer. Chem. Soc.*, **80,** 4431
145. Shoppee, C. W. (1964). *Chemistry of the Steroids* 296 (London: Butterworths)
146. Cuatrecasas, P. and Anfinsen, C. B. (1971). *Methods in Enzymology*, **22,** 345
147. Sica, V., Nola, E., Parikh, I., Puca, G. A. and Cuatrecasas, P. (1973). *Nature New Biology*, **244,** 36

148. Truong, H., Geynet, C., Millet, C., Soulignac, O., Bucourt, R., Vignau, M., Torelli, V. and Baulieu, E.-E. (1973). *FEBS Letters* 00034
149. Ludens, H., de Vries, J. R. and Fanestil, D. D. (1972). *J. Biol. Chem.*, **247**, 7533
150. Jensen, E. V., DeSombre, E. R. and Jungblut, P. W. (1967). In: *Hormonal Steroids* (L. Martini, F. Frascini and M. Motta, editors) p. 492. (Excerpta Med. Found., Amsterdam)
151. Vonderhaar, B. and Mueller, G. C. (1969). *Biochim. Biophys. Acta*, **176**, 626
152. Puca, G. A., Nola, E., Sica, V. and Bresciani, F. (1971). *Biochemistry*, **10**, 3769
153. Baxter, J. O., Rousseau, G. C., Benson, M. L., Garcea, P. L., Ito, J. and Tomkins, G. M. (1972). *P.N.A.S.*, **69**, 1892
154. DeSombre, E. R., Mohla, S. and Jensen, E. V. (1972). *B.B.R.C.*, **48**, 1601
155. Milgrom, E., Atger, M. and Baulieu, E. E. *Biochemistry*, submitted for publication

9
Mode of Action of Plant Hormones

R. CLELAND
University of Washington, Seattle

9.1	INTRODUCTION	284
9.2	AUXINS	284
	9.2.1 *The hormone and the range of its activities*	284
	9.2.2 *Auxin and the control of cell elongation*	286
	9.2.2.1 *The test system*	286
	9.2.2.2 *The basic response*	286
	9.2.2.3 *The site of auxin action*	287
	9.2.2.4 *The effect of auxin on RNA synthesis*	287
	9.2.2.5 *The proton pump hypothesis*	289
9.3	GIBBERELLINS	290
	9.3.1 *The hormone and its activities*	290
	9.3.2 *The test system*	291
	9.3.3 *The basic response and mechanism of GA action*	292
9.4	CYTOKININS	294
	9.4.1 *The hormone and its activities*	294
	9.4.2 *The mechanism of action*	296
	9.4.2.1 *The test system: callus growth*	296
	9.4.2.2 *Cytokinins and t-RNA*	296
	9.4.2.3 *Cytokinins and c-AMP*	298
9.5	OTHER PLANT HORMONES	298
	9.5.1 *Abscisic acid*	298
	9.5.2 *Ethylene*	299
	9.5.3 *Florigen*	300

9.1 INTRODUCTION

Plants are the product of an integrated series of events involving continued cell division, cell enlargement and cell differentiation[1]. In plants, as in animals, the integration of these events is carried out to no small extent by hormones. As these hormones are not the same ones that are found in animals, the study of the mode of action of plant hormones is a separate and distinct area of research.

In this article only three classes of plant hormones will be discussed in depth: the auxins, the gibberellins and the cytokinins. For each of these, their chemical nature, their location in plants and the scope of their activities will first be given. Then the mode of action of each will be discussed by examining the information which has been obtained from a single selected test system. Finally, there will be a brief discussion of some of the other diverse types of plant hormones.

9.2 AUXINS

9.2.1 The hormone and the range of its activities

Auxins first attracted attention because of their role in the control of cell elongation in stems and coleoptiles. Charles Darwin carried out some of the pioneering studies on auxin in the 1880s, but it was not until 1928 that an auxin was first isolated by Went, and the name auxin was given to this group of hormones[2]. Within a few years the chemical structure of one natural auxin, indoleacetic acid (IAA), had been determined (Figure 9.1).

Auxins have been found in all higher plants in which they have been sought,

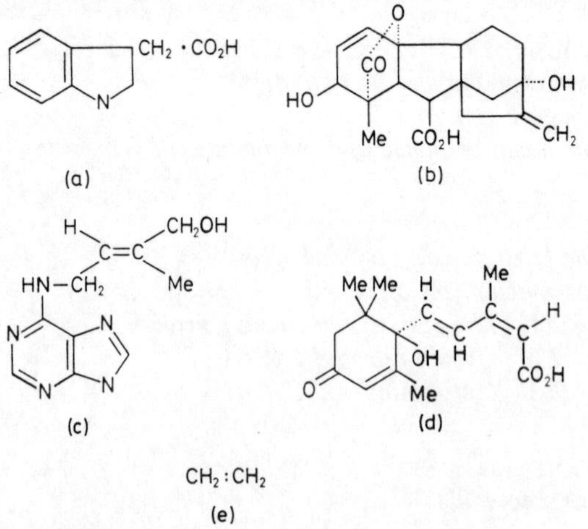

Figure 9.1 Structures of plant hormones. (a) indoleacetic acid (auxin), (b) gibberellic acid (gibberellin), (c) *trans*zeatin (cytokinin), (d) abscisic acid, (e) ethylene

and are present in many lower plants and bacteria as well[3]. It now appears that plants contain a single major auxin, IAA, although a variety of closely-related auxins such as indoleacetonitrile and indolepyruvic acid can be found in certain plants as well[4]. There have been many reports that non-indolic auxins exist as well, but to date no such auxin has been isolated and characterised chemically. (The original report of auxin a and b seems to have been in error.) It should be pointed out, however, that the job of identifying natural auxins is hampered by the low level of auxins which exist in plants. For example, the tip of the oat coleoptile is one of the richest sources of auxin, but 50 000 tips contain only *ca.* 1 µg of auxin[5]. A consequence of this paucity of auxin is that chemical determinations of the structure of auxins have been few and far between.

In addition to the natural auxins, there are many synthetic auxins. These include the herbicide 2,4-D (2,4-dichlorophenoxyacetic acid) and the now infamous 2,4,5-T (2,4,5-trichlorophenoxyacetic acid)[6]. Studies of these synthetic auxins have led to an understanding of the molecular properties which confer auxin activity on a molecule, but not to their mode of action[4].

Auxins can be isolated from any part of a higher plant, although they are in highest concentration in the areas where they are made; i.e. the tips of stems, roots and coleoptiles, the young leaves and developing seeds[7]. It should be noted that these areas are also the centres of cell division (meristems) in plants. Auxins them move from these meristematic areas to the slightly older elongating and differentiating cells. An unusual aspect of this movement is that it is polar[8]; the cells of stems and coleoptiles can transport auxins more readily in one direction (away from the tip) than the other (towards the tip). This polar transport is rapid, requires metabolism, and is not restricted to any particular cell type. The auxins are eventually removed from the system by a combination of destruction by peroxidase-type degradative enzymes (IAA-oxidase) or by conjugation into some inactive form such as indoleacetylaspartic acid[4].

The range of processes which require auxin or are regulated by it is nothing short of remarkable (see Ref. 4 for further references). Plant cell division requires auxin, at least in tissue cultures. Cell elongation in stems is regulated by the auxin level. The differentiation of protoxylem cells into tracheids requires auxin, as does the differentiation of protophloem cells into sieve-elements. The formation of lateral roots in intact plants is promoted by auxin while the formation of buds in callus cultures is suppressed. Auxin can induce the formation of specific enzymes such as the enzymes involved in ethylene biosynthesis, or it can modify the isozymic composition of already existing enzymes such as peroxidase. Finally, the integration between organs of a plant can be influenced by auxin; e.g. the inhibition of the growth of lateral buds is due to the auxin produced by the rapidly-growing apical buds.

Obviously, any hormone which regulates this variety of different processes can hardly be acting as a specific regulator in each case. For example, the control of cell division and organ formation in tissue cultures depends on the balance between auxin and at least one other hormone, the cytokinins. Likewise the differentiation of tracheids and sieve-elements depends on the interaction between auxin and certain nutrients such as the monosaccharides. But in one process, the control of cell elongation, auxin appears to act as a specific regulating substance. As a result, the bulk of studies on the mode of action of auxin have focused on the question as to how auxin induces cell elongation.

9.2.2 Auxin and the control of cell elongation

9.2.2.1 The test system

The most popular system for the study of auxin-induced cell elongation makes use of the oat (*Avena*) coleoptile[9,10]. When an oat seed germinates, a stem-like mesocotyl surmounted by the first leaf starts to grow upward. This first leaf is protected from abrasion during its passage through the soil by a hollow sheath-like organ, the coleoptile. The coleoptile grows rapidly for 4–5 days but, when a length of 4–6 cm is reached, growth ceases. An unusual characteristic of this growth is that after the coleoptile has reached a length of 1 cm, further growth is entirely by cell elongation. This rapid elongation requires the hormone auxin which in intact coleoptiles is supplied by the coleoptile tip. If a subapical section is cut from a coleoptile it will quickly be depleted of auxin and will therefore elongate only very slowly. But upon reapplication of auxin, growth will resume with only a short lag. This almost complete dependence of coleoptile growth on auxin coupled with the rapid growth rate, the ability of the tissue to respond rapidly to auxin and the lack of cell division makes the coleoptile a particularly suitable material for use in studies on the mode of action of auxin.

The standard test system consists of sections (usually 5–15 mm in length) cut from the region starting 2–3 mm below the tip of 20–30 cm long oat coleoptiles. Sections are usually incubated in a dilute buffer (pH 4.7–7.0) which contains sucrose whenever growth is to be followed over more than a few hours. The desired concentration of auxin is then added and elongation growth is followed with a continuous recording apparatus or simply by measuring the lengths of sections at suitable intervals.

9.2.2.2 The basic response

Treatment of an auxin-depleted coleoptile section with auxin results in an increase in the rate of cell elongation after a lag of only 8–10 min[11,12] (Figure 9.2). The rate then continues to increase over the next 10–20 min until a new steady-state rate is reached. At optimal auxin levels (10^{-6}–10^{-4} M) this rate is five-tenfold greater than the endogenous growth rate[10]. If the auxin concentration remains constant, and if a supply of osmoticum (sucrose) is maintained, the auxin-induced growth rate can persist at a nearly constant rate for up to 24 h. Also needed for this response are respiration[12], protein[13] and nucleic acid synthesis[14]; inhibition of any of these causes the growth rate to fall after a short lag.

Theoretical considerations indicate that the size of a plant cell, like that of a balloon, is primarily a function of the hydrostatic pressure (turgor pressure) and the extensibility of the encasing cell wall[15]. Auxin might cause

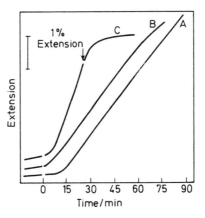

Figure 9.2 Comparison of *Avena* coleoptile cell elongations induced *in vivo* by auxin (A) and hydrogen ions (B) and *in vitro* by hydrogen ions (C). Incubation conditions: A, 5 mm sections in 0.01 M K–Pi buffer, pH 6.0, with IAA (2 ug ml^{-1}) added at time 0; B, 5 mm sections incubated in 0.01 M K–Pi buffer, pH 6.0, and transferred to 0.01 M K–citrate buffer, pH 3.0 at time 0; C, frozen-thawed sections incubated in 0.01 M K–Pi buffer, pH 7.0 +15 g imposed force and solution changed to 0.01 M K–citrate, pH 3.0 at time 0. At the arrow the solution was replaced with pH 7.0 buffer. Note that removal of hydrogen ions stops the elongation response

cells to enlarge through an increase in turgor pressure, but this does not seem to be the case[16]. Instead, a variety of tests have shown that the ultimate effect of auxin action is an increase in the ability of the cell walls to undergo a visco-elastic extension[17], although these tests have given no hint as to the nature of the changes which bring about this increase in wall extensibility.

9.2.2.3 The site of auxin action

The cell wall might seem to be the logical place to look for the site of auxin action, but two considerations have suggested that the site must be located somewhere in the protoplast. First, there is the negative evidence that no one has been able to demonstrate any effect of auxin on cell walls unless the protoplast is intact. Secondly, the requirement for a variety of aspects of cellular metabolism in all auxin responses has been taken as evidence for an interaction of auxin with some intracellular component[18].

Attempts to find auxin binding sites have proved frustrating. The problem is that, while auxin readily enters all parts of plant cells, it can be washed out almost as readily[19]. It has simply not been possible to find any evidence for covalent bonding of auxin to any cell constituent under physiological conditions (although linkage to macromolecules can occur under non-physiological conditions[20]). Apparently the interaction between auxin and its receptor only involves weak bonds. Recently, the first indications of a breakthrough were obtained from the reports that auxin-binding sites had been found in the plasma membrane[21, 22] and in the nucleoplasm[23]. The location of these binding sites is of particular interest since the nucleus and the plasma membrane are the proposed sites of auxin action in two of the current theories of auxin action; the 'gene activation' and the 'proton-pump' theories.

9.2.2.4 The effect of auxin on RNA synthesis

The possibility that auxin affects RNA synthesis in some manner so as to alter gene transcription has received much attention in recent years, and there is considerable evidence to support such an idea[24]. For example, auxin can induce the formation of enzymes such as benzoylmalate synthetase[25] and alter the isozyme composition of others such as peroxidase[26]. RNA synthesis is often, although not always, enhanced by auxin[24]. This enhancement has been reported to occur within 10 min of application of auxin[27], although the data are only barely significant, at best. Finally, inhibition of D-RNA synthesis (by actinomycin D) but not inhibition of r-RNA (by 5-fluorouracil) renders plant tissues incapable of undergoing auxin-induced cell elongation[14]. These data have been interpreted to mean that auxin directly or indirectly induced the formation of new species of m-RNA[14, 24].

Four groups of investigators have now isolated proteins which show some affinity for auxin and which alter the rate of *in vitro* RNA synthesis. But the details differ so greatly that it would appear that no two of these proteins are identical. For example, Matthysee and Phillips[28] obtained a protein from pea seedlings which in the presence of auxin stimulated RNA synthesis with pea chromatin, but not pea DNA. Since this stimulation was evident both when endogenous RNA polymerase and saturating amounts of *E. coli* polymerase were used, they suggested that it altered the amount of template available for transcription. On the other hand, Venis[29] isolated an auxin-binding protein from pea seedlings by affinity chromatography which stimulated RNA synthesis using DNA as the template and *E. coli* polymerase. In this case auxin was not necessary except for the isolation of the factor. A third factor, which has been obtained from soybean hypocotyls, also is active in the absence of added auxin, but in this case only on chromatin[30]. Furthermore, since it was far more active with native polymerase than with *E. coli* polymerase, Hardin *et al.*[31] suggested that the factor was altering the native polymerase. Recently this factor has been localised as existing in the plasma membrane; when fractions enriched in plasma membrane are treated with auxin this factor is released into the supernatant[22]. Finally, Mondal *et al.*[23] have obtained a protein factor from the nucleoplasm of coconut endosperm nuclei which binds to added auxin, interacts with coconut DNA and stimulates RNA synthesis. In this case gel electrophoresis and DNA–RNA hybridisation were used to show that the increased RNA represents, at least in part, the formation of new species of RNA.

These results leave more questions unanswered than they answer. While it is apparent that auxin-binding proteins exist which can modify RNA synthesis, it is not at all clear whether a cell possesses one or many such proteins, whether the protein needs auxin for its activity, whether it affects the polymerase or the template activity and whether it alters the types of RNA which are being produced.

Even if auxin can modify RNA synthesis, there are good reasons to believe that the nucleus is not the site of auxin action in a number of responses such as cell elongation. In the first place, the already short lag before the onset of rapid cell elongation can be reduced to *ca.* 1 min by several tricks, including the use of the methyl ester of IAA rather than of the free IAA[31]; such a short lag makes it very unlikely that gene activation is a necessary prerequisite for the growth response. Furthermore, no differences in RNA species could be detected by competitive DNA–RNA hybridisation between RNA species isolated from auxin-treated and control pea stem sections[32].

Finally, in several tissues at least a limited amount of auxin-induced growth can occur even when protein synthesis is essentially completely abolished by cycloheximide[13, 33]. These facts have led many investigators to focus their attention on some other part of the cell as a site for auxin action.

9.2.2.5 The proton-pump hypothesis

If the site of auxin action is somewhere in the protoplast, while the ultimate effect is a modification of the extracellular wall (a wall loosening) there must be some means of communication between the protoplast and the wall. This has led to the suggestion that protoplasts excrete a wall-loosening factor (WLF) in response to auxin[17]. If the nature of this hypothetical WLF were known, it might shed some light on where and how auxin acted. A number of candidates for the WLF have been suggested including the cell wall degrading enzymes cellulase[34], β-1,3-glucanase[35] and dextranase[36]. New polysaccharides which are intercalated into the wall have also been proposed as the WLF[37]. But in each case there exists a paucity of direct evidence, and none of the suggested WLFs have been shown to alter the extensibility of isolated cell walls in a manner analogous to that which occurs *in vivo* with auxin.

Another candidate for the WLF is hydrogen ions. It has long been known that acidic solutions (e.g. pH 3) will substitute for auxin in promoting rapid cell elongation for short periods, but it was believed that the acid simply causes the release of bound auxin in cells[38]. Recent work clearly shows that hydrogen ions are themselves active, as they induce cell elongation in cycloheximide-treated tissues which are insensitive to auxin[39]. This growth promotion begins within one minute (Figure 9.2) and quickly reaches the same maximum rate as is induced by optimal concentrations of auxin[39].

Hydrogen ions are also able to act on isolated cell walls[39, 40]. If coleoptiles are first frozen and thawed to disrupt the protoplast, and the walls are then subjected to a constant applied force of 20 g while incubated in a pH 7 buffer, the walls extend viscoelastically for a short time and then reach a nearly constant length. Addition of auxin, sucrose or potassium ions causes no further elongation. But if treated with hydrogen ions (i.e. the pH is lowered to 3) the walls begin to extend after a short lag, then elongate at a constant, high rate for 4–8 h and can increase in length by up to 50% over a 12 h period (Figure 9.2). It appears that hydrogen ions, added to isolated walls which are under tension, cause an elongation which resembles in kinetics and duration the elongation produced in live cells by auxin. The optimum pH for this *in vitro* elongation is *ca.* 3, but it now appears that a much lower concentration of hydrogen ions is actually needed in the cell wall. Coleoptiles are surrounded by a cuticle which is at least a partial barrier to hydrogen ions. If the cuticle and its adhering outer epidermal wall are stripped from the coleoptile before freezing, a maximal elongation response can be elicited at pH 5.0 in either the remaining coleoptile or in the epidermal strip[41]. Apparently, whenever the pH surrounding cell walls drop below *ca.* 5.5, the walls undergo some change resulting in an increased extensibility. If the walls are under tension, they will then extend.

If hydrogen ions are the WLF in auxin-induced growth, one should expect that auxin would induce plant cells to excrete hydrogen ions. Such an effect has now been demonstrated for *Avena* coleoptile[42], *Helianthus* hypocotyl[43], and sycamore callus cells[44]. In at least the first tissue the H^+ excretion can be detected within 20 min, requires respiration and can be blocked by cycloheximide[42]. These results provide strong evidence that one of the primary actions of auxin is to set into motion proton-pumps at the plasma membrane.

In 1970 Hager *et al.*[45] proposed that auxin activates aleady-existing ATPases in the plasma membrane and that these serve as the proton pumps. In support of this is the fact that the K^+–Na^+ activated ATPases of plants have been shown to be concentrated in the plasma membrane[46]. Furthermore, Hertel *et al.*[21] have used a pelleting-binding technique to show that auxin binding sites are concentrated at the plasma membrane as well. But as yet there has been no direct demonstration that auxin can activate any plasma membrane ATPase, and the fact that the proton-pumps in sycamore[44] and *Avena* coleoptiles[42] are stimulated by Ca^{2+} and not by K^+ suggest that K^+–Na^+ activated ATPases are not the ATPases which are involved.

Is there any connection between an auxin-activated proton-pump and an auxin-binding protein involved in RNA synthesis? When protein synthesis of auxin-treated tissues is blocked with cycloheximide, the growth rate begins to fall within 8–10 min[13,33]. This suggests that auxin-induced elongation requires some as yet unidentified labile growth-limiting protein (GLP). The GLP would appear to have a functional life of only *ca.* 25 min, with the result that continued synthesis of this protein would be necessary if rapid elongation is to persist for more than short periods of time. Furthermore, in *Avena* coleoptile[13] and lupin hypocotyls[33] the pool of GLP is low in the absence of auxin, but rapidly expands following addition of auxin; this suggests that auxin may influence transcription or translation of this protein, either directly or indirectly. The following possibility would seem to fit the available facts. Auxin binds to an ATPase in the plasma membrane, with two consequences. First, it activates the ATPase with the result that protons are excreted, the wall becomes loosened and the cell starts to extend. Secondly, it causes the release of a protein factor which affects RNA synthesis and results in the production of more ATPase. These possibilities are within our capabilities to test.

9.3 GIBBERELLINS

9.3.1 The hormone and its activities

At virtually the same time that the auxins were first being isolated in Europe, a second group of plant hormones was being discovered in Japan[47]. These hormones, called gibberellins after the fungus *Gibberella fujikuroi* from which they were isolated, were found to be particularly active in promoting stem elongation. Although extensively studied by the Japanese, the gibberellins remained unknown to the rest of the scientific world until after the second World War when the first articles appeared in English. By 1954 the structure of one gibberellin, gibberellic acid (Figure 9.1), was identified[48] and 2 years later gibberellins were first isolated from higher plants[49]. Since then interest in this group of plant hormones has grown year by year.

It has become apparent that there are a large number of different gibberellins, all possessing the four-ring gibbane backbone, and differing in the substituents attached to their rings[50]. This caused a crisis in nomenclature which was solved by the brilliant idea of simply identifying each chemically-characterised gibberellin by the prefix GA and a number[51]; thus gibberellic acid became GA_3. When GA is used without a subscript number it stands for the gibberellins as a class. To date 38 numbers have been assigned and there is every reason to believe that many additional gibberellins will ultimately be identified.

Gibberellins have been found in all higher plants in which they have been carefully sought, and are present in some lower plants as well[50]. As with auxins, the actual amounts of GA in a plant are small (e.g. 1–10 ug kg^{-1} fresh weight). No plant contains all of the gibberellins, but most plants possess at least several of them. For example, *Echinocystis* seeds have yielded GA_1, GA_3, GA_4, GA_7 and at least one other unidentified one[52]. The reason for such a variety of gibberellins in a plant is not clear, for unlike the steroids, to which they are related biosynthetically, gibberellins possess a remarkably similar spectrum of biological activities[53].

While all organs of a plant contain GA, the largest amounts are found in seeds. Roots appear to produce GA which is transported in the xylem to the stem, while the young leaves and the apical bud are rich in GA which can move down the stem, either by diffusion or in the phloem[54].

Gibberellins, like the auxins, exhibit a variety of effects in plants[54]. One of the major effects is to activate the subapical meristems of stems[55]. Many plants exist as both dwarf and tall varieties; the difference between the two is that in the tall plant there is a region of intense cell division activity just below the apical bud which contributes the bulk of the cells to the stem, while in the dwarf varieties this area is dormant. Treatment of dwarf plants with GA converts them into tall plants, while treatment of tall plants with chemicals which block GA synthesis (e.g. AMO–1618) causes them to revert to the dwarf form. A second effect of GA is to induce certain long-day plants to flower regardless of the day length[56]. In some plants GA will alter the sexuality of the resulting flowers, promoting maleness. Treatment of unfertilised tomato flowers with GA results in the development of parthenocarpic fruits. Finally, the synthesis of a variety of enzymes, including α-amylase, is induced in grain seeds by GA.

Many of the gibberellin responses are complex (e.g. stem elongation) or involve long time periods (e.g. flowering). As a result, much of the work on the mechanism of GA action has centred around the α-amylase induction system. These studies will be described, now, in more detail.

9.3.2 The test system

Grain seeds contain at one end an embryo from which the new plant will arise; the remainder of the seed is occupied by a mass of dead starch-containing endosperm cells surrounded by a couple of layers of live aleurone cells. When such a seed is wetted, the following sequence of events is set into motion. First, the embryo secretes gibberellin which diffuses to the aleurone cells. There the synthesis of α-amylase is initiated. The α-amylase is then secreted into the endosperm where it hydrolyses the starch into sugars. When the sugars reach the embryo they allow it to grow and the seed germinates.

In 1960 Paleg showed[56] that if a dry grain seed is bisected so as to separate the embryo from the bulk of the endosperm, the endosperm half will produce little or no α-amylase unless treated with GA. Seeds of barley, wheat, oats or rice can be used to demonstrate this effect, but as the largest and most reproducible effects are obtained with barley seeds (especially the Himalaya variety), most workers have concentrated their attention on the barley system.

The standard system is as follows[57]. Barley seeds are bisected and the portion containing the embryo is discarded. Endosperm halves are placed on sterile sand and are allowed to imbibe water for 3 days. At this time virtually no α-amylase has been produced, but the endosperm can be detached and removed so that only the aleurone layers remain. These aleurone layers are then incubated in buffer, with or without GA, for varying times, and the amount of α-amylase in the medium and in the aleurone cells is then determined.

9.3.3 The basic response and mechanism of GA action

After treatment of aleurone cells with GA, no α-amylase appears for about 8 h[58]. Then the level of α-amylase begins to rise and reaches a maximum after 48 h. At first there was some question as to whether α-amylase is synthesised *de novo* or is only activated by GA, but Varner and co-workers[59] used density-labelling techniques to show that the enzyme is made *de novo*. As would be expected, cycloheximide prevents the appearance of the enzyme if added during the lag, and terminates its synthesis if added during the period of synthesis[60].

When actinomycin D was added during the lag, the synthesis of α-amylase was blocked, while when added after the lag, it was without effect. This led to the suggestion[58] that the lag represented the period of time necessary for the production of the m-RNA for α-amylase, and that synthesis of the α-amylase commenced only when the synthesis of the messenger had been completed. Furthermore, it seemed possible that GA might be acting on this system as a specific inducer of the α-amylase messenger.

It soon became apparent, however, that the situation was more complex. In the first place, GA was needed during the period of synthesis as well as during the lag period even though RNA synthesis was not needed during this same period[60]. Secondly, GA was not specific for α-amylase but induced the synthesis of other hydrolytic enzymes such as protease[61], although it did not affect the synthesis of all enzymes produced by aleurone cells. Finally, there are reports that other agents can substitute for GA in this system; amino acids are active in *Avena fatua* seeds[62] while adenosine 3′,5′-cyclic monophosphate (c-AMP) has been reported[63] to substitute for GA in the barley system (some difficulties have been encountered in repeating this observation, however). It would appear, then, that GA is not a specific inducer for a single enzyme, a-amylase, but that it is required in some manner for the chain of events leading to the synthesis of the class of enzymes, the hydrolases.

It was apparent that the critical period for GA action is during the 8 h lag preceding enzyme synthesis. In this period GA must set into motion events leading to the synthesis of the hydrolases. The first clue as to the nature of these events came from the study by Jones[64] of the ultrastructural changes which precede a-amylase synthesis. One of the most noticeable events is a large increase in the amount of rough endoplasmic reticulum (RER) in GA-treated cells. Correlated with this is an increase in polysomes, starting 4 h after GA treatment[65]. A change in the pattern of RNA synthesis also occurs after 4 h. Although it is difficult with this tissue to obtain enough RNA for suitable analysis, Zwar and Jacobsen[66] solved the problem by taking advantage of the sensitivity of the double-labelled ratio method, coupled with gel electrophoresis. The synthesis of heterodisperse RNA (presumably messenger), but not r-RNA or t-RNA was found to be enhanced by GA and this stimulation was found to be inhibited by those agents (actinomycin D and abscisic acid) which prevent a-amylase synthesis. RNA synthesis, polysome formation and amount of RER are all found to reach a maximum at 12–16 h, the period of maximum a-amylase synthesis.

Is the a-amylase synthesised on the RER? It is known that in mammalian cells different types of protein can be formed on RER and free polysomes[67], RER being the site of synthesis of extracellular enzymes, while intracellular enzymes are made on free polysomes. Since the hydrolases are all extracellular enzymes, their synthesis might be expected to occur on RER, although this has not been proven. The role of GA might simply be to cause the formation of some constituent of the RER.

But what constituent limits the formation of RER? It is apparently not ribosomes, since the numbers of ribosomes increase to a much smaller extent than the increase in polysomes after GA treatment[66]. One possibility is the availability of membrane to which the ribosomes can attach. Supporting this idea is the fact that choline incorporation into membranes is stimulated by GA, starting at 3 h, or just prior to the increase in RER and polysomes[68]. Johnson and Kende[69] have pursued this idea by examining the effects of GA on three enzymes involved in the incorporation of choline into lecithin. The first enzyme, choline kinase, is present in equal amounts in the presence or absence of GA. But the other two enzymes, phosphorylcholine–cytidyl transferase and phosphorylcholine–glyceride transferase, increase markedly in the presence of GA. These increases are apparent within 2 h and the enzyme levels reach their peak *ca.* 12 h after the start of the GA treatment. In addition, the incorporation of ^{32}P into CTP is stimulated by GA within 30 min[70]. These results are all consistent with the idea that the primary effect of GA is on phospholipid synthesis.

The exact mode and site of GA action remains a mystery. The fact that the GA-induced increases in the two lecithin-pathway enzymes are blocked by either cycloheximide or actinomycin D[69] might suggest that these enzymes are synthesised in response to GA and that GA alters either the transcription or translation processes. It is known from other systems that GA can bind to DNA[71] and it has been reported that treatment of isolated pea nuclei with GA causes an increase and a change in the RNAs that are produced[72]. But it is still premature to conclude that GA acts at the gene level. Furthermore, there is evidence to suggest another mode of action. In barley aleurone cells GA not only induces the formation of a-amylase, but also stimulates the subsequent excretion of the enzyme[73]. The enzyme β-glucanase, whose synthesis is independent of GA, is also secreted in response to GA[74]. In a model system consisting of phospholipid-steroid micelles, GA markedly increased the permeability of the micelles to glucose and chromate ions[75]. The possibility that GA acts at a membrane to alter its permeability properties deserves further consideration.

Obviously GA-binding sites must exist in plant cells. If we knew where these were located we might have a clue as to the mode of action of GA. Unfortunately these binding sites have so far defied detection. Until they are found, a solution to the problem of how GA acts is likely to elude us.

9.4 CYTOKININS

9.4.1 The hormones and its activities

The presence in plants of factors that regulate cell division was first demonstrated by Haberlandt[76] in 1913, when he showed that diffusates of phloem tissues could induce cell division in potato parenchyma. Over the next four decades there were additional reports of division-inducing substances, but it never proved possible to isolate and identify chemically the active agent. Then in 1955 Skoog and co-workers[77] separated from autoclaved yeast DNA a division-inducing factor which they called kinetin. When added to parenchyma cells (such as tobacco pith) along with an auxin and suitable nutrients, kinetin would cause massive and continued cell division, while the result that the parenchyma cells would give rise to a callus tissue. Within a few years closely related compounds had been isolated from a variety of plant sources and in 1963 the structure of one, zeatin (Figure 9.1) was published[78]. The usual crisis in nomenclature ensued and a number of names for these compounds were proposed and rejected before the generic name of cytokinins was finally adopted.

Although cytokinins were first discovered because of their ability to induce cell division, it soon became apparent that they influence a number of other processes in plants as well (for references see Ref. 79). For example, the ability of callus to differentiate to form buds and roots is controlled, at least in part, by the cytokinin/auxin ratio; a high ratio favours bud formation while a low ratio promotes root formation. Another effect of cytokinins is to retard the senescence of detached leaves; in fact, the control of the lifespan of attached leaves may also be regulated by the cytokinins. In addition, leaf cells treated with a cytokinin possess the ability to attract metabolites from other regions of the leaf or the plant. Cytokinins have also been implicated in the control of xylem differentiation in some tissues such as the radish root and in the induction of some enzymes, such as nitrate reductase in *Agrostemma* embryos.

Cytokinins have been found in two forms in cells: free, or as part of t-RNA[80]. The principal free cytokinins are *trans*-zeatin (Figure 9.1) and its riboside and ribotide relatives, but dihydrozeatin has also been found free in plants. Four cytokinins have been isolated from t-RNA zeatin, (principally the *cis*-isomer), isopentenyladenine (IPA) and their respective CH_3S-derivatives. Neither kinetin nor the other commonly employed synthetic cytokinin, benzylaminopurine (BAP) has ever been found to occur naturally.

Cytokinin-containing t-RNAs have been obtained from bacteria, yeast and mammalian cells as well as from a variety of higher plant sources[79,80]. It seems likely that they exist in all organisms. When the t-RNA was fractionated, only certain of the t-RNAs were found to contain cytokinin bases; cytokinin activity is restricted to those t-RNAs which recognise codons beginning with the letter U. Thus the cytokinins are only found in phenylalanine, leucine, serine, tyrosine and cysteine t-RNA. In the case of the leucyl-t-RNAs, only those species recognising the U codons contained cytokinins, while the others had no cytokinin bases.

The amount of cytokinin bases in t-RNA is small, with only a single cytokinin base being present in any t-RNA molecule[80]. In all cases where the location of this base has been determined, it has been located adjacent to the 3'-end of the anticodon (e.g. Ref. 81). It apparently plays a vital function in the binding of the anticodon to the messenger RNA, as the chemical removal of the side chain has been found to render the t-RNA incapable of binding to the m-RNA–ribosome complex, even though it can still be charged with the appropriate amino acid[82]. It is interesting to note that certain of the t-RNAs which recognise codons beginning with A also have a modified base next to the codon, *N*-(purin-6-ylcarbamoyl)threonine. Again, this base is necessary for binding activity, but it has no cytokinin activity.

Two groups have recently raised questions as to whether the known cytokinins are actually the biologically-active compounds. Wood and Braun[83] have isolated two cell-division promoting compounds from tumourous *Vinca rosea* cells or from normal *Vinca* callus cells induced to divide with a cytokinin. These compounds, which are reported to be nicotinamide derivatives and to contain glucose, are believed to be produced in response to the cytokinins and to be the actual factors which regulate cell division. Fox and his co-workers[84] have reported that when BAP is fed to tobacco callus cells, it is rapidly metabolised, first to the nucleotide and then to 6-benzyl-amino-7-glucosylpurine. This compound is far more stable than BAP in tobacco cells, possesses cytokinin activity and might be the actual division-inducing substance. Only time and further experiments will tell.

Free cytokinins have been isolated from all parts of a plant, but some parts contain more cytokinins than others[79]. Particularly rich sources are seeds, young fruits and the bleeding sap obtained from roots. The site of biosynthesis of free cytokinins is not certain, but there are two pieces of evidence which suggest that they are produced in the root tips and are then transported in the xylem to the shoot portion of the plant. First, cytokinins continue to appear in the exudate of roots in undiminished amounts for at least 4 days after removal of the stems[85]. Secondly, when Holm and Key[86] compared the growth of rooted and derooted soybean seedlings, they found that the lack of roots inhibited the hypocotyl growth, but that this could be overcome by addition of cytokinins and gibberellins.

9.4.2 The mechanism of action

9.4.2.1 The test system; callus growth

Much of the work on the mechanism of action of cytokinins has made use of some variation of the tobacco callus system. The basic procedure is as follows[87]. Sterile pieces of parenchyma are excised from the pith of tobacco stems and are placed on a sterile agar medium. This medium contains mineral salts, carbohydrate (usually sucrose), selected organic compounds such as thiamin and an auxin. There are a number of different recipes which are used commonly, but most popular are the formulations of White[88] and of Murashige and Skoog[87]. Without a cytokinin the parenchyma cells will enlarge, but not divide to any significant extent, but if a cytokinin is also present in the medium, extensive division will take place and the callus will grow in size. After a callus has formed, it can be maintained indefinitely by subculturing it on fresh cytokinin-containing medium.

9.4.2.2 Cytokinins and t-RNA

The work on the mechanism of cytokinin action has centred around the relation between free cytokinins and t-RNA. This is not surprising, since the occurrence of the same molecule both as a free hormone and as an integral bound, necessary part of t-RNA must either be a remarkable coincidence or must indicate some connection between free and bound cytokinins. Furthermore, free cytokinins are reported to stimulate protein synthesis in several systems[79] and t-RNA is an essential part of such a process. A theory has evolved, then, which states that free cytokinins are required in cytokinin-dependent tissues for the appearance of those t-RNA species which contain bound cytokinin and that without these t-RNA species, certain essential messages cannot be translated[80, 89]. Furthermore, the theory states that the lack of these cytokinin-containing t-RNAs is due to an inability of the cells to produce these modified bases and that free cytokinins overcome this deficiency by being incorporated directly into t-RNA. If this theory is correct, certain predictions can be made; namely, that treatment of cytokinin-requiring tissues with a cytokinin will result in the appearance of new t-RNA species, that free cytokinins will be incorporated directly into t-RNA, and that cytokinin-requiring cells will not be able to make t-RNA-bound cytokinin. There is now considerable, but conflicting evidence on these points.

Cherry and his co-workers[90] have examined the amount of the different leucyl–t-RNA species in soybean cotyledons after treatment with BAP or water. In each case six different leucyl–t-RNAs could be found, but in the cytokinin-treated tissues two of the species showed two–three-fold increase. It is not known, however, whether these are the cytokinin-containing species. Furthermore, since this tissue does not have an obligate cytokinin requirement, the experiment does not answer the question as to whether free cytokinin is needed for the appearance of specific t-RNA species.

A number of attempts have been made to demonstrate direct incorporation of cytokinins into t-RNA, with uncertain results. McCalla et al.[91] fed BAP[8-^{14}C] to cocklebur leaves and recovered from RNA hydrolysates a small amount of the radioactivity which cochromatographed with BAP. Fox repeated these experiments[89] using tobacco and soybean callus and detected a sizeable incorporation into what was believed to be t-RNA on the basis of its elution from benzoylated DEAE–cellulose columns. However, Richmond et al.[92] found that the t-RNA peak was contaminated by a second component, and that the bulk of the radioactivity was in the non-t-RNA fraction. But recently Burrows et al.[93] have provided convincing evidence that at least some incorporation of cytokinin can occur. Large amounts of tobacco callus, which requires added cytokinin for growth, were grown in the presence of BAP, and the t-RNA was then isolated and hydrolysed. Among the bases recovered from the hydrolysate was BAP. But even if a small amount of cytokinin finds its way into t-RNA, is this its mode of action?

Another approach has been to utilise BAP which is substituted in the 9-position. These compounds are biologically active, but they cannot be incorporated into t-RNA unless the substituent at the 9-position is cleaved off in the cell. When Fox et al.[94] administered to tobacco callus 9-methyl-BAP labelled in both the methyl group and in the ring, they could detect hydrolysis of the methyl group within 10 min. On the other hand Kende and Tavaras[95] could detect no incorporation from [7-^{14}C]-9methyl-BAP into t-RNA, even though the starting material was of sufficient specific activity that they should have been able to detect an incorporation of one cytokinin molecule per t-RNA molecule. Undoubtedly the last word on this subject has not yet been printed.

Perhaps the most illuminating experiments have been concerned with the mode of synthesis of the RNA-bound cytokinin bases. In a series of elegant experiments Hall and his co-workers[96] showed that the IPA is formed by attachment of the sidechain to adenine which is already present in preformed t-RNA. t-RNA from yeast, bacteria or mammalian cells were first given a mild permanganate treatment to remove already existing isopentenyl sidechains. When these modified t-RNAs were incubated with mevalonate or isopentenyl pyrophosphate and a crude enzyme preparation, the side chains were reformed. Only the adenine which had originally possessed an isopentenyl sidechain received a new one, indicating that the enzymes involved are able to recognise specific adenine bases in t-RNA.

Chen and Hall[97] then examined the question as to whether a cytokinin-requiring plant tissue can make t-RNA bound cytokinins in the absence of free cytokinins. Tobacco pith callus was maintained on a medium which had no cytokinin and labelled mevalonic acid was supplied. When the t-RNA was isolated and the bases were recovered, radioactive IPA was detected. Apparently these tissues possessed the enzymes needed to form IPA from mevalonic acid and adenine even in the absence of free cytokinin. It should be noted, too, that when Burrows et al.[93] grew tobacco callus in the presence of a single cytokinin, BAP, the bulk of the t-RNA–bound cytokinin proved to be zeatin and IPA, not BAP. These results suggest that most of the RNA-bound cytokinins in cytokinin-requiring tissues come from the normal synthetic pathways and not by direct incorporation of free cytokinin into t-RNA.

The evidence, then, casts doubt on the possibility that free cytokinins exert their biological effects simply by providing an essential ingredient for t-RNA synthesis. But this by no means rules out some other kind of cytokinin–t-RNA interaction. For example, cytokinins might bind to the m-RNA–ribosome complex and alter the codon specificity so as to compensate for the deficiency of a particular t-RNA. A reversible binding of cytokinins to ribosomes has been detected by Berridge et al.[98]; up to 1.34 molecules of BAP were bound per ribosome under optimal conditions, and the amount of binding of various cytokinins was proportional to their biological activity. But neither these workers nor anyone else has been able to demonstrate any effect of cytokinins on protein synthesis in an *in vitro* system. Perhaps when the system for *in vitro* protein synthesis in plants has been perfected these results will change.

9.4.2.3 Cytokins and c-AMP

At this point one must at least consider the possibility that cytokinins act in some way other than via t-RNA. Recently Wood and Braun[99] proposed that cytokinins inhibit c-AMP phosphodiesterases, thereby retarding the degradation of c-AMP, and that c-AMP actually is the substance that mediates all the cytokinin effects. In support of this idea they note that both plant and animal diesterases are inhibited by the natural *Vinca* cytokinin[100], and that a slowly-degraded form of c-AMP, 8-bromo-c-AMP, has definite cytokinin activity[99]. Furthermore, a similar cytokinin–c-AMP interaction would seem to exist in animal cells. The ability of cytokinins to inhibit the growth of certain tumourous animal cells is well documented[101]. Gallo et al.[102] have now shown that dibutyryl–c-AMP can mimic the effect of the cytokinin IPA in inhibiting transformation of phytohemagglutinin-treated human lymphocytes.

9.5 OTHER PLANT HORMONES

The control of plant development is by no means the exclusive province of the three hormones already discussed, as a variety of other hormones have been implicated in different aspects of plant development. For several of these we now possess considerable information about their natural history (e.g. site and pathways of biosynthesis) and their role in the plant, but since we know discouragingly little about their mode of action, an extensive discussion of any of them does not seem warranted. However, three hormones, abscisic acid (ABA), ethylene and florigen, each have some novel features and do deserve a brief mention.

9.5.1 Abscisic acid

Many plants can grow almost indefinitely, but because they cannot escape their environment, there are times when it is to a plant's benefit to be dormant rather than to grow. For example, when winter approaches, a plant must cease growing and convert all the growing areas into resistant dormant buds. In theory this could be accomplished by negative control, i.e. by a withdrawal of the growth-promoting hormones. But instead, plants use inhibitory hormones; as winter approaches, abscisic acid (Figure 9.1) builds up in the growing regions and inhibits cell division and cell elongation[103]. In the spring ABA levels drop and the buds begin to grow again. In addition to its role in bud dormancy, ABA participates in a number of other developmental responses including abscission of leaves, dormancy of seeds and the closure of stomates in wilted plants (for references see Ref. 104).

Studies on the mode of action of ABA have tended to centre around the inhibitory effects of ABA on protein and RNA synthesis[104]. There are a number of reports showing that ABA interferes with these processes, but it is not at all clear how ABA works, since the hormone causes selective inhibitions. For example, in barley endosperm ABA inhibits the synthesis of α-amylase but not nitrate reductase[60]. In tobacco leaves ABA inhibits r-RNA but not t-RNA formation[105]. ABA cannot, therefore, simply be a general inhibitor of either RNA or protein synthesis.

There are two clues which may lead to an understanding as to how ABA works. First, in a number of cases ABA counteracts events induced by GA. In the α-amylase system each of the steps which is induced by GA can be inhibited by ABA[60, 68, 69], while in bud dormancy GA can overcome the inhibitory effects of ABA[103]. When the mode of action of GA is better understood, it may open the door to work on the mode of action of ABA.

The other clue comes from studies on the control of stomate opening in leaves. In the light K^+ is pumped into guard cells, the turgor pressure increases, and the stomates open[106]. In the dark the pump is inactive, the K^+ leaks out, and the stomates close. Anything that interferes with the K^+-pump of guard cells will cause the closure of the stomates. Treatment of guard cells with ABA causes closure[107] which can be detected within 5 min[180]. Horton and Moran[109] have now shown that ABA prevents the entry of K^+ into guard cells. It would appear, then, that ABA must interfere in some way with the K^+ pump, and that a study of the interaction between ABA and this K^+ pump may provide a real clue as to how ABA works.

9.5.2 Ethylene

Plants are unusual in possessing a gaseous hormone, ethylene. Although ethylene probably exerts its primary effects on the cells in which it is produced, it qualifies as a hormone in that it can diffuse to other cells and influence them as well. Like the other plant hormones, ethylene induces a variety of processes (for references see Refs. 110 and 111). Best known of these is fruit ripening, where ethylene plays a dual role; it not only causes the biochemical changes which lead to a ripened fruit, but it also induces the formation of more ethylene so that the process of ripening escalates. Among its other effects are inhibition of cell division and cell elongation in stems and roots[111, 112], a change in the polarity of cell growth from a longitudinal to a transverse direction[113], induction of flowering in pineapples[114] and an acceleration of the breakdown of the cells in the abscission zone in leaves and fruits[115].

In many cases ethylene acts as a 'second messenger' between the hormone auxin and the physiological response. An example of this is root growth where auxin has long been known to inhibit cell elongation[111,112]. It is now known that the auxin simply induces the formation of the ethylene-producing enzymes and that the resulting ethylene actually inhibits the root growth.

Ethylene influences a variety of biochemical processes in plants, including protein and RNA synthesis[115], but there is as yet no evidence that clearly indicates either its site or mode of action. Since many of the ethylene responses can be blocked by actinomycin D or cycloheximide[110], and since chromatin of ethylene-treated soybeans shows changes in activity in parallel to changes which occur in growth and RNA synthesis[116], there has been some temptation to believe that ethylene acts at the level of chromatin, but there is really no direct evidence to support such an idea.

9.5.3 Florigen

Certainly the most frustrating of all plant hormones is the flowering hormone, florigen. In vegetative plants there are no cells set aside and reserved for reproduction. When a plant flowers, cells which would otherwise be involved in the production of leaves and buds are diverted into producing modified leaves, the flowers. This change occurs in many plants in response to photoperiod; plants in general can be divided into short-day, long-day and day-neutral groups.

The part of the plant which perceives the correct photoperiod is the leaf, yet the part which undergoes the transition from vegetative to reproductive activity is the bud. Clearly there must be some communication between these two and a hormone, florigen, has been proposed[117]. Grafting experiments between short-day and long-day, or between photoperiod-requiring and day-neutral plants have led to the conclusion that there exists a single florigen, which is common to all flowering plants[118]. The problem is that no one has been able to isolate and identify this hormone. Over the years there have been reports of crude extracts which contain flower-inducing activity (e.g. Refs. 119 and 120), but as yet no substance has been purified from them and some workers in the area have come to question whether florigen actually exists[121] despite the wealth of indirect evidence in its favour. As such a hormone would have tremendous practical value, as well as value in basic research, the inability to find florigen is frustrating.

References

1. Torrey, J. G. (1967). *Development in flowering plants* (New York: Macmillan)
2. Went, F. W. (1928). *Rec. Trav. Bot. Neerl.*, **25,** 1
3. Bentley, J. A. (1961). *Encycl. Plant Physiol.*, Vol. 14, 485 (H. Burström, editor) (Berlin: Springer-Verlag)
4. Thimann, K. V. (1969). *The Physiology of Plant Growth and Development*, 1 (M. B. Wilkins, editor) (London: McGraw-Hill)
5. Haagen-Smit, A. J. (1951). *Plant Growth Substances*, 1 (F. Skoog, editor) (Madison: Univ. Wisconsin Press)

6. Veldstra, H. (1953). *Annu. Rev. Plant Physiol.*, **4**, 151
7. Söding, H. (1961). *Encycl. Plant Physiol.*, Vol. 14, 583 (H. Burström, editor) (Berlin: Springer-Verlag)
8. Goldsmith, M. H. M. (1968). *Annu. Rev. Plant Physiol.*, **19**, 347
9. Larsen, P. (1961). *Encycl. Plant Physiol.*, Vol. 14, 521 (H. Burström, editor) (Berlin: Springer-Verlag)
10. Cleland, R. (1972). *Planta*, **104**, 1
11. Ray, P M. and Ruesink, A. W. (1962). *Devel. Biol.*, **4**, 377
12. Evans, M. L. and Ray, P. M. (1969). *J. Gen. Physiol.*, **53**, 1
13. Cleland, R. (1971). *Planta*, **99**, 1
14. Key, J. L. and Ingle, J. (1964). *Proc. Nat. Acad. Sci. USA*, **52**, 1382
15. Lockhart, J. A. (1965). *J. Theoret. Biol.*, **8**, 264
16. Ray, P. M. and Ruesink, A. W. (1963). *J. Gen. Physiol.*, **47**, 83
17. Cleland, R. (1971). *Annu. Rev. Plant. Physiol.*, **22**, 197
18. Ray, P. M. (1969). *Communication in Development*, 172 (A. Lang, editor) (New York: Academic Press)
19. Johnson, M. P. and Bonner, J. (1956). *Physiol. Plant.*, **9**, 102
20. Davies, P. J. and Galston, A. W. (1971). *Plant Physiol.*, **47**, 435
21. Hertel, R., Thompson, K. S. and Russo, V. E. A. (1972). *Planta*, **107**, 325
22. Hardin, J. W., Cherry, J. E., Morré, D. J. and Lembi, C. A. (1972). *Proc. Nat. Acad. Sci. (USA)*, **69**, 3146
23. Mondal, H., Mandel, R. K. and Biswas, B. B. (1972). *Nature New Biol.*, **240**, 111
24. Key, J. L. (1969). *Annu Rev. Plant Physiol.*, **20**, 449
25. Venis, M. A. (1972). *Plant Physiol.*, **49**, 24
26. Ockerse, R., Siegel, B. Z. and Galston, A. W. (1966). *Science*, **151**, 452
27. Masuda, Y. and Kamisaka, S. (1969). *Plant and Cell Physiol.*, **10**, 79
28. Matthysse, A. G. and Phillips, C. (1969). *Proc. Nat. Acad. Sci. (USA)*, **63**, 897
29. Venis, M. A. (1971). *Proc. Nat. Acad. Sci. (USA)*, **68**, 1824
30. Hardin, J W., O'Brien, T. J. and Cherry, J. H. (1970). *Biochim. Biophys. Acta*, **224**, 667
31. Rayle, D. L., Evans, M. L. and Hertel, R. (1970). *Proc. Nat. Acad. Sci. (USA)*, **65**, 184
32. Thompson, W. F. and Cleland, R. (1971). *Plant Physiol.*, **48**, 663
33. Penny, P. (1971). *Plant Physiol.*, **48**, 720
34. Fan, D. F. and Maclachlan, G. A. (1966). *Can. J. Bot.*, **44**, 1025
35. Masuda, Y. and Yamamoto, R. (1970). *Develop. Growth Diff.*, **11**, 287
36. Heyn, A. N. J. (1969). *Science*, **167**, 874
37. Ray, P. M. (1967). *J. Cell Biol.*, **35**, 659
38. Bonner, J. (1934). *Protoplasma*, **21**, 406
39. Rayle, D. L. and Cleland, R. (1972). *Planta*, **104**, 282
40. Rayle, D. L., Haughton, P. M. and Cleland, R. (1970). *Proc. Nat. Acad. Sci. (USA)*, **67**, 1814
41. Rayle, D. L., and Cleland, R. Unpublished data
42. Cleland, R. (1973). *Plant Physiol.*, **51**, 52
43. Ilan, I. (1973). *Physiol. Plant.*, **28**, 146
44. Fisher, M. L. and Albersheim, P. (1973). *Plant Physiol.*, **51**, 52
45. Hager, A., Menzel, H. and Krauss, A. (1971). *Planta*, **100**, 47
46. Hodges, T. K., Leonard, R. T., Bracker, C. E. and Keenan, T. W. (1972). *Proc. Nat. Acad. Sci. (USA)*, **69**, 3307
47. Stowe, B. B., Stodola, F. H., Hayashi, T., and Brian, P. W. (1961). *Plant Growth Regulation*, 465 (R. M. Klein, editor) (Ames: Iowa, State Univ.)
48. Cross, B. E. (1954). *J. Chem. Soc.*, 4670
49. Phinney, B. O., West, C. A., Ritzel, M. and Neely, P. M. (1957). *Proc. Nat. Acad. Sci. (USA)*, **43**, 398
50. Lang, A. (1970). *Annu. Rev. Plant Physiol.*, **21**, 537
51. MacMillan, J. and Takahashi, N. (1968). *Nature (London)*, **217**, 170
52. Elson, G. W., Jones, D. R., MacMillan, J. and Suter, P. J. (1964). *Phytochem.*, **3**, 93
53. Cleland, R. (1969). *The Physiology of Plant Growth and Development*, 49 (M. B. Wilkins, editor) (London: McGraw-Hill)
54. Sach, R. M., Bretz, C. F. and Lang, A. (1959). *Amer. J. Bot.*, **46**, 376
55. Lang, A. (1957). *Proc. Nat. Acad. Sci. (USA)*, **43**, 709
56. Paleg, L. (1960). *Plant Physiol.*, **35**, 293

57. Jones, R. L. and Varner, J. E. (1967). *Planta*, **72**, 155
58. Varner, J. and Ram Chandra, G. (1964). *Proc. Nat. Acad. Sci. (USA)*, **52**, 100
59. Filner, P. and Varner, J. E. (1967). *Proc. Nat. Acad. Sci. (USA)*, **58**, 1520
60. Chrispeels, M. J. and Varner, J. E. (1967). *Plant Physiol.*, **42**, 1008
61. Jacobsen, J. V. and Varner, J. E. (1967). *Plant Physiol.*, **42**, 1596
62. Naylor, J. M. (1966). *Can. J. Bot.*, **44**, 19
63. Galsky, A. G. and Lippincott, J. A. (1969). *Plant and Cell Physiol.*, **10**, 607
64. Jones, R. L. (1969). *Planta*, **87**, 117
65. Evins, W. H. (1971). *Biochemistry*, **10**, 4295
66. Zwar, J. A. and Jacobsen, J. V. (1972). *Plant Physiol.*, **49**, 1000
67. Ganoza, M. C. and Williams, C. A. (1969). *Proc. Nat. Acad. Sci. (USA)*, **63**, 1370
68. Evins, W. H. and Varner, J. E. (1971). *Proc. Nat. Acad. Sci. (USA)*, **68**, 1631
69. Johnson, K. D. and Kende, H. (1971). *Proc. Nat. Acad. Sci. (USA)*, **68**, 2674
70. Collins, G. G., Jenner, C. F. and Paleg, L. (1972). *Plant Physiol.*, **49**, 404
71. Kessler, B. and Snir, I. (1969). *Biochim. Biophys. Acta*, **195**, 207
72. Johri, M. M. and Varner, J. E. (1968) *Proc. Nat. Acad. Sci. (USA)*, **59**, 269
73. Chrispeels, M. J. and Varner, J. E. (1967). *Plant Physiol.*, **42**, 398
74. Taiz, L. and Jones, R. L. (1970). *Planta*, **92**, 73
75. Wood, A. and Paleg, L. G. (1972). *Plant Physiol.*, **49**, 103
76. Haberlandt, G. (1913). *Ber. K. Preuss. Akad. Wiss.*, **1913**, 318
77. Miller, C. O., Skoog, F., von Saltza, M. H. and Strong, F. M. (1955). *J. Amer. Chem. Soc.*, **77**, 1392
78. Letham, D. S. (1963). *N. Z. J. Bot.* **1**, 336
79. Kende, H. (1971). *Int. Rev. Cytol.*, **31**, 301
80. Skoog, F. and Armstrong, D. J. (1970). *Annu. Rev. Plant. Physiol.*, **21**, 359
81. Madison, J. T. and Kung, H. K. (1967). *J. Biol. Chem.*, **242**, 1324
82. Fittler, F. and Hall, R. H. (1966). *Biochem. Biophys. Res. Commun.*, **25**, 441
83. Wood, H. N. and Braun, A. C. (1967). *Ann. N.Y. Acad. Sci.*, **144**, 244
84. Deleuze, G. G., McChesney, J. D. and Fox, J. E. (1972). *Biochem. Biophys. Res. Commun.*, **48**, 1426
85. Kende, H. (1965). *Proc. Nat. Acad. Sci. (USA)*, **53**, 1302
86. Holm, R. E. and Key, J. L. (1969). *Plant Physiol.*, **44**, 1295
87. Murashige, T. and Skoog, F. (1962). *Physiol. Plant.*, **15**, 473
88. White, P. R. (1963). *The Cultivation of Animal and Plant Cells*, 2nd edn. (New York: Ronald Press)
89. Fox, J. E. and Chen, C. M. (1967). *J. Biol. Chem.*, **242**, 4490
90. Cherry, J. H. and Anderson, M. B. (1972). *Plant Growth Substances 1970*, 181 (D. J. Carr, editor) (Berlin: Springer-Verlag)
91. McCalla, D. R., Morré, D. J. and Osborne, D. J. (1962). *Biochim. Biophys. Acta*, **55**, 522
92. Richmond, A., Back, A. and Sachs, B. (1970). *Planta*, **90**, 57
93. Burrows, W. J., Skoog, F. and Leonard, N. J. (1971). *Biochemistry*, **10**, 2189
94. Fox, J. E., Sood, C. K., Buckwalter, B. and McChesney, J. D. (1971). *Plant Physiol.*, **47**, 275
95. Kende, H. and Tavares, J. E. (1968). *Plant Physiol.*, **43**, 1244
96. Kline, L. K., Fittler, F. and Hall, R. H. (1969). *Biochemistry*, **8**, 4361
97. Chen, C. M. and Hall, R. H. (1969). *Phytochem.*, **8**, 1687
98. Berridge, M. V., Ralph, R. K. and Letham, D. S. (1972). *Plant Growth Substances 1970*, 248 (D. J. Carr, editor) (New York: Springer-Verlag)
99. Wood, H. N. and Braun, A. C. (1973). *Proc. Nat. Acad. Sci. (USA)*, **70**, 447
100. Wood, H. N., Lin, M. C. and Braun, A. C. (1972). *Proc. Nat. Acad. Sci. (USA)*, **69**, 403
101. Suk, D., Simpson, C. L. and Mihich, E. (1968). *Proc. Amer. Assoc. Cancer Res.*, **8**, 23
102. Gallo, R. C., Hecht, S. M., Whang-Peng, J. and O'Hopp, S. (1972). *Biochem. Biophys. Acta*, **281**, 488
103. Eagles, C. F. and Wareing, P. F. (1964). *Physiol. Plant.*, **17**, 697
104. Addicott, F. T. and Lyon, J. L. (1969). *Annu. Rev. Plant Physiol.*, **20**, 139
105. Leshem, Y. and Schwarz, L. (1972). *Physiol. Plant.*, **26**, 328
106. Fischer, R. A. and Hsiao, T. C. (1968). *Plant Physiol.*, **43**, 1953
107. Mittelheuser, C. J. and Van Steveninck, R. F. M. (1969). *Nature (London)*, **221**, 281

108. Cummins, W. R., Kende, H. and Raschke, K. (1971). *Planta*, **99**, 347
109. Horton, R. F. and Moran, L. (1972). *Zeit. Pflanzenphysiol.*, **66**, 193
110. Abeles, F. B. (1972). *Annu. Rev. Plant Physiol.*, **23**, 259
111. Burg, S. P. (1973). *Proc. Nat. Acad. Sci. (USA)*, **70**, 591
112. Chadwick, A. V. and Burg, S. P. (1970). *Plant Physiol.*, **45**, 192
113. Eisinger, W. and Burg, S. P. (1972). *Plant Physiol.*, **50**, 510
114. Burg, S. P. and Burg, E. A. (1966). *Science*, **152**, 1269
115. Abeles, F. B. and Holm, R. E. (1966). *Plant Physiol.*, **41**, 1337
116. Holm, R. E., O'Brien, T. J., Key, J. L. and Cherry, J. H. (1970). *Plant Physiol.*, **45**, 41
117. Chailakyan, M-Kh. (1937). *Comp. Rend. Acad. Sci. USSR*, **16**, 227
118. Lang, A. (1965). *Encyc. Plant Physiol.*, **15**(1), 1380
119. Lincoln, R. G., Mayfield, D. L. and Cunningham, A. (1961). *Science*, **133**, 756
120. Hodson, H. K. and Hamner, K. C. (1970). *Science*, **167**, 384
121. Evans, L. T. (1971). *Annu. Rev. Plant Physiol.*, **22**, 365

10
Cell Culture in the Study of the Mechanism of Hormone Action

E. H. MACINTYRE
National Jewish Hospital and Research Center, Denver, Colorado

10.1	INTRODUCTION	306
	10.1.1 *General comments*	306
	10.1.2 *Historical background*	307
	10.1.3 *Current status of the art*	308
10.2	CELL AND ORGAN CULTURE	308
	10.2.1 *Definition of terms*	308
	10.2.2 *Culture techniques*	309
	10.2.2.1 *Disaggregation*	309
	10.2.3 *Constituents of growth media*	310
	10.2.3.1 *General review*	310
	10.2.3.2 *Supplementation with serum*	312
	10.2.3.3 *The effect of different media on cell functions*	313
	10.2.3.4 *Contamination*	313
	10.2.4 *Culture apparatus*	314
	10.2.5 *Procedures for culture and selection of cells*	314
	10.2.5.1 *Primary culture of cells*	314
	(a) *Viruses in cell cultures*	314
	(b) *Organ cultures*	315
	(c) *Dispersed cultures*	315
	10.2.6 *Hormone producing, hormone-dependent, and hormone-responsive cultures*	315
	10.2.6.1 *Sources of primary tissue*	315
	10.2.6.2 *Cell culture techniques for selecting hormone-dependent cultures*	316

	10.2.6.3	*The selection and potentiation of specialised cells in culture*	317
	10.2.6.4	*The growth of normal hormone-responsive cells in culture*	317
	10.2.6.5	*Routine testing of specific cell function* in vitro	317

10.3 PROCEDURES OF POTENTIAL VALUE FOR THE *in vitro* STUDY OF HORMONES — 318

10.4 STUDIES OF HORMONE SYNTHESIS AND ACTION *in vitro* — 318
    10.4.1 *Early studies of hormone synthesis* in vitro — 318
    10.4.2 *The mammary gland-differentiation and cancer* — 318
        10.4.2.1 *Hormone dependence for differentiation of normal breast tissue* — 319
        10.4.2.2 *Cell differentiation and the expression of mammary tumour virus (MTV)* — 319
        10.4.2.3 *Autonomy of breast cancer and synthesis of hormones by breast cancer cells* — 320
    10.4.3 *Culture of the pituitary gland* — 320
    10.4.4 *Culture of the adrenal gland* — 321
    10.4.5 *Culture of the pineal gland* — 322
        10.4.5.1 *Historical review* — 322
        10.4.5.2 *Aspects of the biochemistry of the pineal gland* — 322
            (a) In vitro *experiments* — 322
            (b) In vivo *experiments* — 322
        10.4.5.3 *The use of organ cultures of the pineal in the study of the mechanism of drug action* — 323
    10.4.6 *Culture of cells from the central nervous system* — 323
    10.4.7 *Miscellany* — 324
    10.4.8 *Cultures of cells from insects* — 324

10.5 FUTURE PROSPECTS FOR THE STUDY OF HORMONES *in vitro* — 324
    10.5.1 *General comments* — 324

ACKNOWLEDGMENTS — 325

10.1 INTRODUCTION

10.1.1 General comments

The past 15 years have seen the emergence of cell culture as a routine laboratory tool, accepted by most workers as a valuable technique for certain investigations. The potential of this method has increased with the greater sophistication of culture techniques, the standardisation of much of the methodology and the availability of commercial sources for basic needs, such as sera, media and culture vessels. The demonstration that differentiated cells from diverse tissues could be grown in culture and retained their differentiated functions even when isolated from their normal milieu in the body[1-40], revolutionised the attitude of many investigators who hitherto had regarded 'dedifferentiation' as an inevitable consequence of cell culture. Such cells offer the means of studying the response of a specific cell type to a wide range of manipulations. Changes, not only in protein synthesis or in cell numbers[41], but also in the characteristic product of the cell can now be measured.

Methods of selection derived in many cases from the area of bacterial genetics[42] are now being applied to mammalian cells[3,27,43-46]. Such procedures have already yielded many established cell lines which produce hormones autonomously *in vitro*[3,10,15,22,26-28,32-36,40]; hormone-dependent cultures are also available[17,21,27,38,47]. The existence of such cells offers new and exciting possibilities for the workers interested in the synthesis and mode of action of hormones. Our knowledge in this area should expand considerably, not just with respect to the mechanism of hormone synthesis and action, but also with respect to the role played by hormones in processes as seemingly diverse as tissue differentiation[5,17,38,47-51], virus production[52], and disease syndromes such as Albright's[53].

This review will deal primarily with work on mammalian cells. Considerable space will be devoted to discussion of the pitfalls inherent in the use of cell culture for hormone studies. This is essential because technical problems associated with work in this young field are not yet summarised in one review, but are scattered throughout the literature. Foreknowledge of their existence can prevent serious blunders. There will be no attempt to supply an exhaustive list of every cell type or every experiment relating to this field. Examples will rather serve to indicate the variety of investigations for which cell culture is already, or may be, of value. If the reader is stimulated to follow the development of this field and perhaps to use such techniques as are applicable to his research, the writer's goal will have been accomplished.

10.1.2 Historical background

There are two milestones in the history of cell culture—the demonstration early this century that cells completely isolated from the body[54] could proliferate, and, somewhat later, the finding that specialised function (such as the production of cartilage[9], melanin[7], or the secretion of gondatrophin[11]) could be retained over a lengthy period by cells proliferating in culture. These findings are basic to all subsequent developments.

The difficulties which beset early workers in cell culture were formidable. Many were technical—the nutritional requirements of cells for amino acids, vitamins, sources of energy, salts, etc. were unknown; every facet of the technique from preparing constituents of growth media to the cleaning of culture vessels had to be handled in the laboratory. Contamination by bacteria was a major problem.

Other problems associated with cell culture were conceptual. There were many who regarded cells in culture as laboratory peculiarities, bearing little relationship to the tissue of origin.

10.1.3 Current status of the art

This attitude persisted until the early sixties when a concerted effort was made to culture cells which either produced hormones or were hormone-dependent. Their characteristics were defined not only on the basis of morphology but also by their specific products. The latter denoted and typified the tissue of origin of the cells. For example, adrenocorticotropic hormone (ACTH), prolactin and somatotropic hormone (SH) were synthesised by cultures of anterior pituitary cells[3,10,26,27,40]. In many cases, the cells continued to make specific substances after many generations in culture[1-4,6,8,10,13-20,22-29,32-40]. 'Dedifferentiation' in culture is now attributed either to an explicable change in behaviour by a specialised cell (such as is shown by hormone-dependent cultures after hormone deprivation[17,27,38,47]), or to overgrowth and replacement by rapidly proliferating cells such as fibroblasts[55].

10.2 CELL AND ORGAN CULTURE

10.2.1 Definition of terms

In this text, the term 'specialised cell' denotes a cell which is capable of producing a specific product(s) in culture under a prescribed set of conditions. Such a cell is also to some extent, at least, a differentiated cell. The capability to synthesise collagen, protocollagen or proline hydroxylase will not be considered evidence of specialisation here, since most cells are capable of such syntheses to some extent[56,57], and we are seeking products which distinguish one cell type from another. (The astrocyte appears to be an exception to the general statement about collagen production[58].) The present range of specialised cells includes lines[59] which produce hormones[3,10,15,22,26-28,32-36,40] or are hormone-dependent[17,21,27,38,47]. Among other specialised products are cartilage[6,30], bone[30], myosin[12,16,19,39], melanin[4,14], casein[38], lactose[38], bilirubin[25], liver-specific enzymes[23,37], and histamine[28]. Specialised products which are not hormones are of interest in this text only in so far as their synthesis or excretion can be manipulated by hormone action.

The terms 'cell culture' and 'tissue culture' are used interchangeably in the literature; the former term is preferred here, since 'tissue culture' implies some supra-cellular organisation. 'Organ culture' is used in this text where organised pieces of tissue, though not necessarily a whole organ, form the culture[30,31]. This distinction becomes important where proliferating and non-proliferating cells of different types coexist in culture and the cell type responsible for the synthesis of a specific product must be identified[55].

'Cell line' denotes a cell population which has an apparently infinite life span *in vitro*; 'strain' designates a population having a finite span[59]. Equivalent terms are 'established cell line' and 'primary cell line', respectively[60]. Finally, the tradition will be honoured which equates 'in culture' with '*in vitro*', reserving '*in vivo*' for situations where intact animals are involved.

Specific aspects of cell and organ culture will be covered in separate sections which will deal with problems peculiar to the method of, and pertinent to, hormone studies in culture.

10.2.2 Culture techniques

10.2.2.1 Disaggregation

The three main techniques for the disaggregation of tissues are the use of proteolytic enzymes, the use of chelating agents, and mechanical separation. The choice of agent is dictated both by the tissue and by the procedure planned after disaggregation.

The proteolytic enzyme trypsin[61] is the most commonly used disaggregating agent[60]. Crude preparations of trypsin contain impurities, such as elastase in small amounts[62], as well as bacterial and fungal spores, and different batches are of variable potency. Collagen-rich tissues are resistant to its action. These minor problems are solved by filtration, by standardisation of batches, and by substitution of another agent in the case of resistant tissues[60].

A more serious problem is its effect in destroying certain receptors located at the cell surface[63,64]. Trypsin, for example, ablates the insulin receptor site(s) of lipocytes; this renders these cells insensitive to insulin[63]. In this case the substitution of collagenase for trypsin solves the problem[65]. Collagenase is ineffective, however, in other tissues, such as adrenal, pituitary and corpus luteum[64], where trypsin also destroys surface receptors. Fortunately, tissue disaggregation by trypsin precedes receptor destruction and the judicious addition of lima bean trypsin inhibitor[60] at an appropriate time protects the receptors[64]. This was demonstrated experimentally when fresh isolates of cells from the cortex of the adrenal gland and similar isolates stimulated *in vitro* by the tropic hormone ACTH were compared for levels of hormone production[64]. By use of the precautions listed above cells which were viable, physiologically equivalent to cells of the intact organ, and retaining the trypsin-sensitive surface receptors were obtained. Serum, which also inactivates trypsin[60], was unsuitable for such adrenal cortex isolates[64], because it contains (a) ACTH-binding substances[51,64], (b) an indefinite assortment of hormones[48], and (c) other interfering factors[48,51,64]. Certain workers have used the mixture of proteolytic enzymes, viokinase, for disaggregating adrenal and pituitary cultures[3,27].

The contamination (even of crystalline preparations of trypsin) by ribonuclease[66] is a hazard if cell fractionation (and therefore cell disruption), following trypsinisation, is to be undertaken. In such cases, appropriate quantities of sodium dodecyl sulphate (SDS) serve to inactivate the ribonuclease[67].

Chelating agents, such as ethylene diamine tetra-acetate (versene, EDTA) act by binding divalent cations such as Ca^{2+} and Mg^{2+} at the cell surface; these ions are essential for the adhesion of certain cells. The chelating agents are chemically stable and withstand autoclave temperatures. Unlike proteolytic agents, they do not destroy surface receptors. Unfortunately, they act on a limited range of cells and are occasionally cytotoxic. They are useful where a culture contains cells with differential sensitivity to chelating agents[68].

The third form of disaggregation—mechanical separation—has none of the above disadvantages, but it may cause cell damage, with release of toxic intracellular substances, such as lysosomal hydrolases[69,70]. It has also been shown[64] that dead cells adsorb many substances (apparently non-specifically) including ACTH[64].

Primary insect cultures often carry a chitinous external coat; this coat is digested by a group of enzymes including chitinase[60]. For secondary cultures[60], trypsin is satisfactory. Culture methods for insect cells are improving rapidly[71].

In summary, the method of disaggregation and the type of damage that may result from it merit attention.

10.2.3 Constituents of growth media

10.2.3.1 General review

The growth of cells in culture, especially in the case of primary cultures and normal cells[60], demands an environment resembling that of the tissue of origin. In the early days, lymph clot, and later, plasma clot with embryo extract were used[7,9,11,54]. These are now generally superseded by liquid synthetic media, of standardised composition, and usually supplemented by serum. Three such media often used by investigators studying hormones *in vitro* are Eagle's Minimum Essential Medium (MEM)[72], Ham's F10[73] and Medium 199[74]. Their constituents are detailed below (Table 10.1); they are

Table 10.1 The constituents of Minimum Essential Medium (MEM), F-10 and 199 (From Paul[60], by courtesy of E & S Livingstone.)

Compounds	mg/l		
	MEM	F–10	199
L-Alanine		8.91	25.00
L-Arginine.HCl	126.98	211.00	70.00
L-Asparagine.H$_2$O		15.00	
L-Aspartic acid		13.30	30.00
L-Cysteine.HCl.H$_2$O		35.09	0.11
L-Cystine	24.00		20.00
L-Glutamic acid		14.70	66.82
L-Glutamine	292.00	146.20	100.00
Glycine		7.51	50.00
L-Histidine.HCl.H$_2$O	41.88	22.97	21.88
L-Hydroxyproline			10.00
L-Isoleucine	52.00	2.60	20.00
L-Leucine	52.00	13.10	60.00
L-Lysine.HCl	72.48	29.30	70.00
L-Methionine	15.00	4.48	15.00
L-Phenylalanine	32.00	4.96	25.00
L-Proline		11.50	40.00
L-Serine		10.50	25.00
L-Threonine	48.00	3.57	30.00
L-Tryptophan	10.00	0.60	10.00
L-Tyrosine	36.00	1.81	40.00
L-Valine	46.00	3.50	25.00

Component				
p-Aminobenzoic acid				0.05
Ascorbic acid				0.05
d-Biotin			0.024	0.01
Calciferol				0.10
Cholesterol				0.20
Choline chloride		1.00	0.698	0.05
Folic acid		1.00	1.32	0.10
i-Inositol		2.00	0.541	0.05
Menadione				0.01
Niacin				0.025
Niacinamide			0.615	0.025
Nicotinamide		1.00		
d-Pantothenate.Ca		1.00	0.715	0.01
Pyridoxal.HCl		1.00		0.025
Pyridoxine.HCl			0.206	0.025
Riboflavin		0.10	0.376	0.01
Thiamine.HCl		1.00	1.012	0.01
α-Tocopherol phosphate.Na$_2$				0.01
Tween 80				5.00
Vitamin A acetate				0.1147
Vitamin B$_{12}$			1.36	
Adenine sulphate				10.00
Adenosine triphosphate.Na$_2$				10.00
Adenylic acid				0.20
Cupric sulphate.5H$_2$O			0.0025	
Deoxyribose				0.50
Dextrose		1 000.00	1 100.00	1 000.00
Ferric nitrate.9H$_2$O				0.10
Ferrous sulphate.7H$_2$O			0.834	
Glutathione				0.05
Guanine.HCl				0.30
Hypoxanthine			4.08	0.30
Lipoic acid			0.20	
*Phenol red		10.00	1.20	20.00
d-Ribose				0.50
Sodium acetate.anhydrous				50.00
Sodium pyruvate			110.00	
Thymidine			0.727	
Thymine				0.30
Uracil				0.30
Xanthine				0.30
Zinc sulphate.7H$_2$O			0.0288	

* Phenol red is omitted in certain analyses where its emission spectra would cause interference[31,45].

combined with a balanced salt solution to maintain pH, osmotic pressure and to supply inorganic ions[60]. The simplest medium, MEM, was developed from Eagle's Basal Medium to supply minimal requirements for the growth of mouse and human cell lines with respect to essential amino acids and vitamins[75]. The other two media probably contain an excess of some constituents as well as some unnecessary ingredients. The continuing search for a a chemically defined medium capable of permitting cell growth without supplementation by serum has produced some very complicated mixtures, such as NCTC 109 (Table 10.2)[76]; several cell lines are now adapted to grow

Table 10.2 Constituents of NCTC 109 (From Paul[60], by courtesy of E & S Livingstone.)

Amino acids	mg l^{-1}	Other components	mg l^{-1}
L-Alanine	31.5	Coenzyme A	2.5
L-a-Amino butyric acid	5.5	Deoxyadenosine	10.0
L-Arginine HCl	31.2	Deoxycytidine HCl	10.0
L-Asparagine	8.0	Deoxyguanosine	10.0
L-Aspartic acid	9.9	Dextrose	1 000.0
L-Cysteine HCl	259.9	Diphosphopyridine nucleotide,	
L-Cystine	10.5	tetrahydrate (DPN.4H$_2$O)	7.0
L-Glutamic acid	8.3	Flavin adenine dinucleotide	
L-Glutamine	136.0	(FAD)	1.0
L-Glycine	13.5	D-Glucosamine	3.2
L-Histidine HCl.H$_2$O	26.7	D-Glucuronolactone	1.8
L-Hydroxyproline	4.1	L-Glutathione	10.1
L-Isoleucine	18.0	5-Methylcytosine	0.1
L-Leucine	20.4	Phenol Red	20.0
L-Lysine HCl	38.4	Sodium acetate.3H$_2$O	83.0
L-Methionine	4.4	Sodium glucuronate	1.8
L-Ornithine	7.4	Thymidine	10.0
L-Phenylalanine	16.5	Triphosphopyridine nucleotide	
L-Proline	6.1	monosodium salt (TPN)	1.0
L-Serine	10.8	Uridine-5'triphosphate, tetrasodium,	
L-Taurine	4.2	tetrahydrate (UTP)	1.0
L-Threonine	18.9		
L-Tryptophan	17.5		
L-Tyrosine	16.4		
L-Valine	25.0		
Vitamins			
p-Aminobenzoic acid	0.125	Nicotinamide	0.063
Ascorbic acid	49.900	Nicotinic acid	0.063
D-Biotin	0.025	Pyridoxal HCl	0.063
Calciferol	0.250	Pyridoxine HCl	0.063
D-Ca-Pantothenate	0.025	DL-a-Tocopherolphosphate Na$_2$	0.025
Choline chloride	1.250	Thiamine HCl	0.025
Cocarboxylase	1.000	Tween 80	12.500
Folic acid	0.025	Riboflavin	0.025
i-Inositol	0.125	Vitamin A	0.250
Menadione	0.025	Vitamin B$_{12}$	10.000

without serum in this medium[76,77]. So far, however, no medium has been devised which will permit ordinary cells or even the majority of cell lines to grow without serum.

10.2.3.2 Supplementation with serum

This section will concentrate upon problems inherent in the supplementation of growth media with serum.

Serum is a potential and actual source of many forms of contamination—bacteria, mycoplasma and viruses being the most obvious. The gravity of this problem for users of commercial sera is shown by several recent surveys[69,78-80].

High levels of fatty acids in serum (resulting from release of lysosomal lipases[70] from white cells included in serum as a result of poor fractionation) are cytotoxic[69]. Mycoplasma (PPLO) interfere with growth, cause chromosomal abnormalities in infected cells[81] (possibly due to the action of the associated viroid[82]) and have a requirement for arginine greatly in excess of that of the cells[60,83]. Mycoplasmal infections are often difficult to detect and next to impossible to cure[60,84]; their presence has vitiated many biochemical investigations[60], and might seriously hamper any *in vitro* studies on hormones. A simple chromatographic technique for detecting the high levels of uridine phosphorylase found in cultures contaminated by mycoplasma has recently been reported[84a]. This appears to offer a much better detection method than any which have been available to date. Serum also contains variable quantities of hormones, hormone-binding proteins, labile amino acids, etc.[48,50,51,64,85]. Glutamic acid in serum, for example, can serve as a course of glutamine[48,86]. Serum is also very difficult to remove from the surface of cells[48,64,85].

10.2.3.3 The effect of different media on cell function

Many workers using different cell lines or strains have noted that the composition of the media employed affected the functions expressed by the cells[17,48,85-87a]. In one experiment[86], the L-M mouse fibroblast line[87,88], which had been carried in peptone 199 medium, was transferred to 2 × E (Eagle's Basal Medium containing twice the usual amount of essential amino acids, vitamins and glutamine). Although their utilisation of essential amino acids was similar to that of L-M cells which had been carried continuously in 2 × E, their synthesis and re-utilisation of non-essential amino acids differed consistently and changed as the cells adapted to 2 × E[86]. The rate of proliferation of L-M cells adapted to 2 × E medium is actually decreased by the addition of non-essential amino acids to the medium[86].

All this argues for stringent standardisation of all substances used in cell culture.

10.2.3.4 Contamination

The danger of contamination inherent in the use of serum has been discussed. Details of procedures for determining if contamination exists and for eliminating or preventing it are detailed in general textbooks on cell and organ culture[60].

Cross-contamination between cells from different culture types is a continuing hazard. For example, a recent report[89] warns workers in the field that the widely used CHB glial cell line[90] is not of human, but of rat origin, despite the fact that this line was cultured originally from a human tumour. There are now several methods for detecting such contamination[60,89,91-93]; they are cumbersome, but reliable. Infection of the workers by mammalian tumour viruses is a potential hazard[19,94].

10.2.4 Culture apparatus

Details of the different types of culture apparatus in use are given in standard textbooks. A great convenience has been the development, in the past few years, of sterile plasticware, usually polystyrene, for use in cell culture. This plasticware does suffer irradiation damage upon exposure to ultraviolet light, which makes the plastic cytotoxic; sometimes the plasticware itself is cytotoxic. (It is also soluble in organic esters, and to some extent in dimethylsulphoxide (DMSO); the latter is used in the preservation of cells by freezing.) Microbiological plasticware cannot be substituted for the culture of mammalian cells. Cells attach poorly to it, apparently because it has the wrong surface charge. Microbiological plasticware was used to advantage in the selection of rat pituitary cells capable of growing in suspension culture[95]. The resultant line retained the ability to synthesise growth hormone and prolactin[95]. The above instances of cytotoxicity can easily give completely spurious findings.

Teflon and polyethylene are used for many containers, as well as to cover magnetic stirrers. Teflon is preferable because polyethylene has greater surface-binding affinities for steroids such as ACTH[64].

10.2.5 Procedures for culture and selection of cells

10.2.5.1 Primary culture of cells

The *in vitro* cultivation of cells from primary tissue, especially cells of the more fastidious types, is as yet more an art than a science. For those who wish to work with human primary tissue, access to suitable material can be a major problem. There are many advantages in working with animal sources. The problems of access and to some extent, of standardisation are minimal; the availability of inbred animals makes possible subsequent re-injection of cultured cells without histocompatibility rejection problems[96]. The carcinogenic potential of cells grown in culture can easily be tested by their injection into autologous hosts or into immunologically incompetent, usually newborn, animals[60].

(a) *Viruses in cell cultures*—Another aspect, however, which should be considered is the extent to which the tissue to be used harbours viruses, tumourigenic or otherwise. It is now accepted, for example, that virtually all murine cells carry tumour-virus information. Other species carry some tumour viruses, though not to the extent seen in the mouse; the same may hold true for human tissues[19,94,97]. The transplantable C1300 neuroblastoma of mouse origin[1], from which Sato derived his neuroblastoma lines, produces many viruses[98] resembling the RNA murine tumour virus group. This in no way interfered with their value as models for studying the production of enzymes which synthesise and degrade neurotransmitters[1,99], their membrane potentials[100], or excitability[100]; their usefulness as models of RNA synthesis by neuroblastoma would be quite limited, however.

The worker with mammalian cultures should remember that he may be a potential host[19,94,97] for the tumour viruses which infect the cells upon which he experiments.

(b) *Organ cultures*—There are two types of primary cultures: (i) organ (organoid) cultures where architectural integrity is preserved[60,101]; and (ii) dispersed cultures[60]. It is technically difficult to supply adequate nutrition to all parts of an organ culture. They have, however, served to delineate some of the functions of whole organs, such as the rat pineal. Here two groups of investigators have used quite different culture techniques. One group, for example, kept the bisected pineals sealed in tubes containing rich medium on a roller drum; the pineal sections were attached to the tube by chick embryo extract and plasma[31,102-104]. The pineal cells were histologically and functionally intact throughout the experimental period (up to 10 days). Another group of investigators[105,106] continuously perfused the organ culture, set on a grid, with a mixture of CO_2 and O_2; albumin fraction V in minimal quantities was used instead of serum. This maintained the nutritional equilibrium and gaseous exchange very well. Non-specific adsorption by serum and embryo extract of, for example, steroids and other hormones was avoided[48,51,64,85,86]. Organ cultures of oestrogenised chick oviduct have been used to demonstrate the induction by progesterone of avidin, the organ-specific protein of the goblet cells[47,107].

(c) *Dispersed cultures*—This is the more usual type of primary culture. Problems caused by dispersing agents and growth media have been considered above. Another common hazard is the overgrowth of specialised cells by fibroblasts; this can be prevented by alternate animal passage and *in vitro* culture passage—a technique developed by Sato[1,3,26,27,55] and considered below. A different method has also proved effective in holding overgrowth by fibroblasts in check[17]; thus collagenase was used both in the initial primary cultures of breast and at intervals for 1–3 day periods during subsequent cultures. Collagenase is not inactivated by serum or divalent cations[17], and destroys the collagen produced by the fibroblasts in the rich growth medium (MEM with 20% foetal calf serum, insulin, prolactin and cortisol)[17]. The epithelial cells of the breast then outgrow the fibroblasts. By this means established lines of functional, hormone-dependent mammary epithelium have been established from mice, rats and hamsters[17]. Pure collagenase must be used, as the impure form contains toxic substances[17].

10.2.6 Hormone-producing, hormone-dependent, and hormone-responsive cultures

10.2.6.1 Sources of primary tissue

Reference has already been made to organ culture; the growth of hormone-producing and hormone-dependent cells *in vitro* was greatly facilitated by the tumour 'bank' created by Furth[108-110]. Normal endocrine tissue is difficult to maintain in culture without attendant loss of specific function[60,109]. Furth demonstrated that continued stimulation of a specific target organ by excessive amounts of its tropic hormone(s) was an effective means of inducing hormone-dependent tumours of that organ[108-110]; minimal doses of carcinogen were sometimes used in addition[109,110]. This principle could be applied to many different tissues. For example, mammotropic and growth hormones, supplemented by subthreshold carcinogen, induced hormone-dependent breast carcinoma of the rat[110]. The induction process could be very slow—thyroid cancers developed after years of overstimulation by thyrotropic hormone[109]. A well known application, developed from the technique of Biskind and Biskind[111], gives a reliable method for producing hormone-dependent tumours of the gonads. In this method, ovary or testis of the newborn was transplanted into the spleen of adult, castrated female or male rats, respectively[5,27,108,109]. Gonadotrophic hormone from the host pituitary caused the transplant to grow and to produce hormones; these were inactivated in the liver. Interruption of the feed-back mechanism to hypothalamus and pituitary caused an increase in the levels of gonadotrophic hormone releasing factor (GRF) and of gonadotrophic hormone, and gonadal tumours developed[5,108,111]. All such tumours were initially hormone-dependent in the host, but often escaped from this dependence[109] while continuing to produce organ-specific products. Amongst the tumours induced by the Furth technique were pituitary tumours; their cells produce thyroid stimulating hormone (TSH)[109] and are inhibited by thyroxin; the cells are being adapted to *in vitro* culture[27]. There were also steroid-dependent, androgen and or oestrogen-dependent tumours in hamsters[112]. Tumours induced by this method provided the primary tissue for Sato's adrenal, pituitary and ovarian lines[3,5].

10.2.6.2 Cell culture techniques for selecting hormone-dependent cultures

In cultures from 'Furth' tumours the loss of hormone-dependence has two main causes and leads to two different types of culture. The cause may be (1) the overgrowth of fibroblasts[3,55] or (2) the emergence of a mutant cell independent of tropic hormone, but still synthesising its organ-specific product(s)[27]. The former problem is elegantly solved by virtue of the fact that normal fibroblasts will not grow after injection into the animal whereas tumour cells do[3]. Several cycles of injecting cell cultures (containing a mixture of fibroblasts and tumour cells) into the inbred strain of animals from which the cells were derived, and subsequent reculture of the tumour which grew in the animal usually give a pure culture of specialised tumour cells[3,26,27,55].

A procedure to select against hormone-independent cells is based on the penicillin selection technique for bacterial auxotrophs, as adapted for mammalian cells by Puck and Kao[46,113]. Bromodeoxyuridine (BUdR) is added to the medium and incorporated by proliferating cells into DNA in place of thymidine. The carbon–bromine linkage is sensitive to blue light[46,113]; and cells which have incorporated BUdR are killed by irradiation at appropriate wavelengths. A mixed culture of hormone-dependent and hormone-independent cells is first placed in minimal medium devoid of the specific tropic hormone(s). When all traces of the tropic hormone(s) are gone, hormone-dependent cells will not synthesise DNA, but do not die; hormone-independent cells continue to proliferate. At this stage BUdR is added to the medium for a period of one to two cell doubling times; the hormone-independent cells incorporate BUdR and die following exposure to blue light. The cultures are then returned to normal growth medium supplemented with appropriate hormones and, ideally, a pure culture of hormone-dependent cells is obtained. Subsequent cloning of known specialised cells using established single cell cloning methods, but without collagen or tropocollagen layers[114], is usually successful[27].

10.2.6.3 The selection and potentiation of specialised cells in culture

The method of alternating animal passage with *in vitro* culture of animal tumours was also used to select the faster growing cells of such cultures from adrenal and pituitary[3,115]. Cultures were harvested as soon as they reached confluence and were injected into appropriate animals. Resultant tumours were removed as soon as they became palpable and were immediately re-cultured. This cycle was repeated several times. The amount of organ-specific hormones which were synthesised increased with each cycle of injection and re-culture; eventually a steady level of synthesis was attained. This corresponded in many instances to that level which is considered to be maximal for the respective pituitary or adrenal cells in the intact organ[27,40,115]. This also holds true for the pituitary line which has been adapted to grow in suspension culture[95].

10.2.6.4 The growth of normal hormone-responsive cells in culture

Here, I wish to dwell briefly upon the selection techniques that led to the establishment of normal liver parenchymal cell strains[116,117]—a notable achievement. Recent work[118] suggests the possibility that cancer of the parenchymal cells of liver (hepatoma) may be causally related to long-term therapy with androgenic-anabolic steroids. It will now be possible to test this hypothesis in cultures of normal liver cells.

Both parenchymal and stromal cells of liver require arginine for growth. The former can, in some instances, derive arginine from ornithine; stromal cells cannot synthesise arginine from ornithine and die in its absence. This fact was used to select for parenchymal cells in primary cultures of normal liver in arginine-free, but ornithine-supplemented medium[116,117].

10.2.6.5 Routine testing of specific cell function *in vitro*

Tissue-specific function of cultured cells has been shown by assays of their products in *in vivo* as well as *in vitro*[3,27,40]; specific antibodies to hormones have been developed for immunoassays. The work of Sato and others has ushered in an era where the properties of the cell in culture will indicate not only the *source* of the primary tissue but also the specific *function* of the cell. Neuroblastoma cells[1], pituitary and adrenal lines secreting the cognate hormones[3,27,40,85,115], as well as glial and ovarian lines[2,5,18] are now available.

10.3 PROCEDURES OF POTENTIAL VALUE FOR THE *IN VITRO* STUDY OF HORMONES

This section will indicate several areas where recent work with cultured cells has provided the means for studying effects of gene dosage, complementation, and the complex of interactions described as control mechanisms. That these bear upon future studies of hormone expression is self-evident.

(1) Several sources of somatic cell mutants are now available—'spontaneous'[119-122], nutritional[45,86,113,123], and drug resistant[43,46,113,124].

(2) The general application of somatic cell hydridisation began with the Littlefield procedure[125], in which two mutant parents, each having a different drug resistance marker are used; this was followed by the Davidson procedure[126], where one parent is normal and only one is a drug-resistant mutant; the technique was next simplified in the Harris and Watkins method[127,128] by the use of the fusing virus Sendai; two further modifications by Sato[27] permitted the use of normal parent cells.

These techniques prepare the way for extensive genetic analysis of mammalian cells. The cells from many phyla and from hundreds of species are now in culture. A great variety of cells is available in the American Type Tissue Culture Collection[129] as well as from commercial sources[60].

10.4 STUDIES OF HORMONE SYNTHESIS AND ACTION *IN VITRO*

10.4.1 Early studies of hormone synthesis *in vitro*

The use of cell culture to study hormone synthesis began with the work of Gey[11] in the late thirties. He showed that human placenta and placental tumour in long-term culture secreted a substance which induced ovulation in rats[11]; the response to the substance was identical with that obtained from injecting the urine of pregnant women. Gey identified the substance as prolan (now chorionic gonadotrophin, CGTH) and the cell of synthesis as the cytotrophoblast of placenta. He also presented preliminary evidence that anterior pituitary cells in culture produced a similar substance[11]. Subsequent studies have confirmed the accuracy of his observations[22,130-134].

10.4.2 The mammary gland-differentiation and cancer

Cancer of the breast is among the most common of human malignancies[135,136,136a]; the high resultant mortality and morbidity, even in cases of early diagnosis, demonstrated the inadequacy of our understanding of this cancer, or rather, of this complex of cancers[135-137]. There is evidence that hormones play an integral part in the induction of experimental breast cancer[138,139]; this fact was used by Furth[109,110] to induce hormone-dependent breast cancer (see above). The association between cancer of the mouse breast and tumour viruses is well established[140,141]; recent work suggests that a similar relationship may obtain for certain human breast cancers[142,143].

Furthermore, there is evidence that the degree of differentiation of breast tissue determines which type of mammary tumour virus may proliferate[52]. The contribution which cell culture will make to understanding the difference between normal and cancerous breast tissue, and between hormone-dependent and autonomous cells from breast cancer, may well be a very significant practical application of the culture of specialised cells *in vitro*.

Cell culture has already contributed substantially to our understanding of the role played by hormones in the differentiation of the breast[17,38,136a,137,144], and of the products characterising the cells at different stages of development[38,144]. In the following paragraphs, cell culture work shall be discussed in the context of hormone-dependent breast differentiation, the relationship of cell differentiation and the expression of mature virus. Brief mention will be made of the autonomy of breast cancer and hormone production by cancer cells.

10.4.2.1 Hormone dependence for differentiation of normal breast tissue

Topper[38] studied the differentiation of adult mouse breast using organ cultures from female mice in midterm of pregnancy. He showed that insulin, hydrocortisone and prolactin are essential for the formation of milk in such cultures of breast, and defined the quantity and the temporal sequence in which these hormones were required for the induction of lactation *in vitro*. Changes in morphology and the synthesis of organ-specific products (such as casein, the A- and B-proteins of lactose synthetase and a galactosyl transferase) were monitored. All three hormones were required to obtain maximal secretion of these products and the hormones were effective only in the sequence—insulin, hydrocortisone and prolactin. Differentiation to the lactational state took place in three phases. Insulin was mitogenic, but caused no ultrastructural changes. Hydrocortisone induced extensive development of granular endoplasmic reticulum (site of formation of proteins and carbohydrates) only after prior exposure of the culture to insulin. The synthesis of specific products, such as casein, required the continued presence of insulin as well as the addition of prolactin. This final stage of differentiation was characterised ultrastructurally by major changes in the position of intracellular structures and the development of secretory granules containing the specific substances.

This form of organ culture has also provided a very sensitive assay for prolactin[145]; it can be used to detect 10–20 μg/ml (0.28–0.56 milliunits/ml) of prolactin and has been employed to assay, for example, prolactin synthesis from cultures of anterior pituitary cells. Other studies of the induction of differentiation in cultures of breast have confirmed and extended Topper's findings[52,137,144].

10.4.2.2 Cell differentiation and the expression of mammary tumour virus (MTV)

The interaction of three factors—genetically susceptible tissue, hormones, and mammary tumour virus (MTV)—is known to induce adenocarcinoma of the breast in mice[140,141]. McGrath[52] studied the effect of insulin and hydrocortisone on short-term dispersed cultures of BALB/C$_3$H mouse breast tissue. He observed the formation of three-dimensional structures, forming domes in culture. Prolactin had no effect on these structures, which resembled clusters of alveolar cells. The development of these domes was associated with the appearance of mature MTV of B-type. These results supported previous observations that the replication of MTV of B-type from organ cultures of breast, known to be infected with MTV, depended upon the physiology of the tissue[146].

10.4.2.3 Autonomy of breast cancer and synthesis of hormones by breast cancer cells

The investigation of these aspects of breast cancer has just started. The growth of rat mammary cancer R323OAC is resistant to ovariectomy of the host. The cells lack cytoplasmic 17-β-oestradiol-binding protein (EBP)[147], which normally complexes with the oestrogen taken up by the cell and which carries it to the nucleus, where the oestrogen combines with the chromatin. (The chromatin of the R323OAC mammary cancer cells can bind oestrogen if the specific EBP component is supplied).

Other facets awaiting study in culture are the ability of breast cancer cells to synthesise enzyme systems not expressed in ordinary breast tissue, and often found only in endocrine organs. It is reported that some cancer cells can convert steroids such as dehydroepiandrosterone sulphate (from the adrenal) to the metabolically more active androstenedione[148]; that they can synthesise oestrogens from testosterone[148,149]; that they can form pregnenolone, the common intermediate[149a], from cholesterol[148]; and that they can form 5α-dihydrotestosterone from testosterone[150], a reaction catalysed by the enzyme 5-α-reductase (perhaps also found in normal breast and benign tumours of breast[151]).

The availability of techniques for growing mammary tissue in differentiated form as organ cultures[38,144] and as epithelial sheets, attached to the surface of the culture vessel[17,52] will facilitate the study of hormone–tissue interactions, which are important in the development of cancer of the breast. It is clear that mammary cancer stems from many different causes and can rise in different types of cells. It is probably realistic to expect that it will soon be possible to develop an effective assay system for selecting the therapy best for an individual case, somewhat analogous to the routine testing of pathogenic organisms for sensitivity to antibiotics.

10.4.3 Culture of the pituitary gland

The development of functional, cloned lines of cells from tumours of mammalian anterior pituitary[3,10,26,95,113] has facilitated the study of hormone production. The availability of such lines has simplified studies of the site of action of the tropic hormone. Adrenocorticotropic hormone (ACTH), gonadotropic hormone (GTH), and thyrotropic or thyroid stimulating hormone (TSH) appear to be synthesised by different cell types[3,26,95,115]. Whether growth hormone (somatotropic hormone, GH) and prolactin (mammotropic hormone) are produced in the same cell is not yet certain[95,115,152]. It has been claimed that GH and prolactin activity represent different functions of the same molecule[115]; another view is that there are in fact two differrent hormones[152].

There seems to be a complicated relationship, in at least some instances, between synthesis of GH and repression of prolactin activity and vice versa. This has been studied in cultures of pituitary lines[115]. Tashjian has shown that hydrocortisone stimulates GH production sevenfold and decreases prolactin production three- to five-fold in cloned bifunctional pituitary tumour cells; hydrocortisone exerts no effect on cell doubling time[115]. (Cycloheximide and puromycin suppressed the production of GH as well as the stimulatory response to hydrocortisone in this system[115]). Conversely, extracts from several tissues including hypothalamus increased the production of prolactin six to nine times in this same system, and decreased that of GH by 50%[115]. Cultures of normal pituitary and of some tumours of pituitary show a changing pattern of GH synthesis and release with time[153]. Also the amount of prolactin and growth hormone in the normal pituitary varies enormously depending upon the age, sex and physiological state of the animal. Prolactin levels, for example, are very high at certain stages of the oestrous cycle, in lactation, and after section of the pituitary stalk[154].

Recent work has demonstrated that the hypothalamus synthesises several tropic hormone releasing factors[155-157] as well as inhibitory factors preventing secretion of hormones[158]; preliminary work with corticotropin releasing factor (CRF) shows that it is effective with suspensions of pituitary cells[64]. The use of such factors may resolve the problem of the multifunctional capability of pituitary cells for GH and prolactin synthesis. They may also be used to test whether cloned cells which appear to produce only one hormone can be induced to synthesise other pituitary hormones. (See Chapter 3 by Schreiber for further details about hypophyseal–pituitary hormones).

10.4.4 Culture of the adrenal gland

The adaptation of ACTH-sensitive adrenal tumours of mice to culture, resulting in the establishment of functional, cloned lines of cells[3,26,40,159,159a] has facilitated greatly the study of the effects of ACTH on the adrenal cortical cell. *In vitro* studies have shown that ACTH interacts with the cell surface exclusively[159]. The evidence for this is that an eicosapeptide analogue of ACTH, covalently coupled to cellulose, induced steroidogenesis in cultures of adrenal cells. The cellulose–eicosapeptide ACTH complex was too large to be engulfed by the target cells; control experiments demonstrated that filtrates of the insoluble hormone–cellulose complex could not promote steroidogenesis; it was therefore deduced that the complex remained intact during the experiment[159]. Subsequent investigations in culture demonstrated that the steroidogenic effect of ACTH could be mimicked by the intracellular 'second messenger'[160,161] cyclic AMP or its analogue, dibutyryl cyclic AMP[162,163]. Kowal claims that ACTH can cause certain alterations in the metabolic activity of the adrenal cell which are independent of the adenyl cyclase system[164].

10.4.5 Culture of the pineal gland

10.4.5.1 Historical review

The function of the pineal gland has been a source of controversy for many years. One school argued that the pineal was a vestigial remnant devoid of function and another attributed to it many functions, amongst which the capacity to induce either gonadal atrophy or hyperplasia is perhaps the most striking. The arguments are reviewed by Reiter and Fraschini[165].

10.4.5.2 Aspects of the biochemistry of the pineal gland

The pineal is a readily accessible, easily separable organ, and a major storehouse[102-106] of the neurotransmitter, serotonin[166-170]. It is the unique site of synthesis of melatonin[102-106] (which has antigonadotropic activity[171,172]) and of the enzyme, hydroxyindole-*O*-methyltransferase[102-104]. Fluctuations in the levels of serotonin and of melatonin may be important in certain diseases and behaviour disorders[173-177]. The pineal gland is innervated solely by the sympathetic neurones of the superior cervical ganglia[165,178,179]; these are light-sensitive and release norepinephrine upon stimulation[102]. It has been suggested that norepinephrine controls the respective levels of serotonin and melatonin by affecting the activity of the enzyme, *N*-acetyl methyltransferase[179,180]. The biosynthetic pathway (in the intact animal) is tryptophan → 5-OH tryptophan → serotonin (5-OH tryptamine) → *N*-acetylserotonin (catalysed by *N*-acetyl methyltransferase). The last step in the pathway which yields melatonin is catalysed by hydroxyindole-*O*-methyltransferase[102-104]. The pineal can perform this complete biosynthesis in organ culture. It has thus been possible to correlate *in vitro* and *in vivo* studies.

(a) In vitro *experiments*—The addition of norepinephrine to organ cultures of pineal glands (removed from the animal during daylight hours) stimulated the uptake of ^{14}C tryptophan, the synthesis of serotonin, of large amounts of *N*-acetyltransferase, of hydroxyindole-*O*-methyltransferase and of melatonin, as well as the incorporation of labelled amino acids into protein[102]. The substitution of dibutyryl cyclic AMP (DBcAMP) for norepinephrine induced synthesis of the first four products, but there was no significant accumulation of labelled protein[103]. β-adrenergic, but not α-adrenergic blocking agents prevented the effects of norepinephrine (a β-adrenergic neurotransmitter); neither agent prevented the DBcAMP effect[104]. This suggested that some, but not all, of the effects of norepinephrine on the pineal were mediated through adenyl cyclase.

(b) In vivo *experiments*—*In vivo* experiments showed very high levels of serotonin and low levels of the melatonin-synthesising enzymes as well as of melatonin during daylight hours. A rise in *N*-acetyltransferase levels to 15–70 times that of daylight levels[106,178,180] with a corresponding rise in melatonin and a drop in serotonin levels occurred during darkness; i.e. there was an inverse relationship between the levels of serotonin and melatonin in intact animals.

The discrepancy between the *in vitro* and the *in vivo* findings has been partly explained by experiments with pineals from dark-adapted rats, which had been exposed immediately before sacrifice to short periods of bright light (0.25–10 min). The high nocturnal levels of N-acetyltransferase dropped sharply (in 10 min from a level of 5 000 to 400 units) during this short period[179]; the mechanism controlling this rapid change is not as yet known. Thus, combining the *in vivo* and the *in vitro* studies helped clarify one aspect of the function of the pineal.

10.4.5.3 The use of organ cultures of the pineal in the study of the mechanism of drug action

The pineal system may be of value in studying the mode of action of certain drugs. *In vitro* studies with organ cultures indicate that the hallucinogen, harmine, stimulates the production of melatonin, and decreases the level of serotonin, by increasing the rate of synthesis of N-acetyltransferase[106,181].

Lysergic acid diethylamide displaces serotonin effectively from its specific receptor sites[167]; this has not yet been studied in the pineal system. The morphine-cocaine group of drugs is known to affect serotonin levels; the experimental findings have been summarised recently[182]; different workers report conflicting observations; the use of the pineal system could be of value in resolving this conflict.

10.4.6 Culture of cells from the central nervous system

The successful establishment of lines of mouse, rat, and human brain cells has permitted determination of some of the specific functions which distinguish astrocytes and neuroblasts from each other as well as from other cell types[1,2,18,99,100,183,184]. Although only a limited number of lines of cells derived from the central nervous system is available at present, this represents a great advance over the methods of histological and histochemical examination of a decade ago[185]. The neuroblasts have probably been the most extensively studied. Recent work defines three distinct genetic types[99] derived from the neuroblastoma lines of Sato's[1]. These include clones which are (1) cholinergic, (2) adrenergic, and (3) both cholinergic and adrenergic. All clones produce large quantities of cholinesterase; some also synthesise an amount of histamine equivalent to that found in whole brain extracts. Other work with these neuroblastoma cells has demonstrated membrane excitability[100].

Morphologic differentiation in culture can be modulated by certain substances. Thus DBcAMP can induce axon development of murine neuroblasts *in vitro*[186], whereas colcemid represses it. DBcAMP has been reported to over-ride the colcemid repression[187]; it has been suggested on the basis of this work that DBcAMP may effect reassembly of microtubules dissociated by colcemid. However, other studies using human neuroblasts[188], have shown that DBcAMP induces a remarkable, irreversible hyperplasia of intracellular microfilaments, but that it exerts no effect on microtubules in this system.

Astrocyte cultures *in vitro*[2] have been shown to produce S100 protein, a low molecular weight compound[189], which is present in all species examined, ranging from fish to man, and which is reported to be unique to brain[190]. Astrocytes also synthesise glycerol phosphate dehydrogenase in response to hydrocortisone[184], and carry specific surface transplantation antigens (TSTA)[128,191]. Their content of cAMP is increased after stimulation by catechols such as norepinephrine and by histamine[183]. The response to norepinephrine is maximal in log phase, and prevented by β-adrenergic blocking agents; the response to histamine is the same in all phases of growth and is not blocked by such compounds. It has therefore been suggested that astrocytes carry different surface receptors for these compounds[183].

10.4.7 Miscellany

Several compounds produced by fungi or plants affect the integrity of mammalian membranes and prevent the expression of differentiated function. An example is cytochalasin B, a fungal product[192]. Cytochalasin B interferes with the release of growth hormone from pituitary slices[193], with lymphotoxin production by antigen-activated lymphocytes[194], with specific sorting of embryonic cells in process of reaggregation[195], and with the dispersion of pigment granules[196].

10.4.8 Cultures of cells from insects

Hormones are clearly implicated in the differentiation and sexual maturation of insects[71,197]; many workers have studied these aspects with a view to planning insect control. In the past few years, several lines of insect cells have been developed[198-202] and some standardisation of media and of hormone preparations has been accomplished[71,197-202]. *In vitro* work has extended the information available about hormones and differentiation in insects[71,196]. Insect virsuses have been identified[203,204]. The study of hormone–cell interactions in cultures of insect cells will be of great value in designing, for example, safer methods of insect control[205] than DDT or the eldrin group of pesticides.

10.5 FUTURE PROSPECTS FOR THE STUDY OF HORMONES IN VITRO

10.5.1 General comments

The use of cell culture as a vehicle to study the mechanism of hormone action is in its early stages. The basic techniques needed to develop mutants of mammalian cells or hormone-dependent cultures are now clearly defined and the range of such cultures available should increase very soon. This, in turn, will improve our understanding of cell physiology and of the control mechanisms of the cell, and in consequence, our treatment of disease processes.

The shift of emphasis in molecular biology to the study of mammalian cells should increase the application of investigative techniques which have proven their value in studies of bacterial genetics. Hormone-producing cultures may serve as a source for hormones which are produced in small amounts or are species-specific.

Cell culture will be a very useful tool for studying the mechanism of action of hormones and of control mechanisms associated with hormones.

Acknowledgements

The author would like to thank Mrs M. Goren, Miss D. H. Forster, and Mr N. G. Eig for their painstaking assistance with secretarial and library matters. This work was partly supported by U.S.P.H.S. grant CA-14362 from the National Cancer Institute.

References

1. Augusti-Tocco, G. and Sato, G. (1969). *Proc. Nat. Acad. Sci. USA*, **64**, 311
2. Benda, P., Lightbody, J., Sato, G., Levine, L. and Sweet, W. (1968). *Science*, **161**, 370
3. Buonassisi, V., Sato, G. and Cohen, A. I. (1962). *Proc. Nat. Acad. Sci. USA*, **48**, 1184
4. Cahn, R. D. and Cahn, M. B. (1966). *Proc. Nat. Acad. Sci. USA*, **55**, 106
5. Clark, J. L., Jones, K. L., Gospodarowicz, D. and Sato, G. H. (1972). *Nature New Biol.*, **236**, 180
6. Coon, H. G. (1966). *Proc. Nat. Acad. Sci. USA*, **55**, 66
7. Ebeling, A. H. (1924). *C. R. Soc. Biol.*, **90**, 562
8. Finch, B. W. and Ephrussi, B. (1967). *Proc. Nat. Acad. Sci. USA*, **57**, 615
9. Fischer, A. (1922). *J. Exp. Med.*, **36**, 379
10. Gala, R. R. (1971). *J. Endocrinol.*, **50**, 637
11. Gey, G. O., Seegar, G. E. and Hellman, L. M. (1938). *Science*, **88**, 306
12. Harary, I. and Farley, B. (1960). *Science*, **131**, 1674
13. Harris, A. W. (1970). *Exp. Cell. Res.*, **60**, 341
14. Hu, F. (1965). *Tex. Rep. Biol. Med.*, **23** (*Suppl*), 308
15. Kodani, M. and Kodani, K. (1966). *Proc. Nat. Acad. Sci. USA*, **56**, 1200
16. Konigsberg, I. R. (1963). *Science*, **140**, 1273
17. Lasfargues, E. Y. and Moore, D. H. (1971). *In Vitro*, **7**, 21
18. Pontén, J. and Macintyre, E. H. (1968). *Acta Pathol. Microbiol. Scand.*, **74**, 465
19. McAllister, R. M., Nicolson, M., Gardner, M. B., Rongey, R. W., Rasheed, S., Sarma, P. S., Huebner, R. J., Hatanaka, M., Oroszlan, S., Gilden, R. V., Kabigting, A. and Vernon, L. (1972). *Nature New Biol.*, **235**, 3
20. Metcalf, D., Bradley, T. R. and Robinson, W. (1967). *J. Cell. Physiol.*, **69**, 93
21. Orgebin-Crist, M. C. and Tichenor, P. L. (1972). *Nature New Biol.*, **239**, 227
22. Pattillo, R. A. and Gey, G. O. (1968). *Cancer Res.*, **28**, 1231
23. Pitot, H. C., Peraino, C., Morse, P. A., Jr. and Potter, V. R. (1964). *Nat. Cancer Inst. Monogr.*, **13**, 229
24. Rosenthal, M. D., Wishnow, R. and Sato, G. (1970). *J. Nat. Cancer Inst.*, **44**, 1001
25. Rugstad, H. E., Robinson, S. H., Yannoni, C. and Tashjian, A. H., Jr. (1970). *J. Cell. Biol.*, **47**, 703
26. Sato, G. and Buonassisi, V. (1964). *Nat. Cancer Inst. Monogr.*, **13**, 81
27. Sato, G., Clark, J., Posner, M., Leffert, H., Paul, D., Morgan, M. and Colby, C. (1971). *Acata Endocrinol.* (*Copenhagen*) *Suppl*., **153**, 126
28. Schindler, R., Day, M. and Fischer, G. A. (1969). *Cancer Res.*, **19**, 47
29. Schroeder, F. H., Sato, G. and Gittes, R. F. (1971). *J. Urol.*, **106**, 734

30. Schwartz, P. L., Wettenhall, R. E. and Bornstein, J. (1968). *J. Exp. Zool.*, **168**, 517
31. Shein, H. M., Wurtman, R. J. and Axelrod, J. (1967). *Nature (London)*, **213**, 730
32. Shin, S. I. (1967). *Endocrinology*, **81**, 440
33. Siegel, E. (1971). *J. Cell Sci.*, **9**, 49
34. Spooner, B. S. (1970). *J. Cell Physiol.*, **75**, 33
35. Steinberger, E., Steinberger, A. and Ficher, M. (1970). *Rec. Progr. Horm. Res.*, **26**, 547
36. Targovnik, J. H., Rodman, J. S. and Sherwood, L. M. (1971). *Endocrinology*, **88**, 1477
37. Tashjian, A. H., Jr., Bancroft, F. C., Richardson, U. I., Goldlust, M. B., Rommel, F. A. and Ofner, P. (1970). *In Vitro*, **6**, 32
38. Topper, Y. J. (1970). *Rec. Progr. Horm. Res.*, **26**, 287
39. Yaffé, D. (1968). *Proc. Nat. Acad. Sci. USA*, **61**, 477
40. Yasamura, Y., Tashjian, A. H., Jr. and Sato, G. H. (1966). *Science*, **154**, 1186
41. Foley, G. E. and Eagle, H. (1958). *Cancer Res.*, **18**, 1012
42. Hayes, W. (1968). *Genetics of Bacteria and their Viruses; Studies in Basic Genetics and Molecular Biology*, (New York: John Wiley & Sons, Inc.)
43. Kao, F.-T. and Puck, T. T. (1967). *Genetics*, **55**, 513
44. Meiss, H. K. and Basilico, C. (1972). *Nature New Biol.*, **239**, 66
45. Patterson, M. K., Jr., Maxwell, M. D. and Conway, E. (1971). *In Vitro*, **7**, 152
46. Puck, T. and Kao, F. (1967). *Proc. Nat. Acad. Sci. USA*, **58**, 1227
47. O'Malley, B. W. and McGuire, W. L. (1968). *Proc. Nat. Acad. Sci. USA*, **60**, 1527
48. Baulieu, E.-E., Alberga, A., Jung, I., Lebeau, M.-C., Mercier-Bodard, C., Milgrom, E., Raynaud, J.-P., Raynaud-Jammet, C., Rochefort, H., Truong, H. and Robel, P. (1971). *Rec. Progr. Horm. Res.*, **27**, 351
49. Comstock, J. P., Rosenfeld, G. C., O'Malley, B. W. and Means, A. R. (1972). *Proc. Nat. Acad. Sci. USA*, **69**, 2377
50. Hsu, Y.-C. (1971). *Nature (London)*, **231**, 100
51. Hsu, Y.-C. (1972). *Nature (London) New Biol.*, **239**, 200
52. McGrath, C. M. (1971). *J. Nat. Cancer Inst.*, **47**, 455
53. Hall, R. and Warrick, C. (1972). *Lancet*, **1**, 1313
54. Carrel, A. (1913). *J. Exp. Med.*, **17**, 14
55. Sato, G., Zaroff, L. and Mills, S. E. (1960). *Proc. Nat. Acad. Sci. USA*, **46**, 963
56. Green, H., Goldberg, B. and Todaro, G. J. (1966). *Nature (London)*, **212**, 631
57. Udenfriend, S. (1966). *Science*, **152**, 1335
58. Macintyre, E. H., Pontén, J. and Vatter, A. E. (1972). *Acta Pathol. Microbiol. Scand.*, **A80**, 267
59. Hayflick, L. and Moorhead, P. S. (1961). *Exp. Cell Res.*, **25**, 585
60. Paul, J. (1970). *Cell and Tissue Culture*, (Edinburgh and London: E. & S. Livingstone)
61. Hofmann, K. and Bergmann, M. (1941). *J. Biol. Chem.*, **138**, 243
62. Phillips, H. J. (1972). *In Vitro*, **8**, 101
63. Kono, T. (1968). *Fed. Proc. Fed. Amer. Soc. Exp. Biol.*, **27**, 495
64. Sayers, G., Portanova, R., Beall, R. J., Seelig, S. and Malamed, S. (1971). *Acta Endocrinol. (Copenhagen) Suppl.*, **153**, 11
65. Rodbell, M. (1964). *J. Biol. Chem.*, **239**, 375
66. Tunis, M. (1968). *Science*, **162**, 912
67. Spahr, P. F. and Hollingworth, B. R. (1961). *J. Biol. Chem.*, **236**, 823
68. Macintyre, E. H., Grimes, R. A. and Vatter, A. E. (1969). *J. Cell Sci.*, **5**, 583
69. Boone, C. W., Mantel, N., Caruso, T. D., Jr., Kazam, E. and Stevenson, R. E. (1972). *In Vitro*, **7**, 174
70. deDuve, C. and Wattiaux, R. (1966). *Annu. Rev. Physiol.*, **28**, 435
71. Chas, P. and Ball, G. H. (1971). *Curr. Top. Microbiol. Immunol.*, **55**, 288
72. Eagle, H. (1959). *Science*, **130**, 432
73. Ham, R. G. (1963). *Exp. Cell. Res.*, **29**, 515
74. Morgan, J. F., Morton, H. J. and Parker, R. C. (1950). *Proc. Soc. Exp. Biol. Med.*, **73**, 1
75. Eagle, H. (1955). *Science*, **122**, 501
76. Evans, V. J., Bryant, J. C., Kerr, H. A. and Schilling, E. L. (1964). *Exp. Cell Res.*, **36**, 439
77. Evans, V. J., Fioramonti, M. C. and Earle, W. R. (1959). *Amer. J. Hyg.*, **70**, 28
78. Kniazeff, A. J. (1968). *Nat. Cancer Inst. Monogr.*, **29**, 123
79. Fogh, J. and Fogh, H. (1969). *Ann. N.Y. Acad. Sci.*, **172**, 15

80. Molander, C. W., Kniazeff, A. J., Boone, C. W., Paley, A. and Imagawa, D. T. (1971). *In Vitro*, **7**, 168
81. Fogh, J. and Fogh, H. (1965). *Proc. Soc. Exp. Biol. Med.*, **119**, 233
82. Liss, A. and Maniloff J. (1971). *Science*, **173**, 725
83. Stanbridge, E. J., Hayflick, L. and Perkins, F. T. (1971). *Nature New Biol.*, **232**, 242
84. Schafer, T. W., Pascale, A., Shimonaski, G. and Came, P. E. (1972). *Appl. Microbiol.*, **23**, 565
84a. Levine, E. M. (1972). *Exp. Cell Res.*, **74**, 99
85. Kowal, J. (1970). *Rec. Progr. Horm. Res.*, **26**, 623
86. Stoner, G. D. and Merchant, D. J. (1972). *In Vitro*, **7**, 330
87. Hsu, T. C. and Merchant, D. J. (1961). *J. Nat. Cancer Inst.*, **26**, 1075
87a. Spooner, B. S. and Hilfer, S. R. (1971). *J. Cell. Biol.*, **48**, 225
88. Merchant, D. J. and Hellman, K. B. (1962). *Proc. Soc. Exp. Biol. Med.*, **110**, 194
89. Stoolmiller, A. C., Dorfman, A. and Sato, G. H. (1972). *Science*, **178**, 1308
90. Lightbody, J., Pfeiffer, S. E., Kornblith, P. L. and Hershman, H. (1970). *J. Neurobiol.*, **1**, 411
91. Brand, K. G. and Syverton, J. (1962). *J. Nat. Cancer Inst.*, **28**, 147
92. Herrick, P. R., Baumann, G. W., Merchant, D. J. Schearer, M. G., Shipman, C., Jr. and Brackett, R. G. (1970). *In Vitro*, **6**, 143
93. Zimmerman, E. M., Freeman, A. E., Price, P. J., Holbrook, Z. and Uhlendorf, C. P. (1972). *In Vitro*, **8**, 85
94. Stewart, S. E., Kasnic, G., Jr., Draycott, C. and Ben, T. (1972). *Science*, **175**, 198
95. Bancroft, F. C. and Tashjian, A. H. Jr., (1971). *Exp. Cell Res.*, **64**, 125
96. Hellström, K. E., Hellström, I. and Bergheden, C. (1967). *Int. J. Cancer*, **2**, 286
97. Das, M. R., Sadasivan, E., Koshy, R., Vaidya, A. B. and Sirsat, S. M. (1972). *Nature New Biol.*, **239**, 92
98. Schubert, D., Humphreys, S., Baroni, C. and Cohn, M. (1969). *Proc. Nat. Acad. Sci. USA*, **64**, 316
99. Amano, T., Richelson, E. and Nirenberg, M. (1972). *Proc. Nat. Acad. Sci. USA*, **69**, 258
100. Nelson, P., Ruffner, W. and Nirenberg, M. (1969). *Proc. Nat. Acad. Sci. USA*, **64**, 1004
101. Trowell, O. A. (1959). *Exp. Cell. Res.*, **16**, 118
102. Axelrod, J., Shein, H. M. and Wurtman, R. J. (1969). *Proc. Nat. Acad. Sci. USA*, **62**, 544
103. Shein, H. M. and Wurtman, R. J. (1969). *Science*, **166**, 519
104. Wurtman, R. J., Shein, H. M. and Larin, F. (1971). *J. Neurochem.*, **18**, 1683
105. Klein, D. C., Berg, G. R., Weller, J. and Glinsmann, W. (1970). *Science*, **167**, 1738
106. Klein, D. C. and Weller, J. (1970). *In Vitro*, **6**, 197
107. Comstock, J. P., Rosenfield, G. C., O'Malley, B. W. and Means, A. R. (1972). *Proc. Nat. Acad. Sci. USA*, **69**, 2377
108. Furth, J. and Sobel, H. (1947). *J. Nat. Cancer Inst.*, **8**, 7
109. Furth, J. (1968). *Thule Internat. Symp. Cancer And Aging*, 131 (Stockholm: Nordiska Bokanoelns Förlag)
110. Kim, U. and Furth, J. (1960). *Proc. Soc. Exp. Biol. Med.*, **105**, 490
111. Biskind, M. S. and Biskind, G. S. (1944). *Proc. Soc. Exp. Biol. Med.*, **55**, 176
112. Kirkman, H. and Algard, F. T. (1965). *Cancer Res.*, **25**, 141
113. Kao, F.-T. and Puck, T. (1968). *Proc. Nat. Acad. Sci. USA*, **60**, 1275
114. Puck, T. T., Marcus, P. I. and Cieciura, S. J. (1956). *J. Exp. Med.*, **103**, 273
115. Tashjian, A. H., Jr., Bancroft, F. C. and Levine, L. (1970). *J. Cell Biol.*, **47**, 61
116. Leffert, H. L. and Paul, D. (1972). *J. Cell Biol.*, **52**, 559
117. Paul, D., Leffert, H., Sato, G. and Holley, R. W. (1972). *Proc. Nat. Acad. Sci. USA*, **69**, 374
118. Johnson, F. L., Feagler, J. R., Lerner, K. G., Majerus, P. W., Siegel, M., Hartmann, J. R. and Thomas, E. D. (1972). *Lancet*, **2**, 1273
119. Fluharty, A. L., Porter, M. T., Lassila, E. L., Trammell, H., Carrel, R. E. and Kihara, H. (1970). *Biochim. Med.*, **4**, 110
120. Gartler, S. M. (1963). *Proc. Int. Conf. Human Genetics, 2nd*, **1**, 626, (1961: Rome)
121. Seegmiller, J. E., Rosenbloom, F. M. and Kelley, W. M. (1967). *Science*, **155**, 1682
122. deWeerd-Kastelein, E. A., Keijzer, W. and Bootsma, D. (1972). *Nature (London) New Biol.*, **238**, 80

123. Broome, J. D. and Schwartz, J. H. (1967). *Biochim. Biophys. Acta*, **138**, 637
124. Chu, E. H. Y. and Malling, H. V. (1968). *Proc. Nat. Acad. Sci. USA*, **61**, 1306
125. Littlefield, J. W. (1964). *Science*, **145**, 709
126. Davidson, R. L. and Ephrussi, B. (1965). *Nature (London)*, **205**, 1170
127. Harris, H. and Watkins, J. F. (1965). *Nature (London)*, **205**, 640
128. Klein, G. and Harris, H. (1972). *Nature New Biol.*, **237**, 163
129. American Type Culture Collection, Rockville, Maryland
130. Garancis, J. C., Pattillo, R. A., Hussa, R. O., Schultz, J. and Mattingly, R. F. (1970). *Amer. J. Obstet. Gynecol.*, **108**, 1257
131. Hammond, J. M., Bridson, W. E., Kohler, P. O. and Chrambach, A. (1971). *Endocrinology*, **89**, 801
132. Huang, W. Y., Pattillo, R. A., Delfs, E. and Mattingly, R. F. (1969). *Steroids*, **14**, 755
133. Knoth, M., Pattillo, R. A., Garancis, J. C., Gey, G. O., Ruckert, A. C. F. and Mattingly, R. F. (1969). *Amer. J. Path.*, **54**, 479
134. Pattillo, R. A., Hussa, R. O., Delfs, E., Garancis, J., Bernstein, R., Ruckert, A. C. F., Huang, W. Y., Gey, G. O. and Mattingly, R. F. (1970). *In Vitro*, **6**, 205
135. Stewart, F. W. (1950). *Armed Forces Inst. Pathology Manual, Section IX*, **F. 34**, 5
136. Haagensen, C. D. (1971). *Diseases of the Breast* (2nd ed.) (Philadelphia: W. B. Saunders)
136a. Barker, J. R. and Richmond, C. (1971). *Brit. J. Surg.*, **58**, 732
137. Stoll, B. A. (1970). *Cancer*, **25**, 1228
138. Bruni, J. E. and Montemurro, D. G. (1970). *Cancer Res.*, **31**, 854
139. Welsch, C. W., Nagasawa, H. and Meites, J. (1970). *Cancer Res.*, **30**, 2310
140. Bittner, J. J. (1942). *Cancer Res.*, **2**, 710
141. Mühlbock, O. (1956). *Advan. Cancer Res.*, **4**, 371
142. Moore, D. H., Charney, J., Kramarsky, B., Lasfargues, E. Y., Sarkar, N. H., Brennan, M. J., Burrows, J. H., Sirsat, S. M., Paymaster, J. C. and Vaidya, A. B. (1971). *Nature (London)*, **229**, 611
143. Schlom, J., Spiegelman, S. and Moore, D. (1971). *Nature (London)*, **231**, 97
144. Mayne, R. and Barry, J. M. (1970). *J. Endocrinology.*, **46**, 61
145. Frantz, A. G. and Kleinberg, D. L. (1970). *Science*, **170**, 745
146. Lasfargues, E. Y. and Feldman, D. G. (1963). *Cancer Res.*, **23**, 191
147. McGuire, W. L., Huff, K., Jennings, A. and Chamness, G. C. (1972). *Science*, **175**, 335
148. Adams, J. B. and Wong, M. S. (1969). *J. Endocrinol.*, **44**, 69
149. Jones, D., Cameron, E. H. D., Griffiths, K., Gleave, E. N., and Forrest, A. P. M. (1970). *Biochem. J.*, **116**, 919
149a. Burstein, S. and Gut, M. (1971). *Rec. Progr. Horm. Res.*, **27**, 303
150. Jenkins, J. S. and Ash, S. (1972). *Lancet*, **2**, 513
151. Miller, W. R. and Forrest, A. P. M. (1972). *Lancet*, **2**, 714
152. Solomon, I. L., Grant, D. B., Burr, I. M., Kaplan, S. L. and Grumbach, M. M. (1969). *Proc. Soc. Exp. Biol. Med.*, **132**, 505
153. Kohler, P. O., Brisdon, W. E. and Chrambach, A. (1971). *J. Clin. Endocrinol. Metab.*, **32**, 70
154. Turkington, R. W., Underwood, L. E. and Van Wyk, J. J. (1971). *New England J. Med.*, **285**, 707
155. Burgus, R., Dunn, T. F., Desiderio, D., Ward, D. N., Vale, W. and Guillemin R. (1970). *Nature (London)*, **226**, 321
156. Mitnick, M. and Reichlin, S. (1971). *Science*, **172**, 1241
157. Schally, A. V., Arimura, A., Kastin, A. J., Matsuo, H., Baba, Y., Redding, T, W., Nair, R. M. G., Debeljuk, L. and White, W. F. (1971). *Science*, **173**, 1036
158. Brazeau, P., Vale, W., Burgus, R., Ling, N., Butcher, M., Rivier, J. and Guillemin, R. (1973). *Science*, **179**, 77
159. Schimmer, B. P., Ueda, K. and Sato, G. (1968). *Biochem. Biophys. Res. Commun.*, **32**, 806
159a. Schimmer, B. P. (1969). *J. Cell Physiol.*, **74**, 115
160. Grahame-Smith, D. G., Butcher, R. W., Ney, R. L. and Sutherland, E. W. (1967). *J. Biol. Chem.*, **242**, 5535
161. Jost, J.-P. and Rickenberg, H. V. (1971). *Ann. Rev. Biochem.*, **40**, 741
162. Saffran, M., Mattthews, E. K. and Pearlmutter, F. (1971). *Rec. Progr. Horm. Res.* **27**, 607

163. Sayers, G., Ma, R. M. and Giordano, N. D. (1971). *Proc. Soc. Exp. Biol. Med.*, **136**, 619
164. Kowal, J. (1970). *In Vitro*, **6**, 174
165. Reiter, R. J. and Fraschini, F. (1969). *Neuroendocrinology*, **5**, 219
166. Aghajanian, G. K., Graham, A. W. and Sheard, M. H. (1970). *Science*, **169**, 1100
167. Alivisatos, S. G. A., Ungar, F., Seth, P. K., Levitt, L. P., Geroulis, A. J. and Meyer, T. S. (1971). *Science*, **171**, 809
168. Axelrod, J. (1971). *Science*, **173**, 598
169. Cottrell, G. A. (1970). *Nature (London)*, **225**, 1060
170. Gerschenfeld, H. M. (1971). *Science*, **171**, 1252
171. Klein, D. C. and Weller, J. L. (1972). *Science*, **177**, 532
172. Skutsch, G. (1972). *Lancet*, **2**, 1258
173. Calne, D. B. and Sandler, M. (1970). *Nature (London)*, **226**, 21
174. Cotzias, G. C., Tang, L. C., Miller, S. T. and Ginos, J. Z. (1971). *Science*, **173**, 450
175. Goldstein, M. and Frenkel, R. (1971). *Nature New Biol.*, **233**, 179
176. Weiss, B. F., Munro, H. N. and Wurtman, R. J. (1971). *Science*, **173**, 833
177. Wise, C. D., Berger, B. D. and Stein, L. (1972). *Science*, **177**, 180
178. Ellison, N., Weller, J. L. and Klein, D. C. (1972). *J. Neurochem.*, **19**, 1335
179. Klein, D. C. and Weller, J. L. (1972). *Science*, **177**, 532
180. Klein, D. C. and Weller, J. L. (1970). *Science*, **169**, 1093
181. Klein, D. C. and Rowe, J. (1970). *Mol. Pharmacol.*, **6**, 164
182. Knapp, S. and Mandell, A. J. (1972). *Science*, **177**, 1209
183. Clark, R. B. and Perkins, J. P. (1971). *Proc. Nat. Acad. Sci. USA*, **68**, 2757
184. Davidson, R. L. and Benda, P. (1970). *Proc. Nat. Acad. Sci. USA*, **67**, 1870
185. Penfield, W. (1932). *Cytology and Cellular Pathology in the Nervous System*, Vol. 2, (W. Penfield, editor) (New York: Hafner)
186. Prasad, K. N. and Hsie, A. W. (1971). *Nature New Biol.*, **233**, 141
187. Roisen, F. J., Murphy, R. A. and Braden, W. G. (1972). *Science*, **177**, 809
188. Macintyre, E. H., Wintersgill, C. J., Perkins, J. P. and Vatter, A. E. (1972). *J. Cell Sci.*, **11**, 639
189. Kessler, D., Levine, L. and Fasman, G. (1968). *Biochem.*, **7**, 758
190. Moore, B. W. and McGregor, D. (1965). *J. Biol. Chem.*, **240**, 1647
191. Klein, G., Gars, U. and Harris, H. (1970). *Exp. Cell Res.*, **62**, 149
192. Carter, S. B. (1967). *Nature (London)*, **213**, 261
193. Schofield, J. G. (1971). *Nature New Biol.*, **234**, 215
194. Yoshinaga, M., Waksman, B. H. and Malawista, S. E. (1972). *Science*, **176**, 1147
195. Maslow, D. E. and Mayhew, E. (1972). *Science*, **177**, 281
196. McGuire, J. and Moellmann, G. (1972). *Science*, **175**, 642
197. Marks, E. P. (1970). *Gen. Comp. Endocrinol.*, **15**, 289
198. Echalier, G. and Ohanessian, A. (1969). *C. R. H. Acad. Sci. Ser. D.*, **268**, 1771
199. Grace, T. D. C. (1966). *Nature (London)*, **211**, 366
200. Greene, A. E., Charney, J., Nichols, W. W. and Coriell, L. L. (1972). *In Vitro*, **7**, 313
201. Schneider, I. (1969). *J. Cell Biol.*, **42**, 603
202. Singh, K. R. P. (1967). *Curr. Sci.*, **36**, 506
203. l'Héritier, Ph. (1958). *Advan. Virus Res.*, **5**, 195
204. Ohanessian, A. and Echalier, G. (1967). *Nature (London)*, **213**, 1049
205. Wright, J. E. and Spates, G. E. (1972). *Science*, **178**, 1292

Index

Abscisic acid
　characteristics of, 298, 299
　structure, 284
Acceptor factors, role in complex retention, 169
Acceptor sites on nuclear chromatin-DNA, 194
Adrenocorticotrophic hormone
　analogues, activity of, 28, 72, 73
　　structure and activity of, 48–52
　　characterisation of, 49–51
　CRH effect on, 74
　effects on adrenal cortical cell, 321, 322
　modes of action of, 26–60
　characteristics of, 71–77
Actinomycin D
　blocking of transcription, 221
　effect on a-amylase synthesis, 292
　elevation of transaminase activity by, 220
　inhibition, of transaminase activity, 218
　　of transaminase induction, 220
Activity
　of androgens, 155–157
　assessment of, 271
　of steroid analogues, 252–257
　of synthetic steroids, 264, 265
　of auxins, 284–286
　of calcitonins, 129–131
　of insulin, 5, 6
　of parathyroid hormones, 106–108
　of proparathyroid hormone, 110
　structural requirements for, 118, 131
Adaptive increases in hormone secretion, 114, 115
Adenohypophysial hormones, secretion and action of, 62–100
Adenohypophysis, biochemistry and enzymology of, 62–64
Adenosine, effect on lipolysis in fat cells, 9–11
Adenosine monophosphate
　cyclic, accumulation in liver cells, 12, 13
　　action, glucocorticoid requirement for, 232, 233
　　a-amylase induction by, 292

Adenosine monophosphate *continued*
　analogues, activity of, 35, 36
　calcium effect on action of, 54
　corticosterone production in response to, 34, 35
　distribution of, 32–34
　effect on sensitivity, 56
　as effector of alanine transaminase, 223
　and gene expression, 175
　inhibition by, 9–11
　inhibition by adenosine, 9–11
　insulin effect on, 9–13
　interaction with cytokinins, 298
　and lipolysis in fat cells, 9–11
　as mediator of glucocorticoid action, 232, 233
　modes of action of, 37–41
　relationship to ACTH, 321
　role in Ca action on hormones, 113
　role in parathyroid hormone action on bone, 125
　serine dehydratase induction by, 226
　stimulation, by ACTH, 26, 27, 29–37
　　by calcitonin, 138
　　carboxykinase synthesis, 224
　　by parathyroid hormone, 124, 126
　　by TSH, 71
　　of ascorbic acid depletion, 35
　　of glucose oxidation, 35
　　of hydrolysis of cholesterol esters, 35
　turnover, increased by corticoid deficiency, 231
$3',5'$-Adenosine monophosphate phosphodiesterase, stimulation by insulin, 8, 9
Adenyl cyclase
　ACTH activation of, 26, 27, 76, 77
　activation by GH-RH, 85
　bioassays based on activation of, 121, 125
　calcitonin stimulation of, 138
　and CRH action, 74
　inhibition by insulin, 8, 9, 12
　mediation of parathyroid hormone action, 122
　participation in GnRH action, 80

331

INDEX

Adenyl cyclase *continued*
  sensitivity to Ca, 113
  TSH activation of, 71
Adipose tissue, PEP carboxykinase activity in, 225
Adrenal cortex, ACTH stimulation of, 75–77
Adrenal corticoids, synthetic pathways of, 272
Adrenal gland, culture of, 321, 322
Adrenalectomy, effect on PEP carboxykinase activity, 224
Adrenocorticotrophin, characteristics of, 71–77
Adrenodoxin in mixed function oxidases, 45, 46
Affinity of ACTH analogues, 49–51
Alanine, role in gluconeogenesis, 222
Albright's syndrome, role of hormones in, 307
Aleurone cells as test system for gibberellins, 291, 292
Alkaline phosphatase, glucocorticoids stimulation of, 229, 230
α-Amanatin and RNA-polymerase activity, 198
Amino acids
  of ACTH, 48
  of calcitonins, 129, 130
  in GH, 83, 84
  of insulin, 5, 6
  insulin stimulation of uptake of, 15
  of parathyroid hormones, 103–106
  role in enhancing hormone affinity for receptor, 51
Aminoacylated transfer-RNA, accumulation of, 222
Amplification and enhancement of sensitivity, 55, 56
α-Amylase, gibberellins control of synthesis, 292
Anabolic activity
  of androgens, 154, 156–157
  of steroid analogues, 254–257
Anabolic/androgenic ratio of steroid analogues, 252–254
Anabolic effects
  v. androgenic effects, functional comparison of, 245–257
  of glucocorticoids, 235, 236
Anabolic steroids
  analogues, biological activity of, 245–260
  reduction of 4,5-double bond of, 250
Androgens
  actions, not attributed to DHT, 177, 178
    receptor mechanisms in, 154–185
  analogues, biological activity of, 245–260
    in assay for anabolic/androgenic ratio, 253
  -binding proteins, in target cells, 162–165

Androgens *continued*
  -binding site, on β-proteins, 165, 166
  degradation, 'hydroxy' pathway for, 246, 247
    'ketonic' pathway for, 246
  dysfunction, role of DHT in, 176, 177
  receptors, complex, structure of, 166
    cycling, model for, 173
    definition of, 155
    interaction with androgens, 157
  reduction of 4,5-double bond of, 250
  sterility, testosterone induction of, 178
  structure, role in binding to receptor proteins, 165–167
Androphilic proteins
  in blood, 157, 158
  definition of, 155
5α-Androstane, androgenic activity of, 258
Androstenediols, role in prostates, 178
Anti-androgens
  activity of, 175, 176
  of cyproterone acetate, 264
Antifertility activity of progestogens, 261
Anti-inflammatory activity of corticoid analogues, 273–275
Anti-oestrogens, structural similarity to oestrogens, 270
Antisera for parathyroid hormone, 117
Apparatus for cell culture, 314
Aromatisation
  of A-ring of androgens, 249, 250
  protection against, 251, 252
Ascorbic acid
  ACTH analogues effect on, 72
  ACTH effect on, 76
  in the adenohypophysis, 62, 63
  depletion, cyclic-AMP stimulation of, 35
Assays
  for acceptor factors, 169
  of calcitonin, 133
  of parathyroid hormone, 120–122
  for specific cell function, 317
  of vitamin D, 139
Astrocytes, cell culture of, 324
ATPase
  insulin effect on, 19
  role in H^+ excretion, 290
Autoradiographic techniques for study of androgen uptake, 159, 160
Auxin-binding proteins
  isolation of, 288
  role in cell elongation, 290
Auxins, characteristics of, 284–290
Avidin
  induction by progesterone, 189
  synthesis, and DNA-dependent RNA synthesis, 195
  induction by progesterone metabolite, 190, 191

INDEX

Avidin mRNA, progesterone induction of, 195

Barley endosperm, response to gibberellins, 292
Binding
  of ACTH to receptors, 28, 29
  of glucocorticoids, 213, 214
  sites, for auxin, 287
Biossays
  for activity assessment, 264, 265
  for calcitonin, 133
  of parathyroid hormone, 120, 121
Biochemical sequence of events in progesterone action, 192–195
Biochemistry
  of ACTH, 72, 73
  of the adenohypophysis, 62–64
  of GH, 83, 84
  of gonadotrophins, 77, 78
  of pineal gland, 322, 323
  of prolactin, 87
  of TSH, 67, 68
Biological activity
  of cytokinins, 294, 295
  of gibberellins, 291
Biosynthesis
  of calcitonin, 131–133
  of parathyroid hormone, 108, 111
  precursor, of parathyroid hormone, 108–111
  relationship to secretion, 114
  of vitamin D, 139–142
Blocking effect of progesterone, 188
Blood, androphilic proteins in, 157, 158
Bone
  calcitonin action on, 136, 137
  cells, changes in response to parathyroid hormone, 124
  kinetic responses of, 123, 124
  matrix, parathyroid hormone destruction of, 124, 125
  parathyroid hormone and resorption of, 122–125
Bovine parathyroid hormone, structure of, 103–105
Brain nuclei, dihydrotestosterone retention by, 161
Breast cancer
  hormone induction of, 318
  hormone synthesis by, 320
Breast differentiation, hormone dependence for, 319

C 1300 neuroblastoma, viruses from, 314
C-cell, calcitonin from, 128, 131
Calcitonin, characteristics of, 128–139

Calcium
  blocking of hormone release by, 112
  calcitonin response to, 131, 132
  in control of parathyroid hormone secretion, 111, 112
  effect on hormone biosynthesis, 113
  excretion, stimulation by parathyroid hormone, 126
  metabolism, hormonal regulation of, 102–151
  mobilisation, factors affecting, 125
  modes of action of, 52–55
  parathyroid hormone effect on, 123–125
  regulation of vitamin D activation, 143, 144
  transport, parathyroid hormone action on, 127
  stimulation by vitamin D, 141
Calcium binding protein, activation by vitamin D, 144
Callus growth,
  test system for cytokinins, 296
Carbohydrate metabolism, glucocorticoids effect on, 212
Carrier-mediated transport for hexose uptake, 13
Cartilage production by cells in culture, 307
Catabolic effects of glucocorticoids, 234
Catalytic kinase unit of protein kinases, 39
Cation-ligand interaction at parathyroid cells, 112
Cell-free systems, insulin action on, 8, 9
Cells
  culture, in study of hormone action, 306–329
  differentiation and expression of mammary tumour virus, 319, 320
  division, promotion by cyclic-AMP, 37, 40, 41
  role of cytokinins in, 294
  elongation, auxin induction of, 286–290
  cycloheximide effect on, 289
  mRNA role in, 288
  function, growth media effect on, 313
  growth promotion by cyclic-AMP, 37, 40–41
  insulin entry into, 6–8
  line, definition of, 308
  types, in the adenohypophysis, 64, 65
Cellulose, role in cell elongation, 289
Central nervous system, cell culture from, 323, 324
Chelating agents, role in disaggregation, 309
Chemistry
  of calcitonins, 128–131
  of corticoid analogues, 271, 272
  of oestrogens, 265–271
  of parathyroid hormone, 103–108
  of progestational agents, 261–264

Chick oviduct
  oestrogen action and, 202–204
  progesterone action and, 192, 193
Chitinase as disaggregating agent, 310
Cholesterol
  ACTH effect on, 76
  as precursor of corticosteroids, 41–47
  side-chain cleavage of, 46, 47
  sources of, 41
Cholesterol-binding protein in regulation of steroidogenesis, 46, 47
Cholesterol esterase, activation of, 40
Cholesterol esters, hydrolysis, cyclic-AMP stimulation of, 35
Choline, GA stimulation of incorporation of, 293
Chromatin
  acceptor sites on, 194
  ribonucleoprotein binding to, 170
  template activity, oestrogen effect on, 198, 199, 203
Circulating fragment of parathyroid hormone, 117
Circulation
  immunoreactive hormone in, 118
  parathyroid hormone in, 114, 116–119
Cleavage of parathyroid hormone, 118
Coleoptile, study of cell elongation, 286
Collagenase
  as disaggregating agent, 309
  use in dispersed cultures, 315
Competitive binding radioligand assay for parathyroid hormone, 121
Competitive protein binding radioassay of vitamin D, 139
Conjugation of hydroxyl groups, 246
Contamination of cell cultures, 313
Corner and Allen technique for activity assessment, 264
Corticoids
  ACTH stimulation of synthesis of, 76
  analogues, activity of, 271–276
Corticosterone
  ACTH stimulation of secretion of, 76
  calcium effect on production of, 54
  cyclic-AMP accumulation and production of, 30, 31
  production in response to cyclic-AMP, 34, 35
  production in response to dibutyryl cyclic-AMP, 34, 35
  regulation of ACTH secretion, 74, 75
Corticotrophin-releasing hormone
  characteristics of, 73, 74
  corticoids effect on secretion of, 75
  structure of, 73, 74
Cortisol
  ACTH stimulation of secretion of, 76
  activity of, 212
  anti-inflammatory activity of, 274

Cortisol *continued*
  Na^+ retaining activity of, 274
  regulation of ACTH secretion, 74, 75
Culture procedure for cells, 314, 315
Cuticle, barrier to H^+, 289
Cycloheximide
  effect, on α-amylase synthesis, 292
  on cell elongation, 289
Cyproterone, anti-androgenic activity of, 176
Cyproterone acetate, antiandrogen activity of, 264
Cytochalasin B, effect on growth hormone release, 324
Cytochrome P450
  binding of cholesterol, 46, 47
  binding of pregnenolone, 46, 47
  in mixed function oxidases, 45–47
Cytodifferentiation, oestrogen stimulation of, 203
Cytokinins, characteristics of, 294–298
Cytoplasmic polysomes, ovalbumin synthesis on, 206
Cytoplasmic receptors
  progesterone metabolism and, 190
  proteins, for androgens, 173
Cytosol, glucocorticoid receptors in, 215
Cytotoxicity of sterile plasticware, 314

2,4-D, synthetic auxin, 285
D actinomycin and blocking of insulin action, 18
Deciduoma reaction, determination by progesterone, 188
Deciduoma tissue, progesterone metabolism by, 191
Dedifferentiation in cell culture, 307, 308
Definitions
  of anabolic steroids, 245
  of androgens, 245
  of cell culture, 308, 309
  of organ culture, 308, 309
Degradation
  of alkaline phosphatase, 229–230
  rate, of glutamyl alanine transaminase, 222
Dehydrotachysterol, vitamin D analogue, 140
11-Deoxycortisol, inhibition of glucocorticoids, 215
10,19-Desmolase in A-ring aromatisation, 249, 250
Dexamethasone, activity of, 212
DHT
  and androgen dysfunction, 176, 177
  binding proteins, activity of, 175, 176
  protein complex, aggregation of, 162
    nuclear acceptor activity for retention of, 168–170

INDEX

DHT *continued*
    retention by prostate nuclei, 167, 168
    stability of, 164
Diagnostic significance of parathyroid hormone levels, 119
Diaphragm nuclei, dihydrotestosterone retention by, 161
Dibutyryl cyclic-AMP
    calcium effect on, 54
    corticosterone production in response to, 34, 35
Diethylstilboestrol, oestrogenic activity of, 268, 269
Dihydrotestosterone retention by prostate cell nuclei, 160–162
Dimethyl stilboestrol, anti-oestrogenic activity of, 269
Disaggregation, techniques for, 309, 310
Disappearance rate of calcitonin, 134, 135
Dispersed cultures for primary culture of cells, 315
Dissociation
    of anabolic from androgenic activity, 260
    of insulin, 7
Distribution
    of auxins, 285
    of cytokinins, 295
    of gibberellins, 291
    volume, of calcitonin, 134, 135
    of parathyroid hormone, 120
DNA
    -dependent RNA synthesis, avidin synthesis and, 195
    sequence, transcription of, 204
    steroid-receptor complexes binding to, 169, 170
    stimulation of acceptor activity, 169
DNA-RNA competition methods
    for study of nuclear-RNA sequences, 199
    for study of RNA synthesis, 203
Disulphide bonds and insulin binding to receptors, 5
Drugs, pineal system for study of action of, 323
Dwarf plants, role of gibberellins in, 291

Eagle's Minimum Essential Medium, for cell culture, 310, 311
Ectopic secretion of parathyroid hormone, 108
Enzymes
    activity in the adenohypophysis, 63–65
    cleavage by, of prohormone to hormone, 110
    synthesis by, coordinate effect of steroids on, 222
Enzymology of adenohypophysis, 62–64
Established cell line, definition of, 308
Ethylene activity as a hormone, 299, 300
Excitational effects of androgens, 154

Fat cells
    adenosine effect on lipolysis in, 9–11
    cyclic-AMP and lipolysis in, 9–11
    degradation of insulin by, 4
    insulin-agarose binding to, 6–8
    insulin binding to, 3
    insulin-insensitivity of, 5
    insulin stimulation of protein synthesis in, 18
    membranes, insulin effect on glucose release by, 13
    treatment with neuroaminidases, 4
    treatment with phospholipases, 4
Fatty acids, insulin effect on release of, 9
Feed-back
    action of gonadal steroids, 81
    between ACTH and CRH, 75
    regulation, in the adenohypophysis, 66
    of gonadotrophins, 79–81
    of TSH secretion, 70
Female sex steroids, mode of action of, 187–210
Fibroblasts, steroid-receptors in, 215
Fish, insulin activity of, 6
Florigen, role in flowering, 300
Flowering, role of florigen in, 300
Foetal androgen, activity of, 178
Follicle-stimulating hormone
    mechanism of action of, 82
    production by the adenohypophysis, 77
Frog muscle, insulin binding to, 2, 3

Gelfiltration studies of parathyroid hormone, 117
Genes
    expression, receptor mechanisms involved in, 171–175
    repressors, steroids binding to, 174
    transcription products, induction by progesterone, 195
    transcription, progesterone influence on, 193, 194
    regulation by nuclear RNP particles, 173
    role in oestrogen action, 199
Gibberellic acid, structure, 284
Gibberellins, characteristics of, 290–294
Globulin binding testosterone, 158
β-Gluconase
    role in cell elongation, 289
    secretion in response to GA, 294
Glucocorticoids
    definition of, 272
    effect on carboxykinase activity, 224
    mode of action of, 212–241
Gluconeogenesis
    glucocorticoid regulation of, 226, 227
    role of alanine in, 222
    role of alanine transaminase in, 223

Glucose
  insulin inhibition of release from liver, 12, 13
  metabolism, stimulation, by insulin-agarose, 7, 8
    by insulin-dextran-2000, 7, 8
  oxidation stimulation by cyclic-AMP, 35
  transport, glucocorticoid inhibition of, 234
    inhibition by glucocorticoids, 215
    insulin action on, 2, 3
Glucuronide conjugation
  in oestrogen degradation, 266, 267
  in steroids, 246, 247
Glutamine synthetase, glucocorticoid regulation of, 227, 228
Glutamyl alanine transaminase, glucocorticoid regulation of, 222, 223
Glycerol phosphate dehydrogenase, glucocorticoid stimulation of, 228, 229
Glycogen, insulin effect on deposition of, 9, 11–12
Glycogen phosphorylase, ACTH activation of, 37, 38
Glycogen synthase, insulin effect on activity of, 11, 12
Gonadotrophin-releasing hormone
  gonadotrophin secretion, regulation by, 79, 80
  mechanism of action of, 80
Gonadotrophins
  characteristics of, 77–82
  secretion, by cells in culture, 307
  progesterone inhibition of, 189
Gonadal steroids, feed-back action of, 81
Gonads, gonadotrophin action on, 81, 82
4-Gonen-13β-ethyl-17α-ethynyl-17β-ol-3-one as a potent progestogen, 263, 264
Gradient centrifugation for study of DHT-binding proteins, 162, 163
Growth hormone, characteristics of, 82–86
Growth hormone-releasing hormone, structure of, 85
Growth hormone synthesis and prolactin activity, 321
Growth-limiting protein, role in cell elongation, 290
Growth media for cell culture, 310–313
Growth response to auxin, 286, 288–289
Guanosine monophosphate, cyclic stimulation of hormone production by, 36
Guinea-pigs, insulin activity of, 6

Half-life
  of alkaline phosphatase, 229
  of glutamine synthetase, 227
  of glycerol phosphate dehydrogenase, 229
  of transaminase mRNA, 221
  of transaminase template, 218

Ham's F10 Medium for cell culture, 310, 311
Hepatoma cells
  accumulation of transaminase mRNA in, 218, 219
  alanine transaminase activity in, 223
  glucocorticoid receptors in, 213–215
Hepatotoxic effects of steroid analogues, 252
Heterogeneity of parathyroid hormone, 116, 119
Hexoses
  carrier-mediated transport of, 13
  insulin stimulation of transport of, 13
Homosteroids
  activity of, 156
  A-type, activity of, 258
  B-type, activity of, 258
  D-type, activity of, 258
Hooker-Forbes procedure for activity assessment, 264, 265
Hormone-binding proteins in serum, 313
Hormone-dependent cultures
  production of, 315–317
  in study of hormone action, 307
Hormone-dependent mammary epithelium, collagenase in production of, 315
Hormone-membrane interactions, techniques for studying, 14
Hormone-producing cultures, induction of, 315, 316
Hormone-responsive cells, growth in culture, 317
Hormone-secreting cells, morphology of, 41, 43
Hormones
  action, cell culture in study of, 306–329
  Ca metabolism regulation by, 102–151
  in vitro study of, 318
  synthesis, by breast cancer cells, 320
Human parathyroid hormone, structure of, 105, 106
Hybridisable nuclear-RNA, oestrogen induction of, 197, 198
Hydrogen ions
  excretion, in response to auxin, 290
  role in cell elongation, 289
Hydroxylapatite chromatography to study DNA sequences, 203, 204
Hydroxylation
  and degradation of androgens, 245–248
  of steroids, 45, 46, 248, 249
  of vitamin D, 140, 141
Hydroxy steroid dehydrogenases and androgen activity, 156
Hydroxysteroids, activity of, 156
Hyperparathyroidism, radioimmunoassay for differentiation of, 121, 122
Hypothalamic-releasing hormones in secretion of adenohypophysial hormones, 66

INDEX

Hypothalamus, synthesis of tropic hormone releasing factors, 321

Immunochemistry of parathyroid hormone, 110, 111
Immunological activity and insulin amino acids, 6
Immunoreactive hormone in the circulation, 118
Implantation, role of progesterone in, 195, 196
Indoleacetonitrile, occurrence, 285
Indolepyruvic acid, occurrence, 285
Induced protein synthesis, oestrogen stimulation of, 200–202
Informofers, nuclear RNP particles as, 173
Informosomes, nuclear RNP particles as, 173
Inosine monophosphate, cyclic, stimulation of hormone production by, 36
Insects, culture of cells from, 324
Insulin
  -agarose bead complex, and glucose metabolism, 6, 7
  -agarose complex, metabolism stimulation by, 7, 8
  agarose derivatives purification of receptors with, 4, 5
  -dextran-2000 glucose metabolism stimulation by, 7, 8
  inhibition of carboxykinase synthesis, 224, 225
  mode of action of, 1–23
  as repressor of alanine transaminase, 223
—, iodo-
  activity of, 6
  binding to receptors, 3
Insulin-receptor protein, proteins binding to, 7
Intact cells, effects of insulin on, 9–19
Intermedin, secretion in the hypophysis, 72, 73
Intestines
  calcitonin action on, 137, 138
  parathyroid hormone action on, 127
Intracellular cycling of a steroid-receptor complex, 173, 174
Intracellular glucocorticoid receptors and glucocorticoid action, 212–216
Intranuclear progesterone-receptor complex on the chromatin, 194
Intrinsic activity efficiency of ACTH analogues, 49, 50
Iodination and activity of ACTH, 28, 29
Isohormones of parathyroid hormone, 103
Isolation
  of calcitonin, 128, 129
  of parathyroid hormone, 103–106

Isozymes
  of alkaline phosphatase, 230
  of serine dehydratase, 226

K_a value of progesterone-receptor, 191
Ketones,
  substituents, and degradation of androgens, 245–248
Kidney
  calcitonin action on, 137
  parathyroid hormone action on, 125–127
Kinetic responses of bone, 123, 124
Kinetin, isolation of, 294

Labile repressor, role in transaminase m-RNA inhibition, 221
Lag period in response to glucocorticoids, 217, 223, 224, 228, 229
Lipases, activation by protein kinases, 39
Lipolysis
  by ACTH, 76, 77
  gonadotrophins role in, 81
Liver
  cell membranes, insulin receptor from, 3
  cells, cyclic-AMP accumulation in, 12, 13
  insulin binding to, 3
  insulin inhibition of glucose release from, 12, 13
  degradation of proteins in, 17, 18
  effect of glucocorticoids on, 216–227
  enzymes, glucocorticoids effect on, 217, 225–227
  glucocorticoid receptors in, 213–215
  nuclei, dihydrotestosterone retention by, 161
  parenchymal cell strains, selection techniques for, 317
  protein synthesis in, selective action of glucocorticoids on, 216, 217
Luteinising hormone, production by the adenohypophysis, 77
Luteotrophic hormone
  androgens inhibition of secretion of, 80
  characteristics of, 86–90
  mechanism of action of, 82
  production by the adenohyophysis, 77
Lymphocytes, steroid-receptors in, 215
Lymphoid tissues, inhibitory effect of glucocorticoids on, 233–236
Lysosomal hydrolases, release in disaggregation, 310
Lysosomes, stabilisation by insulin, 17

Magnesium
  blocking of hormone release by, 112
  effect on hormone biosynthesis, 113
Male accessory genital organs, androgen actions in, 154

Mammary tumour virus, cell differentiation expression of, 319, 320
McGinty procedure for activity assessment, 265
Medium 199 for cell culture, 310, 311
Melanin production by cells in culture, 307
Melanocyte stimulating hormone, secretion in the hypophysis, 72, 73
Melanophorotropic activity of ACTH, 72, 73
Melatonin level in pineal gland, 322, 323
Membranes
  agents altering structure of, 14
  transport, insulin stimulation of, 13–15
Messengers, calcium as messengers in steroidogenesis, 54
Messenger-RNAs synthesis, oestrogen stimulation of, 174
Metabolism of calcitonin, 134, 135
  corticoid analogues and, 271, 272
  oestrogens and, 265–271
  of parathyroid hormone, 114, 116–120
  progestational agents and, 261–264
  of progesterone, 189–191
  steroid analogues and, 245–252
  steroids and, 251
Metabolites, formation by progesterone, 190, 191
Methionine, role in parathyroid hormone activity, 107, 108
17α-Methylation and steroid resistance to degradation, 251
9-Methyl-BAP for study of cytokinin incorporation, 297
Methylxanthines
  effect on cyclic-AMP accumulation, 33
  inhibition of steroidogenesis by, 36, 37
Milk, parathyroid hormone increase in calcium in, 128
Mineralocorticoids
  definition of, 272
  effect on Na$^+$ retention, 212
Mitochondria, insulin effect on reactions in, 20
Miyake and Pincus procedure for activity assessment, 264
Mixed function oxidases for hydroxylation of steroids, 45, 46
Morphogenetic activity of hormones, 154
Muscles, insulin effect on cyclic-AMP in, 11
Mycoplasma, contamination of cell cultures, 313
Myometric activity of progesterone, 187, 188
Myotrophic effect
  of anabolic steroids, 253
  of androgens, 253
Myotropic activity
  of androgenic steroids, 154
  of steroids, 156

NCTC 109 Medium for cell culture, 311, 312
Neuraminidases, treatment of fat cells with, 4
Neuroblasts, cell culture of, 323, 324
Neurovascular theory for control of the adenohypophysis, 66
Nomenclature of gibberellins, 291
Norepinephrine, effects on pineal gland, 322
19-Nortestosterone
  binding to β-protein, 165, 166
  derivatives, progestational activity of, 261–263
  substitution, and binding affinity, 166
Nuclear acceptor activity for retention of a DHT-protein complex, 168–170
Nuclear acceptor substance(s), structure of, 168
Nuclear receptors, progesterone metabolism and, 190
Nuclear ribonucleoprotein, binding to DHT-protein complex, 170, 171
Nuclear-RNA sequences, oestrogen induction of, 199
Nuclear-RNA synthesis
  oestrogen stimulation of, 201
  progesterone effect on, 195
Nucleic acids, synthesis, effect of glucocorticoids on, 234
Nucleoplasm, auxin-binding sites in, 287
Nucleolar chromatin regions, induced RNA synthesis in, 171
Nucleotides
  cyclic, mimic action of ACTH, 34–36
  sequences, recognition by the steroid-protein complex, 172, 173

Oestradiol
  activity of, 268
  synthesis of, 249
Oestradiol-binding protein in rat uterus, 197
Oestradiol-receptor proteins and oestrogen retention, 159
Oestrogen-induced protein, discovery of, 201
Oestrogen-receptor complex, structure of, 166
Oestrogens
  activity of, 265–271
  definition of, 265
  analogues, activity of, 265–271
  mechanism of action of, 196–207
  non-steroidal synthetic activity of, 268
  plant, activity of, 265
  protein binding of, 158
  synthesis, pathways of, 265, 266
Oestrone, synthesis of, 249, 250
Oral contraceptives, action of, 260–265

Organ cultures
  definition of, 308, 309
  for primary culture of cells, 315
Organ-specific destruction
  of calcitonin, 135
  of parathyroid hormone, 120
Organisational activity of androgens, 154
Ornithine decarboxylase, oestrogen stimulation of, 204
Ovalbumin cell-free synthesis of, 206
Ovalbumin mRNA, activity of, 205, 206
Oviduct translation, oestrogen control of, 202–205

Parathyroid glands, anatomy of, 108
Parathyroid hormone
  Ca metabolism regulation by, 103–128
  calcitonin as antagonist of, 137
  porcine, structure of, 103–105
  structure of, 106
  and vitamin D activation, 143
Penicillin selection technique for hormone-dependent cultures, 316
Peptides
  chains, initiation, insulin stimulation of, 16
  initiation, oestrogen stimulation of, 205
  fragments, of parathyroid hormone, 116
  hormones effect on parathyroid hormone secretion, 112
Permissive effect of glucocorticoids, 230–233
Phosphates
  effect on hormone secretion, 112, 113
  excretion, stimulation by parathyroid hormone, 126
Phosphodiesterase
  effect on cyclic-AMP accumulation, 33
  hydrolysis, of cyclic-AMP by, 35
  of dibutyryl cyclic-AMP by, 35
  inhibition by methylxanthines, 37
Phosphoenol pyruvate carboxykinase, regulation of, 223–225
Phospholipase C, stimulation of glucose metabolism in fat cells, 14
Phospholipases, treatment of fat cells with, 4
Phospholipid synthesis, GA effect on, 293
Physiologic responses to progesterone, 187–189
Pigs, insulin, activity of, 6
Pineal gland, culture of, 322, 323
Pineal system in study of action of drugs, 323
Pituitary cell lines for study of the tropic hormone, 320, 321
Pituitary gland, culture of, 320, 321
pK of induced proteins, 200
Plant hormones, mode of action of, 284–303
Plasma membranes
  auxin-binding sites in, 287

Plasma membranes *continued*
  changes occurring in, 29
  role of calcium in, 52–54
Plastic apparatus, cytotoxicity of, 314
Polymers, activity of ACTH linked to, 28
Polypeptide synthesis, oestrogen initiation of, 200
Polysome profile, oestrogen effect on, 205
Post-transcriptional repressor, function of, 174, 175
Potassium ions
  insulin effect on uptake of, 18–19
  insulin inhibition of release of, 12
Prednisolone, activity of, 212
Pregnancy, progesterone mediation of, 187
5α-Pregnane-3,20-dione
  activity of, 190, 191
  binding to progesterone-receptor, 192
Pregnenolone
  biosynthesis of, 46, 47
  cholesterol conversion to, 44, 45
Primary cell line, definition of, 308
Primary culture of cells, 314, 315
Primary tissue, sources of, 315, 316
Progestational agents, activity of, 260–265
Progestational effect, definition of, 260, 261
Progesterone
  action of, 187–196
  derivatives, antifertility activity of, 261, 262
  discovery of, 260
  inhibition of corticosteroids, 214
Progesterone-binding molecule, shape of, 193
Progesterone-binding protein, as receptor for progesterone, 193
Progesterone-binding sites, in human endometrium, 191
Progesterone-receptor protein, induction of, 189
Prohormone for parathyroid hormone, 108–111
Prolactin
  activity of GH, 83
  activity and growth hormone synthesis, 321
  assay for, 319
  characteristics of, 86–90
  secretion stimulation by TRH, 69, 70
Prolactin-inhibiting hormone, effect on prolactin secretion, 87, 89
Proportionality of control of parathyroid hormone, 111, 112
Prostate microsomal fraction, DHT binding by, 164, 165
Prostate nuclei
  DHT-protein complex retention by, 167, 168
  dihydrotestosterone retention by, 160–162
Protease, induction by gibberellins, 292

Proteolysis
  inhibition by insulin, 16, 17
  insulin inhibition in fat cells, 18
Proteins
  acidic, effect on template activity, 201
    oestrogen effect on, 201
    role in steroid-receptor complex binding, 214
  ACTH effect on biosynthesis of, 76, 77
  breakdown, glucose transport hypothesis in, 235
  cyclic-AMP and synthesis of, 40
  degradation in liver, 17, 18
  insulin effect on metabolism of, 15–18
  insulin stimulation of synthesis of, 15, 16
  metabolism, glucocorticoids effect on, 212
  synthesis, abscisic acid inhibition of, 299
    in adenohypophysis, 62, 63
    effect of glucocorticoids on, 234
    ethylene effect on, 300
    induction of, 203
    glucocorticoids effect on, 216–230
    oestrogen depression of, 199
    oestrogen stimulation of, 200, 204, 205
a-Proteins, DHT binding by, 164
β-Proteins
  androgens binding by, 165–166
  DHT binding by, 164
Protein kinase
  activation by cyclic-AMP, 39, 40
  glucocorticoids role in activation of, 231
  and mechanism of ACTH action, 76, 77
Proteolytic enzymes for disaggregation of tissues, 309
Proton-pump hypothesis for auxin action, 289, 290
Purification
  of androgen-binding proteins, 163, 164
  of DHT-binding proteins, 164
Puromycin, inhibition of steroidogenesis, 38
Pyruvate dehydrogenase, stimulation by insulin, 20

Radioimmunoassays
  for calcitonin, 131, 133
  of parathyroid hormone, 117, 120–122
  of transaminase synthesis, 217
Rat uterus for study of oestrogen action, 196, 197
Rate of disappearance of parathyroid hormone, 119, 120
Receptor-adenylate cyclase complex, effect of calcium on, 53
Receptors
  ACTH binding to, 28, 29, 76
  affinity of hormones for, 51
  in determining sensitivity, 55, 56
  dissociation of insulin from, 4
  effect on sensitivity, 56

Receptors *continued*
  insulin binding to, 2–5
  from liver cell membranes, 3
  mechanisms, in androgen actions, 154–185
  involved in gene expression, 171–175
  molecules, role in tissue specificity, 192
  nature of, 29, 51, 52
  proteins, role of androgen structure in binding to, 165–167
  purification with insulin-agarose derivatives, 4–5
  transport, in the thymus, 215
  TSH binding to, 71
$5a$-Reductase, and DHT-binding protein, 165
Regulation
  of ACTH secretion, 73–75
  of GH secretion, 85, 86
  of gonadotrophin secretion, 78–81
  of prolactin secretion, 87, 89
  of vitamin D, 143
Regulatory protein of protein kinases, 39, 40
Releasing factors in secretion of adenohypophysial hormones, 66
Respiration, stimulation in the adenohypophysis, 62, 63
Ribonuclease, contamination in cell fractionation, 309
Ribosome precursor particle, nuclear RNP particles as, 173
Ribosomes
  cytokinins binding to, 298
  oestrogen stimulation of, 204, 205
  RNA synthesis in, stimulation by growth-promoting hormones, 172
Rings
  contraction, effect on androgenic activity, 258
  expansion, effect on biological activity, 258
RNA
  -bound cytokinin bases, mode of synthesis of, 297
  breakdown, glucose transport hypothesis in, 235
  synthesis, abscisic acid inhibition of, 299
    in action of glucocorticoids, 227
    auxin effect on, 287–289
    blocking by dactinomycin, 18
    changed by gibberellins, 293
    ethylene effect on, 300
    glucocorticoids stimulation of, 235
    progesterone induction of, 195
    role in alanine transaminase activity, 223
    role in glucocorticoid action, 230
    role in oestrogen action, 197
    stimulation by androgens, 171
mRNA
  auxin induction of, 288

mRNA *continued*
  breakdown, inhibition by glucocorticoids, 220
  glucocorticoid stimulation of, 228
  half-life of, 221
  regulation by corticosteroids, 214, 215
  role in α-amylase synthesis, 292
  specific, oestrogen control of, 205–207
    oestrogen induction of, 206, 207
  stability for glutamine synthetase, 227
    of glycerol phosphate dehydrogenase, 229
  synthesis, glucocorticoid effect on rate of, 220
  glucocorticoid, stimulation of, 218
tRNA
  cytokinins in, 295
  incorporation of cytokinins into, 296, 297
  relation with free cytokinins, 296–298
  role of cytokinins in, 295
  unacylated, role in protein degradation, 222
RNA polymerase
  activity of, 172
    oestrogen stimulation of, 197, 198, 201–203
Rough endoplasmic reticulum, increase in response to gibberellins, 293

Saliva, calcium increase by parathyroid hormone, 128
Secretion
  biochemical mechanisms of, 113
  of calcitonin, 131–133
  control of parathyroid hormone, 111–114
  of parathyroid hormone, factors controlling, 112, 113
  of proparathyroid hormone, 110, 111
  regulation of TSH, 67
  relationship to biosynthesis, 114
Sedimentation of DHT-bound proteins, 162, 163
Selection
  of cells, 314, 315
  for normal liver parenchymal cell strains, 317
Sensitivity of adrenal cortex cells, 55, 56
Serine dehydratase, glucocorticoids effect on, 226
Serotonin, level in pineal gland, 322, 323
Serum, supplementation of growth media with, 213, 312
Sialic acid residues and insulin action, 14
Site of action of auxin, 287
Sodium-retaining activity of corticoid analogues, 273–275
Somatic cell mutants for study of hormones, 318
Somatomedin and GH action, 82, 83, 86

Somatotrophic hormone, characteristics of, 82–86
Specialised cells
  characteristics of, 308
  selection of, 317
Steroids
  A-ring, aromatisation of, 249, 250
  integration of structure and function in biosynthesis of, 41, 44, 45
  withdrawal, effect on transaminase activity, 218
Steroid hormones
  ACTH stimulation of secretion of, 26, 27
  analogues, biological activity of, 245–282
  androgenic activities of, 154
  biosynthetic pathways for, 41–47
Steroid-receptor
  complex, binding to DNA sites, 214
  for PEP carboxykinase, 224
Steroidogenesis
  effects of calcium on, 52–54
  regulation of, 46, 47
  role of amino acid sequence in, 48
  stimulation by cyclic-AMP, 37, 39, 40
Sterol-carrier protein in regulation of steroidogenesis, 46, 47
Stilboestrol, oestrogenic activity of, 268
Stomates, abscisic acid effect on, 299
Storage receptors, DHT-binding proteins as, 177
Strain, definition of, 308
Strontium, blocking of hormone release by, 112
Structure-activity relationship
  of corticoid analogues, 273
  and metabolism of steroids, 251, 252
  of parathyroid hormones, 106–108
  of steroid analogues, 258–260
Substitution in steroids, effect on activity, 258–260
Sulphatation factor and GH action, 86
Sulphate conjugation
  in oestrogen degradation, 266, 267
  of steroids, 246, 247
Superinduction
  of glycerol phosphate dehydrogenase, 229
  transaminase activity during, 221
  of tryptophan pyrrolase, 226

2,4,5-T, synthetic auxin, 285
Target cell receptors, progesterone binding to, 191, 192
Target tissues, uptake and retention of androgens by, 159, 160
Teeth, parathyroid hormone effect on development of, 128
Template capacity stimulation of rat uterine chromatin, 198, 199

INDEX

Template RNA
  glucocorticoids stimulation of, 225
  increase by glucocorticoids, 226
Template stability during superinduction, 221
Testicular feminisation syndrome, causes of, 177
Testosterone
  activity of, 178
  analogue of, 245
  in assay for anabolic/androgenic ratio, 253, 254
  blood concentration of, 158
  derivatives, progestational activity of, 261–263
  propionate, in assay for anabolic/androgenic ratio, 253, 254
—, α-methyl-,
  in assay for anabolic/androgenic ratio, 253, 254
Theophylline and cyclic-AMP accumulation, 32, 33
Therapeutic uses of calcitonin, 138, 139
Thiol groups, role in insulin action, 6
Thymocytes, glucocorticoid receptors in, 215, 216
Thymus nuclei, dihydrotestosterone retention by, 161
Thyroid, TSH action on, 70, 71
Thyroprophin, characteristics of, 67–71
Thyrotrophic hormone, characteristics of, 67–71
Thyrotrophin-releasing hormone, action of, 69–70
Thyrotrophin-stimulating hormone
  characteristics of, 67–71
  release stimulation by TRH, 69
Thyroxine, effect on TSH secretion, 70
Time course of cyclic-AMP accumulation, 30, 31
Tissue differentiation, role of hormones in, 307
Tissue-specific function of cultured cells, 317
Tissue uptake of progesterone, 189–191
Transaminase breakdown
  actinomycin D inhibition of, 221
  inhibitors of, 222
Transaminase mRNA, labile repressor of, 221
Transaminase template, half-life of, 218
Transcortin, affinity for progesterone, 193
Transconjugation of steroids, 247, 248
Transcription

Transcription *continued*
  glucocorticoid stimulation of, 220, 225, 229
  oestrogen regulation of, 197–199
Tropic hormone releasing factors, synthesis by hypothalamus, 321
Trypsin
  as disaggregating agent, 309
  inhibition of insulin action, 14, 15
Tryptophan pyrrolase, glucocorticoids effect on, 225, 226
Tubular gland cells, oestrogen effect on differentiation of, 206
Tumour viruses and breast cancer, 318, 319
Tyrosine aminotransferase, insulin effect on synthesis of, 17, 18
Tyrosine transaminase
  glucocorticoids induction of, 213
  induction of, 216
  regulation by glucocorticoids, 217–222

Ultimobranchial body, calcitonin from, 128
Urea, insulin inhibition of production of, 12
Uridine monophosphate
  cyclic, stimulation of hormone production by, 36
Uridine phosphorylase
  detection of mycoplasma with, 313
Uterine chromatin, template capacity stimulation, 198, 199
Uterine polysomes, oestrogen stimulation of, 200
Uterine ribosomes
  capacity, oestrogen regulation of, 200
  synthesis, oestrogen stimulation of, 199
Uterine translation, oestrogen regulation of, 199–201
Uterus, oestrogen effects on, 201, 202

Vasopressin, analogues, corticotrophin-releasing activity of, 74
Viruses in cell culture, 314
Vitamin D, hormone action of, 139–144

Wall extensibility, auxin induction of, 287
Wall-loosening factor, response to auxin, 289

X-ray analysis of insulin structure, 5

trans-Zeatin, structure, 284